AF576064

Assessing Atmospheric Pollution and Its Impacts on the Human Health

Assessing Atmospheric Pollution and Its Impacts on the Human Health

Editor

Alina Barbulescu

MDPI • Basel • Beijing • Wuhan • Barcelona • Belgrade • Manchester • Tokyo • Cluj • Tianjin

Editor
Alina Barbulescu
Civil Engineering
Transilvania University,
Brasov
Brasov
Romania

Editorial Office
MDPI
St. Alban-Anlage 66
4052 Basel, Switzerland

This is a reprint of articles from the Special Issue published online in the open access journal *Atmosphere* (ISSN 2073-4433) (available at: www.mdpi.com/journal/atmosphere/special_issues/Assessing_Atmospheric_Pollution).

For citation purposes, cite each article independently as indicated on the article page online and as indicated below:

LastName, A.A.; LastName, B.B.; LastName, C.C. Article Title. *Journal Name* **Year**, *Volume Number*, Page Range.

ISBN 978-3-0365-6321-3 (Hbk)
ISBN 978-3-0365-6320-6 (PDF)

© 2023 by the authors. Articles in this book are Open Access and distributed under the Creative Commons Attribution (CC BY) license, which allows users to download, copy and build upon published articles, as long as the author and publisher are properly credited, which ensures maximum dissemination and a wider impact of our publications.
The book as a whole is distributed by MDPI under the terms and conditions of the Creative Commons license CC BY-NC-ND.

Contents

About the Editor vii

Preface to "Assessing Atmospheric Pollution and Its Impacts on the Human Health" ix

Alina Bărbulescu, Cristian Ștefan Dumitriu and Nicolae Popescu-Bodorin
Assessing Atmospheric Pollution and Its Impact on the Human Health
Reprinted from: *Atmosphere* **2022**, *13*, 938, doi:10.3390/atmos13060938 1

Alina Bărbulescu
On the Spatio-Temporal Characteristics of Aerosol Optical Depth in the Arabian Gulf Zone
Reprinted from: *Atmosphere* **2022**, *13*, 857, doi:10.3390/atmos13060857 5

Alina Bărbulescu, Cristian Stefan Dumitriu, Iulia Ilie and Sebastian-Barbu Barbeş
Influence of Anomalies on the Models for Nitrogen Oxides and Ozone Series
Reprinted from: *Atmosphere* **2022**, *13*, 558, doi:10.3390/atmos13040558 17

Yi Huang, Li Wang, Xin Cheng, Jinjin Wang, Ting Li and Min He et al.
Characteristics of Particulate Matter at Different Pollution Levels in Chengdu, Southwest of China
Reprinted from: *Atmosphere* **2021**, *12*, 990, doi:10.3390/atmos12080990 43

Yousef Nazzal, Nadine Bou Orm, Alina Barbulescu, Fares Howari, Manish Sharma and Alaa E. Badawi et al.
Study of Atmospheric Pollution and Health Risk Assessment: A Case Study for the Sharjah and Ajman Emirates (UAE)
Reprinted from: *Atmosphere* **2021**, *12*, 1442, doi:10.3390/atmos12111442 57

Joanna Klaudia Buch, Anita Urszula Lewandowska, Marta Staniszewska, Kinga Areta Wiśniewska and Karolina Venessa Bartkowski
The Influence of Transport on PAHs and Other Carbonaceous Species' (OC, EC) Concentration in Aerosols in the Coastal Zone of the Gulf of Gdansk (Gdynia)
Reprinted from: *Atmosphere* **2021**, *12*, 1005, doi:10.3390/atmos12081005 75

Carmen Maftei, Radu Muntean and Ionut Poinareanu
The Impact of Air Pollution on Pulmonary Diseases: A Case Study from Brasov County, Romania
Reprinted from: *Atmosphere* **2022**, *13*, 902, doi:10.3390/atmos13060902 95

Pengcheng Cui, Tao Zhang, Xin Chen and Xiaoyi Yang
Levels, Sources, and Health Damage of Dust in Grain Transportation and Storage: A Case Study of Chinese Grain Storage Companies
Reprinted from: *Atmosphere* **2021**, *12*, 1025, doi:10.3390/atmos12081025 109

Ning Qin, Wei He, Qishuang He, Xiangzhen Kong, Wenxiu Liu and Qin Wang et al.
Multi-Media Exposure to Polycyclic Aromatic Hydrocarbons at Lake Chaohu, the Fifth Largest Fresh Water Lake in China: Residual Levels, Sources and Carcinogenic Risk
Reprinted from: *Atmosphere* **2021**, *12*, 1241, doi:10.3390/atmos12101241 127

Yara de Souza Tadano, Eduardo Tadeu Bacalhau, Luciana Casacio, Erickson Puchta, Thomas Siqueira Pereira and Thiago Antonini Alves et al.
Unorganized Machines to Estimate the Number of Hospital Admissions Due to Respiratory Diseases Caused by PM_{10} Concentration
Reprinted from: *Atmosphere* **2021**, *12*, 1345, doi:10.3390/atmos12101345 143

James D. Johnston, John D. Beard, Emma J. Montague, Seshananda Sanjel, James H. Lu and Haley McBride et al.
Chemical Composition of $PM_{2.5}$ in Wood Fire and LPG Cookstove Homes of Nepali Brick Workers
Reprinted from: *Atmosphere* **2021**, *12*, 911, doi:10.3390/atmos12070911 . . . . **165**

About the Editor

Alina Barbulescu

Dr. Alina Bărbulescu is a Professor in the Department of Civil Engineering at Transilvania University of Brasov, Brasov, Romania. She graduated from the University of Craiova (Romania), Faculty of Mathematics, and from the Petre Andrei University of Iași (Romania), Faculty of Law. After achieving an MSc in Mathematics at Bucharest University (Romania), she achieved a Ph.D. in Mathematics from Al. I. Cuza University of Iasi (Romania), a Ph.D. in Cybernetics and Economic Statistics from the Academy of Economic Studies Bucharest (Romania), and a Ph.D. in Civil Engineering, with Magna cum Laude, at the Technical University of Civil Engineering, Bucharest (Romania). She received a habilitation in Civil Engineering in 2014, with the thesis entitled "Modeling the spatially distributed precipitations", and in Cybernetics and Economics Statistics in 2019, with the thesis entitled "Contribution to statistical analysis and modeling in economics and finance". She is currently a Professor in the Department of Civil Engineering at the Transilvania University of Brașov. Before adopting her current position, she taught at the Ovidius University of Constanța, Romania, Higher Colleges of Technology, UAE, and was a visiting professor at Abu Dhabi University. Her research interests span applied statistics, econometrics, hydrological modeling, water, and air pollution. She has authored over 170 articles and 30 books and book chapters; she was the co-editor of 20 Special Issues in scientific journals and conference proceedings; and she has been an invited speaker at multiple international conferences. She is a member of IAHS, IAENG, and SSMR, and a topic editor at various journals, including *Water* and *Atmosphere*.

Preface to "Assessing Atmospheric Pollution and Its Impacts on the Human Health"

In recent decades, due to industrial expansion and urbanization, atmospheric pollution has become one of the main threats to public health and natural ecosystems. The production of pollutants from different sources can easily be evaluated; however, their dissipation is associated with the mechanisms of diffusion and transport in various atmospheric conditions, with wind speed being one of the main influencing factors. The accumulation of pollutants is a stochastic phenomenon, depending on multiple factors such as atmospheric circulation, turbulence, wind direction and speed, air temperature, humidity, etc., making the assessment and forecasting of this phenomenon difficult. In conditions of atmospheric calm, the particles accumulate near the emission sources, increasing the time that the population is exposed to toxic substances which induce respiratory diseases and, sometimes, irreversible harm to human health. Therefore, assessing air quality is extremely important for maintaining a clean and healthy environment. In this context, the articles included in this collection addressed the following topics:

·Estimating the air quality using statistical and artificial intelligence methods;

·Modeling the extremes of different pollutants over various time series;

·Emphasizing the impact of atmospheric pollution on human health, especially the incidence of pulmonary diseases in highly polluted zones;

·Analyzing the carcinogenic and non-carcinogenic risks of exposure to different pollutants for an extended period.

The topic is of interest to scientists and the general public, addressing problems that we all face.

The editor would like to thank the authors for sharing the results of their research, and the reviewers whose valuable suggestions led to significant improvements to the submitted articles.

Last but not least, we thank the editorial staff who helped to ensure a smooth editorial process.

Alina Barbulescu
Editor

Editorial

Assessing Atmospheric Pollution and Its Impact on the Human Health

Alina Bărbulescu [1], Cristian Ștefan Dumitriu [2,*] and Nicolae Popescu-Bodorin [3]

1 Department of Civil Engineering, Transilvania University of Brașov, 5 Turnului Str., 900152 Brasov, Romania; alina.barbulescu@unitbv.ro
2 SC Utilvavorep SA, 55, Aurel Vlaicu Bd., 900055 Constanța, Romania
3 Department of Engineering and Computer Science, Spiru Haret University, 13, Ion Ghica Str., 030045 Bucharest, Romania; casalot@gmail.com
* Correspondence: cristian.dumitriu@unitbv.ro

Citation: Bărbulescu, A.; Dumitriu, C.Ș.; Popescu-Bodorin, N. Assessing Atmospheric Pollution and Its Impact on the Human Health. *Atmosphere* **2022**, *13*, 938. https://doi.org/10.3390/atmos13060938

Received: 2 June 2022
Accepted: 7 June 2022
Published: 9 June 2022

Publisher's Note: MDPI stays neutral with regard to jurisdictional claims in published maps and institutional affiliations.

Copyright: © 2022 by the authors. Licensee MDPI, Basel, Switzerland. This article is an open access article distributed under the terms and conditions of the Creative Commons Attribution (CC BY) license (https://creativecommons.org/licenses/by/4.0/).

In recent decades, atmospheric pollution has become a major risk for public health and ecosystems. In this era, when industrial development and urbanization are accelerated, decreasing the contamination level from different sources became a must to ensure a friendly and healthy climate for future generations.

The Special Issue "Assessing Atmospheric Pollution and Its Impact on the Human Health" contains articles with the following topics: assessment of the PM10, PM2.5, nitrogen oxides, ozone, and dust on the pollution life using different health indicators and statistical methods, building artificial intelligence method for evaluating the pollution trend and the admission of people in hospital due to pulmonary diseases. It was emphasized that there is insufficient data, and the monitoring network is not uniformly distributed to provide a correct insight into the atmospheric contamination level and its adverse effects on the people. Moreover, the authorities should consider the studies' results and take urgent measures to reduce or eliminate, when possible, the pollution sources.

Given that dust is an important source of pollution in the United Arab Emirates, Nazzal et al. [1] investigated its impact on the health of inhabitants from the Sharjah and Ajman Emirates based on data series collected from April to August 2020, continuing the investigations on the pollution from different sources in the United Arab Emirates [2,3]. They found that the average daily dose (ADD), the hazard quotient (HQ), and the health index (HI) have been used for this aim. The highest concentrations found in the study samples were those of Zn, Ni, and Cu, with anthropogenic origin. The HQ and HI indicated, respectively, an acceptable and negligible non-carcinogenic risk for people's health. Clustering the observation sites based on the original series and those of the health indices found three clusters, one of them formed only by a single location, where the highest concentrations of heavy metals were detected.

Cui et al. [4] investigated the impact of the grain dust on the workers' health using samples of different types of grain collected in six locations in China and developing a probabilistic risk assessment model. Using this approach, the risk to the people's health was transposed into disability-adjusted life years (DALY). It was shown that for the people working in the grain storage and transportation, the mean DALY was greater than 0.4 years, with the values between 0.1 and 3.3 years for the former. The highest DALY corresponds to maize (1.01 years, for in-warehousing), followed by those of rice (0.89 years) and wheat (0.83 years) in the transportation phase.

The article of Maftei et al. [5] addressed the impact of the pollution (with PM10 and nitrogen oxides) on the population's health in the county of Brasov, Romania.

The research tried to correlate the air pollution level with the laboratory analysis results of the patients confirmed with pulmonary malignant tumors. It was shown that most patients suffer from squamous cell carcinoma (76%), the rest of them being diagnosed with pulmonary adenocarcinoma (24%). The disease rate was lower in the rural zones than in the urban ones. In both cases, squamous cell carcinoma has the highest frequency. The limits of this study were the low number of stations recording the atmospheric pollution, their unequal distribution, and the limited database of the medical records.

In their research, Tadano et al. [6] proposed two artificial intelligence models—the echo state networks and the extreme learning machines (ELM) for estimating the impact of the PM10 on the hospital admissions due to respiratory diseases. Other parameters taken into account were air temperature and humidity. The regularization parameter (RP) and the Volterra filter have been used for increasing the model's generalization capability and exploring the nonlinear patterns of the networks' hidden layers. Results show that the ELM better performed in most cases. The research is important for estimating the hospital admission and pointed out the lack of data for other pollutants that could bias the results.

Qin et al. [7] provided the results of the analysis of polycyclic aromatic hydrocarbons (PAHs) in the environment and freshwater fish in the area of Lake Chaohu. First, they identified the atmospheric pollution sources. The exposure to PAHs through water intake, inhalation, and freshwater fish ingestion was evaluated by different techniques, such as the assessment model, probabilistic risk assessment, and Monte Carlo simulation. They showed that the primary source of atmospheric pollution is biomass combustion. The atmospheric transport significantly contributes to the contaminants spreading. Significant differences were found between the samples only for a gaseous BaP equivalent concentration. Among the risk sources, the fish intake and the particles' inhalation occupied the first two places, based on the lifetime average daily dose. The probabilistic cancer risk assessment indicated a potential carcinogenic risk for the population in the neighborhood of Lake Chaohu.

Buch et al. [8] assessed the transport influence on the pollution due to carbon species (elemental—EC and organic—OC) in a zone from the Littoral of the Gdansk Gulf in the periods 13–22 July 2015 (holiday period) and 14–30 September 2015 (holidays and school periods) for two hours in the morning and two in the afternoon. The highest OC (EC) mean concentration in small aerosols was recorded during the holidays (the school period, between 7.00–9. 00 a.m.). Still, the statistical tests rejected the hypothesis that there is a significant difference between the OC concentrations recorded between 7.00–9.00 a.m. and 3.00–5.00 p.m. During the holidays (school period), the EC, sulphate, and nitrate (CO) concentrations were the highest. It was found that the regional wind has an important role in the pollutants' transport.

Huang et al. [9] analyzed the particulate matter (PM) distribution and the trend of heavy metals and water-soluble ions in PM2.5 and PM10 during the haze periods from March 2016 to January 2017 in Chengdu, China, at different pollution levels. It resulted in heavy metals being enriched in fine (PM2.5) particles compared to PM10, and the mobile sources had significant contributions to the haze formation.

Johnston et al. [10] addressed indoor air pollution in the houses of brick workers in the Kathmandu Valley, Nepal, taking into account the type of cooking device used. Higher concentrations of black carbon (349 $\mu g/m^3$) have been detected in the houses using wood fire than where the liquefied petroleum gas cookstoves are used or in outdoor air (5.36 $\mu g/m^3$). Indoor chlorine (potassium) in the first kind of house was 34 (4) times higher than in the second type of residence. Ca, Al, Co, Fe, Ti, and Si concentrations exceeded the allowable limits in all the studied locations. The research pointed out the necessity of the authorities' intervention to improve the region's indoor air quality.

Reliable scenarios or models for atmospheric pollutants dynamics are of high interest for a correct estimation of the pollution impact on the environment and human health. However, the outliers' existence may significantly bias the models' quality and, implicitly, the forecast based on them. In this idea, Bărbulescu et al. [11] studied the existence of outlying values in the daily nitrogen oxides and ozone series collected from 1 January to 8 June 2016 in Timisoara, Romania. Four methods have been employed: the interquartile range (IQR), isolation forest, local outlier factor (LOF), and the generalized extreme studentized deviate (GESD). Three models (ARIMA, GRNN, and ARIMA-GRNN) have been built for the raw series and those without aberrant values. The best one was the hybrid ARIMA-GRNN for the series without aberrants, which can be used for the forecast.

In the article [12], the author analyzed the 387 series of the aerosol optical depth (AOD) collected for 178 months over the Arabian Gulf, continuing the research from [13–15] related to the dust aerosols and storms in the United Arab Emirates. The Principal Component Analysis (PCA) extracted the main data subspace of the temporally indexed and spatially indexed time series (TITS and SITS, respectively). Over 90% of the variance of SITS is explained by the first principal component (PC), and only 60.5% of the variance of TITS by six PCs. Hierarchical clustering applied to SITS indicates that one group contains the locations on the Shamal trajectory, whereas applied to TITS resulted in grouping based on seasonality. The regional and temporal trend series (RTS and TTS, respectively) have been detected using a two-step algorithm, which firstly determined the clusters with the highest number of elements, followed by a mediation process, as presented in [16]. RTS and TTS are trend-stationary, the former being also level-stationary, and fit the data series well.

More research should be done to develop new indices for providing a correct evaluation of the degree of cumulated pollution from different sources and its impact on people's health. At the same time, the decision factors must implement plans to reach a cleaner environment.

Author Contributions: Conceptualization, A.B.; methodology, A.B. and C.Ș.D.; formal analysis, A.B.; writing—N.P.-B. and A.B.; writing—review and editing, A.B. and C.Ș.D.; supervision, A.B.; project administration, A.B. All authors have read and agreed to the published version of the manuscript.

Funding: This research received no external funding.

Conflicts of Interest: The Guest Editor declares that there are no conflict of interest or agreement with private companies that will prevent us working impartially in the editorial process.

References

1. Nazzal, Y.; Orm, N.B.; Barbulescu, A.; Howari, F.; Sharma, M.; Badawi, A.E.; Al-Taani, A.A.; Iqbal, J.; Ktaibi, F.E.; Xavier, C.M.; et al. Study of Atmospheric Pollution and Health Risk Assessment: A Case Study for the Sharjah and Ajman Emirates (UAE). *Atmosphere* **2021**, *12*, 1442. [CrossRef]
2. Al-Taani, A.A.; Nazzal, Y.; Howari, F.M.; Iqbal, J.; Bou Orm, N.; Xavier, C.M.; Bărbulescu, A.; Sharma, M.; Dumitriu, C.-S. Contamination Assessment of Heavy Metals in Agricultural Soil, in the Liwa Area (UAE). *Toxics* **2021**, *9*, 53. [CrossRef] [PubMed]
3. Nazzal, Y.; Bărbulescu, A.; Howari, F.; Al-Taani, A.A.; Iqbal, J.; Xavier, C.M.; Sharma, M.; Dumitriu, C.Ș. Assessment of Metals Concentrations in Soils of Abu Dhabi Emirate Using Pollution Indices and Multivariate Statistics. *Toxics* **2021**, *9*, 95. [CrossRef] [PubMed]
4. Cui, P.; Zhang, T.; Chen, X.; Yang, X. Levels, Sources, and Health Damage of Dust in Grain Transportation and Storage: A Case Study of Chinese Grain Storage Companies. *Atmosphere* **2021**, *12*, 1025. [CrossRef]
5. Maftei, C.; Muntean, R.; Poinăreanu, I. The Impact of Air Pollution on Pulmonary Diseases: A Case Study from Brasov County, Romania. *Atmosphere* **2022**, *13*, 902. [CrossRef]
6. Tadano, Y.d.S.; Bacalhau, E.T.; Casacio, L.; Puchta, E.; Pereira, T.S.; Antonini Alves, T.; Ugaya, C.M.L.; Siqueira, H.V. Unorganized Machines to Estimate the Number of Hospital Admissions Due to Respiratory Diseases Caused by PM_{10} Concentration. *Atmosphere* **2021**, *12*, 1345. [CrossRef]
7. Qin, N.; He, W.; He, Q.; Kong, X.; Liu, W.; Wang, Q.; Xu, F. Multi-Media Exposure to Aromatic Hydrocarbons at Lake Chaohu, the Fifth Largest Fresh Water Lake in China: Residual Levels, Sources and Carcinogenic Risk. *Atmosphere* **2021**, *12*, 1241. [CrossRef]
8. Buch, J.K.; Lewandowska, A.U.; Staniszewska, M.; Wiśniewska, K.A.; Bartkowski, K.V. The Influence of Transport on PAHs and Other Carbonaceous Species' (OC, EC) Concentration in Aerosols in the Coastal Zone of the Gulf of Gdansk (Gdynia). *Atmosphere* **2021**, *12*, 1005. [CrossRef]
9. Huang, Y.; Wang, L.; Cheng, X.; Wang, J.; Li, T.; He, M.; Shi, H.; Zhang, M.; Hughes, S.S.; Ni, S. Characteristics of particulate matter at different pollution levels in Chengdu, southwest of China. *Atmosphere* **2021**, *12*, 990. [CrossRef]
10. Johnston, J.D.; Beard, J.D.; Montague, E.J.; Sanjel, S.; Lu, J.H.; McBride, H.; Weber, F.X.; Chartier, R.T. Chemical composition of $PM_{2.5}$ in wood fire and lpg cookstove homes of Nepali brick workers. *Atmosphere* **2021**, *12*, 911. [CrossRef]
11. Bărbulescu, A.; Dumitriu, C.S.; Ilie, I.; Barbeș, S.B. Influence of Anomalies on the Models for Nitrogen Oxides. *Atmosphere* **2022**, *13*, 558. [CrossRef]
12. Bărbulescu, A. On the spatio-temporal characteristics of aerosol optical depth in the Arabian Gulf zone. *Atmosphere* **2022**, *13*, 857. [CrossRef]
13. Bărbulescu, A.; Nazzal, Y. Statistical analysis of the dust storms in the United Arab Emirates. *Atmos. Res.* **2020**, *231*, 104669. [CrossRef]

14. Bărbulescu, A.; Nazzal, Y.; Howari, F. Statistical analysis and estimation of the regional trend of aerosol size over the Arabian Gulf Region during 2002–2016. *Sci. Rep.* **2018**, *8*, 571. [CrossRef] [PubMed]
15. Nazzal, Y.; Bărbulescu, A.; Howari, F.M.; Yousef, A.; Al-Taani, A.A.; Al Aydaroos, F.; Naseem, M. New insight to dust storm from historical records, UAE. *Arab. J. Geosci.* **2019**, *12*, 396. [CrossRef]
16. Bărbulescu, A.; Postolache, F.; Dumitriu, C.Ș. Estimating the precipitation amount at regional scale using a new tool, Climate Analyzer. *Hydrology* **2021**, *8*, 125. [CrossRef]

MDPI

Article

On the Spatio-Temporal Characteristics of Aerosol Optical Depth in the Arabian Gulf Zone

Alina Bărbulescu 

Department of Civil Engineering, Transilvania University of Brașov, 5 Turnului Street, 900152 Brașov, Romania; alina.barbulescu@unitbv.ro

Abstract: The article investigates some of the available measurements (Terra MODIS satellite data) of the aerosol optical depth (AOD) taken in the Arabian Gulf, a zone traditionally affected by intense sand-related (or even sand-driven) meteorological events. The Principal Component Analysis (PCA) reveals the main subspace of the data. Clustering of the series was performed after selecting the optimal number of groups using 30 different methods, such as the silhouette, gap, Duda, Dunn, Hartigan, Hubert, etc. The AOD regional and temporal tendency detection was completed utilizing an original algorithm based on the dominant cluster found at the previous stage, resulting in the regional time series (RTS) and temporal time series (TTS). It was shown that the spatially-indexed time series (SITS) agglomerates along with the first PC. In contrast, six PCs are responsible for 60.5% of the variance in the case of the temporally-indexed time series (TITS). Both RTS and TTS are stationary in trend and fit the studied data series set well.

Keywords: AOD; classification; dendrogram; PCA

Citation: Bărbulescu, A. On the Spatio-Temporal Characteristics of Aerosol Optical Depth in the Arabian Gulf Zone. *Atmosphere* **2022**, *13*, 857. https://doi.org/10.3390/atmos13060857

Academic Editor: Gabriele Curci

Received: 6 May 2022
Accepted: 23 May 2022
Published: 24 May 2022

Publisher's Note: MDPI stays neutral with regard to jurisdictional claims in published maps and institutional affiliations.

Copyright: © 2022 by the author. Licensee MDPI, Basel, Switzerland. This article is an open access article distributed under the terms and conditions of the Creative Commons Attribution (CC BY) license (https://creativecommons.org/licenses/by/4.0/).

1. Introduction

Dust clouds and storms occur worldwide, especially in the Middle East, southwestern United States, northern China, and the Saharan desert. The essential conditions triggering these phenomena are the existence of huge dust or sand sources, little vegetation, strong surface winds, and an unstable atmosphere [1]. Dust particles primarily enter the lower atmosphere through saltation bombardment, which depends on the meteorological conditions near the surface, the soil texture, and particle size [2–5]. Dust is emitted as hydrophobic particles, relatively ineffective as cloud condensation nuclei. However, during their transport in the atmosphere, due to the interaction with gaseous and particulate air pollutants, their hygroscopicity increases, fortunately enhancing the efficiency of dust removal from the atmosphere through precipitation [6,7]. Haywood et al. [8] indicated that the aerosols cause a strong radiative forcing of climate because of their efficient scattering of solar radiation.

The most abundant aerosol in the atmosphere is dust, composed of oxides (silica, iron oxides), quartz, feldspar, gypsum, and hematite [9]. Ginoux et al. [10] emphasized the anthropogenic and natural dust sources.

Many studies [8,11–13] have already investigated and documented a significant variability of the airborne desert dust during the past decades in the Middle East, Africa, central Asia, and South America, and identified Shamal (the north-westerly wind blowing over Syria, Iraq, and the Arabian Gulf) as the significant natural trigger of dust storm activities across the Arabian Peninsula. Shamal transports the dust lifted from Syria and Iraq to the Arabian Gulf and Peninsula [14–18]. Still, Notaro et al. [16] identified increased dust activity over eastern Saudi Arabia around the Ad Dahna Desert, with dust transported from the Iraqi Desert and local sources. Yu et al. [18] concluded that a strong wind speed determines higher dust activity along the coast of the Persian Gulf in north-central Saudi Arabia, and one has to consider this influence when tracing the phenomenon along and across United Arab Emirates territory.

Different scientists analyzed the aerosol optical depth (AOD) distribution in southeastern Asia and the Middle East in correlation with the seasonal conditions [19–21] and to determine the air quality modifications in China, Sahel, South Africa, and South America [22–25]. They showed that the atmospheric heating rates and the absorption characteristics are linearly dependent, noticing a significant difference between the aerosols in the Indian region and the zones with large deserts and high dunes (such as UAE) contributing to the dust loadings in the atmosphere [26].

Other researchers provided the classification of the aerosol types taking into account different characteristics—fine mode fraction and the aerosol index [27], AOD [4,28,29], refraction index, and Ångström exponent [30–32].

The long-term trend of AOD over the Arabian Peninsula, and eastern and southern Asia for 2002–2009, estimated based on AERONET data, was increasing. The same tendency was observed over most tropical oceans [33]. A long-term positive AOD trend over the Arabian Peninsula occurred, with a higher seasonal tendency during spring and summer (periods when the dust is transported) [26].

Multivariate statistical analysis became one of the most utilized tools for extracting the common characteristics of big data sets issued from environmental sciences [34–38]. The spatio-temporal analysis of AOD is mainly performed by Principal Component Analysis (PCA, also named EOF—Empirical Orthogonal Functions), non-negative matrix factorization (NMF), and combined Principal Component Analysis (CPCA). These tools helped capture the aerosol regimes, the factors influencing the AOD's concentrations, and the trends [20–25].

Recent studies on dust-aerosol in the UAE evaluated the regional distribution of this type of aerosol and the dust storms' intensity [26,39–41]. Using AERONET data collected from 2006 to 2015, Abuelgasim and Farahat [26] found an increasing trend of AOD in summer and spring and a 4.32% mean annual variation of the aerosol loading. They estimated a variation of 11.36% of the mean annual Ångström exponent for the study period. The highest concentration of aerosols was found in summer, while from November to March, an increasing tendency was found during 2011–2016.

Other scientists [40,41] studied the frequency of the dust storms for nine years, using hourly data recorded at eight airports in the UAE. The variation of the aerosol radius was presented in [4], based on monthly series collected at 387 points for 15 years.

Despite the investigations performed to determine the aerosol's characteristics and the effect of meteorological conditions on their loadings and transport in the Arabian Gulf region, many aspects of the aerosol's properties in the UAE remain to be studied.

The AOD time series varies depending on the data structure, aerosol extinction, and surface reflectance [22–24]. Still, here, we shall not analyze the connection between these variables, but the spatial and temporal variation of the AOD series for 178 months. This research continues the attempt to understand the aerosol characteristics in the UAE, aspects that have not been treated in the studies [4,26,40,41].

The main contributions of this research are:

(1) Performing the PCA to extract the principal components that describe the AOD series' characteristics in time and space;
(2) Group the series in clusters (in spatial and temporal dimensions);
(3) Build the 'regional time series' (RTS) and the 'temporal time series' (TTS) of the AOD, employing an original algorithm based on the clusters previously determined;
(4) Compare the RTS and TTS of AOD with those of the 'regional time series' and 'temporal time series' of the aerosol radius (AR) [4] to emphasize the common tendencies.

2. Methods and Data Series

The methods employed at the first stage of our investigations are PCA, also called EOF, and Clustering.

The first one was used to estimate the similarity in terms of linear dependence within the data and eventually to qualify regional/global aspects. The second one was performed to evaluate dissimilarity, the natural tendency of grouping (if present) in data, and identifying aspects and events localized in time and/or space.

PCA is a statistical technique that (linearly) transforms (and deflates) a large set of (possibly correlated) variables into a smaller set of orthogonal uncorrelated variables representing the most significant information of the initial variables set. Initially employed in the study of meteorological series [42], its use became frequent in other fields such as ozone series evaluation [43,44] and isolating aerosols' sources [23] based on AOD retrieved from satellite data.

The PCA method is shortly described in the following.

Let us consider the matrix X, whose columns are the series recorded at each point. If n is the number of series (367, in this case), and m is the number of time units (178 months, here), X is an $m \times n$ matrix. It was shown that X can be written as a product of two matrices, Y $(m \times m)$ and Z $(m \times n)$, of orthogonal functions of position and time.

The equation $X = YZ$ is equivalent to

$$x_{ik} = \sum_{j=1}^{m} y_{ij} z_{jk}, i = \overline{1,m},\ k = \overline{1,n}, \tag{1}$$

so, the vector of the values recorded at a certain point is a linear combination of the Y columns (with different weights), under orthogonality conditions.

Formula (1) is called the PCA analysis.

For a symmetric matrix,

$$A = XX^T = YZZ^TY^T \tag{2}$$

there is a decomposition such as

$$A = Y\Lambda Y^T \tag{3}$$

where Λ is the diagonal matrix formed by A's eigenvalues. Therefore, from (1) and (3) it results that

$$z_{ik} = \sum_{j=1}^{m} y_{ji} x_{jk}, i = \overline{1,m},\ k = \overline{1,n}, \tag{4}$$

because $Z = Y^TX$.

The *j-th* eigenvector has a contribution of $\lambda_j / \sum_{i=1}^{m} \lambda_i$ to X, where $\lambda_i,\ i = \overline{1,m}$ are the eigenvalues [25].

When a small dominant set of principal components exists, the technique detects the common characteristics of the data samples and reveals the regional or temporal aspects. The absence of dominant principal components results in the data series independence, so the phenomenon is localized [45].

Modifications of PCA have been proposed, such as sPCA [46] (that proposes sparse loadings), CPCA [47] (to investigate the pattern of a specific element), Common PCA (to simultaneous reduce the dimensionality in different groups) [48], or Combined PCA (to compare the modes in the AOD decomposition) [23,24,49]. PCA was chosen here because we are interested in the common characteristics of the series.

To determine the number of principal components, the scree plot, the Kaiser criterion, and the proportion of the variance explained by each component may be utilized [50–52]. Here we employed the combined scree plot and variance explained by each component.

Clustering is a method for identifying patterns and similarities within the data and the natural grouping tendency of the similar objects within a data set of interest.

Different scientists introduced various tests for detecting the optimal number of clusters. Although some are more commonly used (because they are well-known or easier to compute), there is no reason to give more credit to one or another, mainly because they rely on different mathematical and computational techniques. Therefore, the strategy for establishing the optimal number of clusters was a multi-criteria decision obeying the majority rule after performing 30 different tests, including silhouette [53], gap, Duda, Dunn, Hartigan, Hubert, and so on [54]. For example, if 15 tests voted for two clusters, eight for three, five for four, and the rest for six, the chosen number is two. This approach assures choosing the number of clusters with the highest probability among the possible number of such groups. In the previous example, the probability associated with two clusters is 15/30 = 1/2, compared to 4/15, 1/6, and 1/15, associated with other choices.

Agglomerative Hierarchical Clustering has been performed using the R software.

The agglomerative coefficient was computed to determine the grouping quality. The higher this coefficient is, the best the clustering is. The dendrogram showing the elements in each cluster was also plotted.

The regional time series (RTS) (temporal time series (TTS)) was built using the spatially-indexed time series (SITS) (temporally-indexed time series (TITS)) and the following algorithm, which is a modified version of *Method II* [4] based on the resulting clusters.

Given k data series each containing n values, consecutively recorded at the same time, denote by $Y = (y_{ji})$ ($j = 1, \ldots, n$, $i = 1, \ldots, k$) the matrix whose column i is formed by the elements of the i-*th* series ($i = 1, \ldots, k$). The steps of the algorithm for determining the representative series are:

(I) Choose the number of clusters, m, and perform the series' clustering, based on a selection algorithm. Here, the choice is made using 30 criteria presented in [54] and the majority principle: m is the number resulting from the highest number of algorithms.
(II) Among the m clusters computed at step (I), choose the one containing the highest number of series (N) and construct a new matrix, Y_1, with the series in this cluster.
(III) Select the representative value in the j-*th line of the matrix* Y_1 *to be the* average of the values in line j.
(IV) Evaluate the error by using the Mean Standard and Mean Absolute Errors (MSE and MAE) corresponding to all the observation sites.
(V) Plot the resulted series.

The novelty of the approach proposed here consists of the following.

1. While selection of the number of clusters in the initial algorithm was left to the user, in this article, an efficient selection procedure is employed.
2. If, in the second step, two clusterings are providing the same maximum number of elements in one of the sets, the chosen one is that which maximizes the distances between the clusters and minimizes those inside the groups.
3. If, after applying the second criterion, there are two clusters with the same number of elements, the computation is performed with each of them. The best result (that gives the minimum mean average error—MAE, mean standard error—MSE, and mean absolute percentage error (%)—MAPE) is reported.

The trend and level stationarity of RTS and TTS has been checked using the KPSS test [55]. The null hypothesis, H_0, was the series stationarity in level (trend), and the alternative one, H_1, was the series nonstationarity in level (trend).

Remember that a time series is stationary if the statistical properties of the process generating it remain unchanged over time. Therefore, the mean, variance, and autocorrelation structure are constant over time.

One of the common causes of the violation of H_0 is the existence of a trend in the mean due to the presence of a unit root or the existence of a deterministic trend. In the first case, the stochastic shocks have persistent effects. In contrast, in the second one, they have only transitory effects after which the variable tends toward a deterministically evolving (non-constant) mean (and the process is called a trend-stationary).

The KPSS test is based on the time series decomposition into a deterministic trend, a random walk, and a stationary error. In the case of stationarity, the series has a fixed element as intercept, or the series is stationary around a fixed level.

The test was performed at the level of significance of 0.05. If the p-value is less than 0.05, the null hypothesis is rejected.

Data used in this study are monthly AOD series retrieved by Terra MODIS (at a wavelength of 412 nm) at 387 points from July 2002 to April 2017 in the Arabian Gulf

Region (Figure 1a), between 24.95–26.25 latitudes and 51.55–55.75 longitudes. The series retrieved at the point of coordinates 26.15 latitude and 51.55 longitude is presented in Figure 1b and the series recorded in January 2003 is shown in Figure 1c.

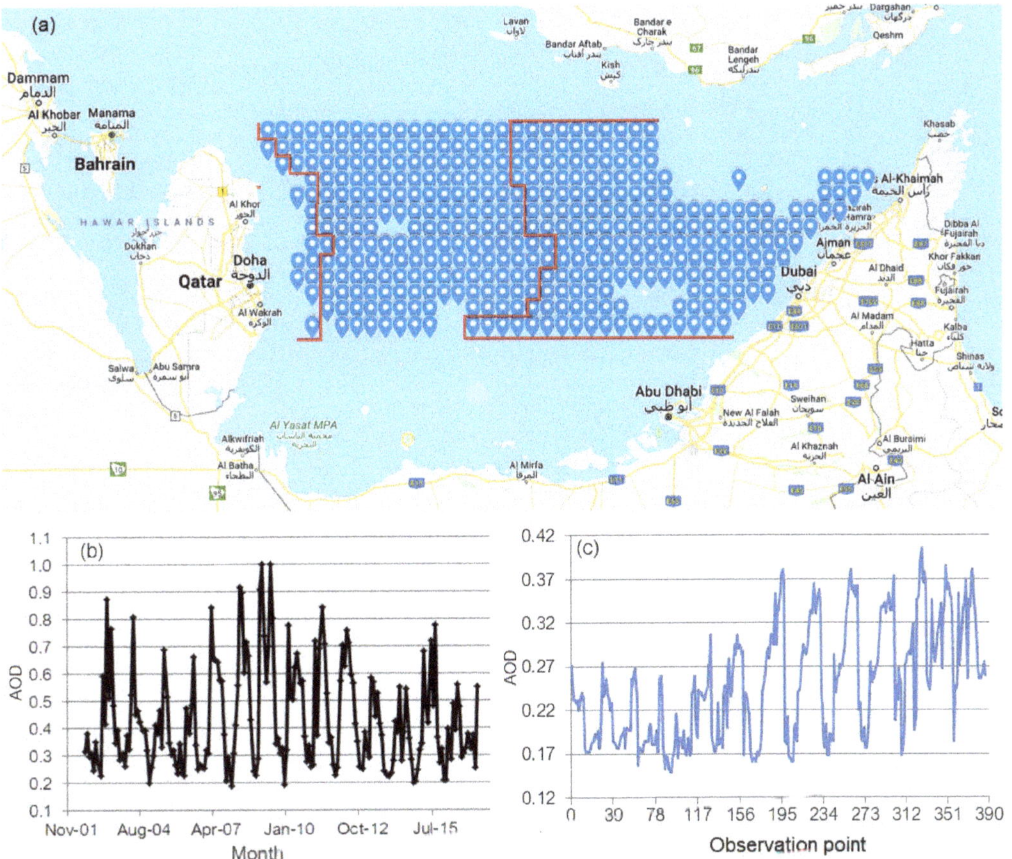

Figure 1. (**a**) Observation area (https://www.google.com/mymaps/, 2022); (**b**) the series retrieved at the point of coordinates 26.15 latitude and 51.55 longitude (SITS); (**c**) series recorded in January 2003 (TITS)

The sites are ordered in increasing order in latitude, and subsequently, by decrease latitude, from the left corner on the map of the studied region to the lower right corner. Details on the study area may be found in [4,40,41]. The coordinates of the sampling points are given in Table S1 in the Supplementary Materials. Data have been organized in a matrix, X, whose columns contain the AOD at each point, and the lines contain the monthly values at the observation points.

3. Results and Discussion

3.1. PCA and Clustering

Table 1 shows the computed eigenvalues greater than 1, the proportion of the variance explained by each component, and the cumulative proportion of the variance explained for SITS.

Table 1. Eigen-analysis of the correlation matrix of SITS.

	PC1	PC2	PC3	PC4	PC5	PC6	PC7	PC8
Eigenvalues	350.98	10.99	4.09	3.79	2.55	1.33	1.17	1.12
Proportion of variance	0.902	0.028	0.011	0.010	0.007	0.003	0.003	0.003
Cumulative proportion	0.902	0.931	0.941	0.951	0.957	0.961	0.964	0.967

Although eight eigenvalues are greater than 1, the first component (PC1) explains 90.20% of the variance within this set. The second component (PC2) explains only 2.8% of the variance within SITS, while the others have even smaller contributions. The first two principal components (PCs) are enough to extract the essential information within the time series set, which proves to be highly PCA compressible. Only 9.80% of the variance within this data is outside the direction of the first dominant PC, and 6.90% is outside the plane determined by PC1 and PC2. These small percentages reveal that the series similarity (linear dependence) is high in this set because the data points agglomerate along with PC1. Therefore, the sand aerosols over the Arabian Gulf have a regional nature, the AOD values being relatively similar (linearly dependent) across the analyzed area.

The optimal number of clusters—two—was selected after running the NbClust package in R. Figure 2 displays the silhouette chart (one of the 30 methods run). The agglomerative clustering has been performed for SITS setting the number of clusters equal to two. The computed agglomerative coefficient was 0.7678, indicating a good partition of the series in two sets. The highest this coefficient is, the better the clustering is.

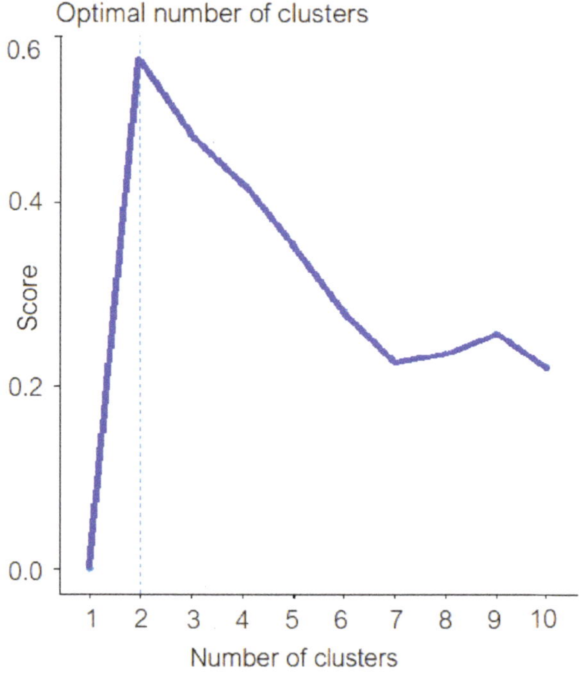

Figure 2. The silhouette chart. The dotted line indicates the optimum number of clusters (two).

The dendrogram displaying the groups of the observation points is presented in Figure 3.

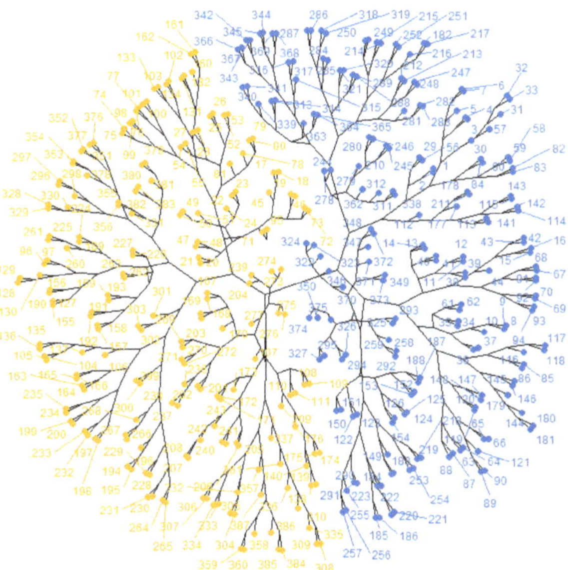

Figure 3. The dendrogram for classification of SITS. The elements in different clusters are in different colors.

Analyzing the elements in the clusters related to the positions of the points on the map (Figure 1a and Supplementary Table S1), results in that one cluster contains the points situated on the eastern side of the region and (very few) near the Qatar shore, while the second one contains the rest of the locations. So, one cluster contains the sites situated in the direction where the Shamal blows—between the red borders in Figure 1.

Figure 4 shows the scree plot of the TITS (the lines of the data matrix). The computation found 25 eigenvalues greater than one, with only six PCs explaining 60.5% of the variance.

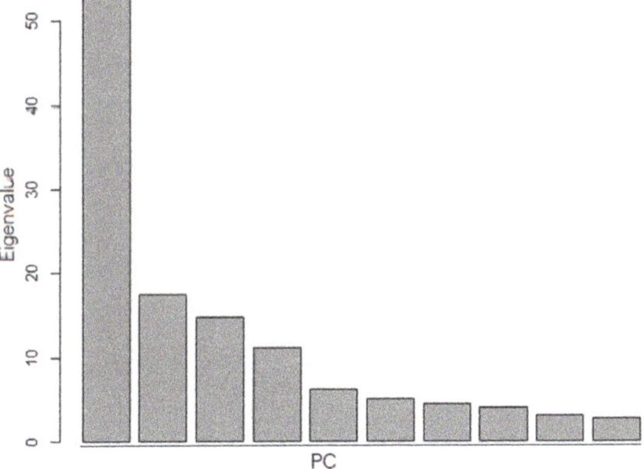

Figure 4. The scree plot for TITS.

The majority principle decided that the number of clusters to classify the series is also two for the TITS. Figure 5 contains the elements in the clusters and the dendrograms for the TITS. The agglomerative clustering provided an agglomerative coefficient of 0.92802, which indicates a strong separation between the groups.

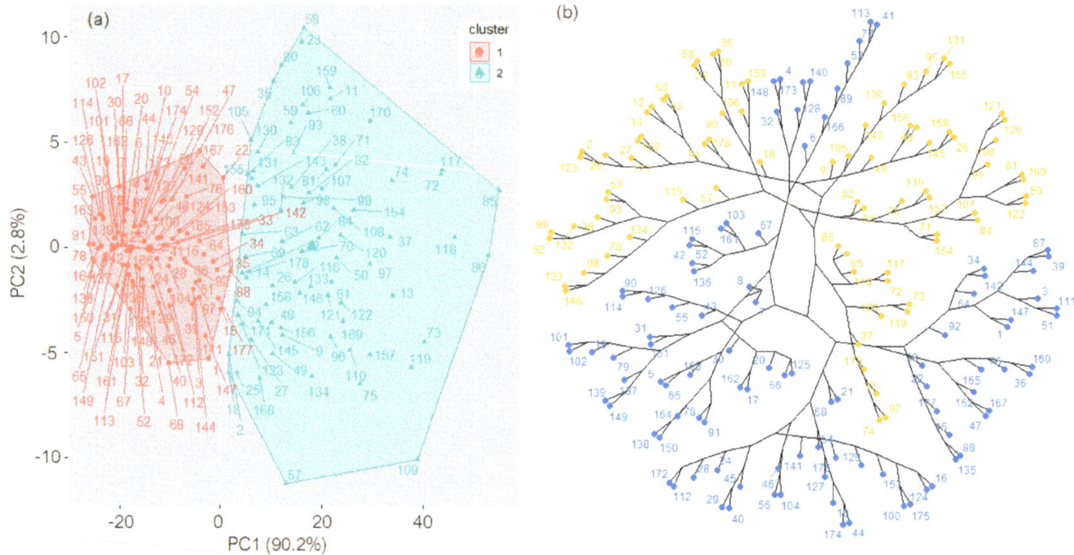

Figure 5. (**a**) Clusters (the values on the axes represent the scores on PC2 and PC1) and (**b**) the dendrogram from the hierarchical clustering of TITS. The elements in different clusters are in different colors.

Comparing the clusters' content, it resulted that one of them mainly contains the series recorded in the summer months (March to August), while the other contains the rest. So, the classification is related to seasonality.

3.2. Estimating the RTS and TTS

Taking into account the results from the previous section, the RTS has been computed by applying the algorithm (I)–(V) described in Section 2 to SITS (columns in the matrix X). The AOD's RTS (as a function of time) is presented in Figure 6a. One can remark on the periodic behavior of this series, whose highest values of the AOD's RTS are primarily recorded in July, while the lowest is in November–January.

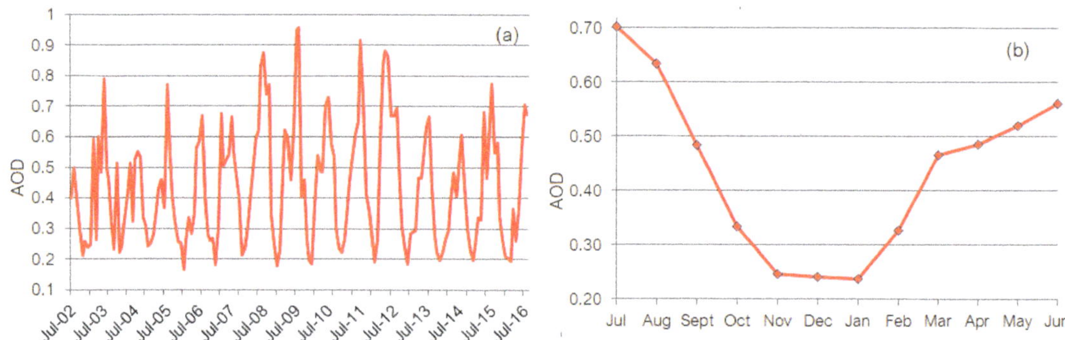

Figure 6. (**a**). The AOD 's RTS; (**b**) AOD's RTS monthly average.

Figure 6b shows that the AOD's RTS monthly average value for July is about three times higher than the corresponding average values for November–January. This result is in concordance with those of Abuelgasim and Farahat [26] and Yoon et al. [56], which indicated a significant increase of AOD over the Gulf Region, especially in summer, related to the dust abundance [39–41,57,58].

The trend and level stationarity hypotheses could not be rejected for the RTS (p-value > 0.1, in both cases) when applying the KPSS test. This means that the RTS does not present a variation in trend or level.

The TTS (Figure 7) was built from the TITS (the transposed matrix, Y^T) and the same algorithm.

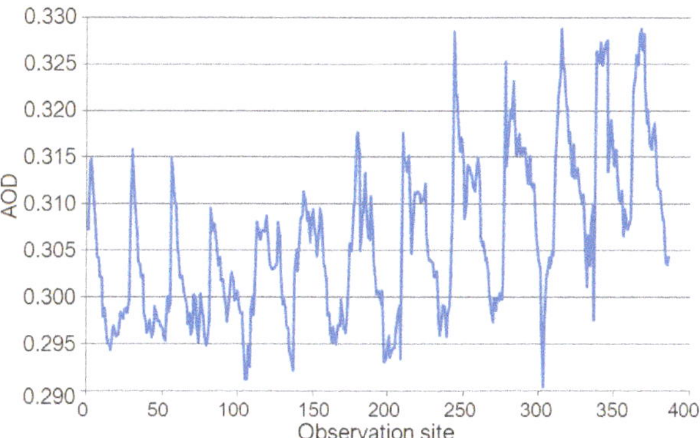

Figure 7. The TTS of AOD obtained running the algorithm from Section 3 for TITS.

The aspect of TTS, with peaks and troughs is related not only by the monthly characteristics of the AOD (higher dust quantities in summer, and lower in winter), but also to the position of the sampling points (Figure 1a) in the direction NW-SE in which Shamal blows.

The non-stationarity in the level and trend stationarity are emphasized by the results of the KPSS test, whose p-values are lower than 0.01 (for level-stationarity) and 0.12376 (for trend-stationarity).

The goodness-of-fit indicators are tools for assessing the fit quality. The lower their values are, the better the fit is. For a correct model quality assessment, using more than one indicator is recommended. For example, MSE can be influenced by values that significantly deviate from the average. Therefore, three goodness-of-fit indicators have been utilized here—MAE, MSE, and MAPE—the last one being non-dimensional, so it is more reliable for comparisons between the two models.

All the values of the goodness-of-fit indicators corresponding to the RTS and TTS are all very low (Table 2). They indicate a better fit for RTS than for TTS when reported to the average indicators. Indeed, the mean values of MAE (MSE and MAPE, respectively) are 0.1724/0.0326 = 5.29 (21.88 and 4.38 times higher, respectively) for TTS than for RTS. The minimum MAE and MSE are also lower for RTS compared to TTS, but the min MAPE is higher for RTS.

Table 2. Goodness of fit indicators of RTS and TTS.

	MAE			MSE			MAPE (%)		
	Min	Mean	Max	Min	Mean	Max	Min	Mean	Max
RTS	0.0207	0.0326	0.0598	0.0008	0.0025	0.0081	4.7925	7.7334	14.1450
TTS	0.0142	0.1724	0.6671	0.0003	0.0547	0.4481	4.3162	33.8819	68.4155

The maximum values of MAE (max MSE and max MAPE, respectively) is 11.15 (55.32 and 4.84, respectively) times higher for RTS than for TTS, indicating a higher variability at the temporal scale than at the spatial one.

This means that RTS better represents the individual series than TTS, so there is a higher homogeneity of the SITS than the TITS. The result is in concordance with the findings related to the seasonal variability of AOD [26].

4. Conclusions

This study extended our previous research on the aerosol radius, using the AOD series collected during the same period at the same sampling points in the Gulf Region. The novelty of the research consists of a dual analysis in time and space and the detection of RTS and TTS that characterizes the AOD behavior over the study zone.

The approach combined the PCA with a new algorithm, building the RTS and TTS series based on the classification provided by the clustering.

It was found that a single principal component explains more than 90% of the variance of SITS, indicating that the series are agglomerated along with PC1. The TITS are scattered (the first six dominant principal components accounting for only 60.5% of the variance in the sets). Still, both RTS and TTS fit data well and are trend stationary.

We intend to extend the research to sets of series with missing data, given that most of the available records present gaps.

Supplementary Materials: The following supporting information can be downloaded at: https://www.mdpi.com/article/10.3390/atmos13060857/s1, Table S1: Data series and the coordinates of the locations.

Funding: This research received funding from the research fund of Transilvania University of Brașov, Romania.

Institutional Review Board Statement: Not applicable.

Informed Consent Statement: Not applicable.

Data Availability Statement: Data can be freely downloaded from the site https://modis.gsfc.nasa.gov/data/ (accessed on 10 July 2017).

Conflicts of Interest: The authors declare no conflict of interest.

References

1. Yassin, M.F.; Almutairi, S.K.; Al-Hemoud, A. Dust storms backward trajectories and source identification over Kuwait. *Atmos. Res.* **2018**, *212*, 158–171. [CrossRef]
2. Grini, A.; Zender, C.S.; Colarco, P. Saltation sandblasting behavior during mineral dust aerosol production. *Geophys. Res. Lett.* **2002**, *29*, 15-1–15-4. [CrossRef]
3. Shao, Y.; Raupach, M.R.; Findlater, P.A. Effect of saltation bombardment on the entrainment of dust by wind. *J. Geophys. Res.* **1993**, *98*, 12719–12726. [CrossRef]
4. Bărbulescu, A.; Nazzal, Y.; Howari, F. Statistical analysis and estimation of the regional trend of aerosol size over the Arabian Gulf Region during 2002–2016. *Sci. Rep.* **2018**, *8*, 9571. [CrossRef] [PubMed]
5. Astitha, M.; Lelieveld, J.; Abdel Kader, M.; Pozzer, A.; de Meij, A. Parameterization of dust emissions in the global atmospheric chemistry-climate model EMAC: Impact of nudging and soil properties. *Atmos. Chem. Phys.* **2012**, *12*, 11057–11083. [CrossRef]
6. Ganor, E.; Osetinsky, I.; Stupp, A.; Alpert, P. Increasing trend of African dust, over 49 years, in the eastern Mediterranean. *J. Geophys. Res.* **2010**, *115*, D07201. [CrossRef]
7. Smoydzin, L.; Fnais, M.; Lelieveld, J. Ozone pollution over the Arabian Gulf—Role of meteorological conditions. *Atmos. Chem. Phys. Discuss.* **2012**, *12*, 6331–6361. [CrossRef]
8. Haywood, J.; Francis, P.; Osborne, S.; Glew, M.; Loeb, N.; Highwood, E.; Tanré, D.; Myhre, G.; Formenti, P.; Hirst, E. Radiative properties and direct radiative effect of Saharan dust measured by the C-130 aircraft during SHADE: 1. Solar spectrum. *J. Geophys. Res.* **2003**, *108*, 8577. [CrossRef]
9. Mahowald, N.; Albani, S.; Kok, J.K.; Engelstaeder, M.; Scanza, R.; Ward, D.S.; Flanner, M.G. The size distribution of desert dust aerosols and its impact on the Earth system. *Aeolian Res.* **2014**, *15*, 53–71. [CrossRef]
10. Ginoux, P.; Prospero, J.M.; Gill, E.T.; Hsu, N.C.; Zhai, M. Global-scale attribution of anthropogenic and natural dust sources and their emission rates based on MODIS Deep Blue aerosol products. *Rev. Geophys.* **2012**, *50*(3), RG3005. [CrossRef]
11. Chen, S.; Huang, J.; Li, J.; Jia, R.; Kang, L.; Ma, X.; Xie, T. Comparison of dust emission, transport, and deposition between the Taklimakan desert and Gobi desert from 2007 to 2011. *Sci. China Earth Sci.* **2017**, *60*, 1338–1355. [CrossRef]
12. Doyle, M.; Dorling, S. Visibility trends in the UK 1950–1997. *Atmos. Environ.* **2002**, *36*, 3161–3172. [CrossRef]
13. Deng, X.; Tie, X.; Wu, D.; Zhou, X.; Bi, X.; Tan, H.; Li, F.; Jiang, C. Long-term trend of visibility and its characterizations in the Pearl River Delta (PRD) region, China. *Atmos. Environ.* **2008**, *42*, 1424–1435. [CrossRef]
14. Goudie, A.S.; Middleton, N.J. *Desert Dust in the Global System*; Springer: Berlin, Germany, 2006.
15. Middleton, N.J. A geography of dust storms in Southwest Asia. *J. Climatol.* **1986**, *6*, 183–196. [CrossRef]

16. Notaro, M.; Alkolibi, F.; Fadda, E.; Bakhrjy, F. Trajectory analysis of Saudi Arabia dust storm. *J. Geophys. Res. Atmos.* **2013**, *118*, 6028–6043. [CrossRef]
17. Shao, Y. A model for mineral dust emission. *J. Geophys. Res. Atmos.* **2001**, *106*, 20239–20254. [CrossRef]
18. Yu, Y.; Notaro, M.; Liu, Z.; Wang, F.; Alkolibi, F.; Fadda, E.; Bakhrjy, F. Climatic controls on the interannual to decadal variability in Saudi Arabian dust activity: Toward the development of a seasonal dust prediction model. *J. Geophys. Res. Atmos.* **2015**, *120*, 1739–1758. [CrossRef]
19. Arfan Ali, M.; Nichol, J.E.; Bilal, M.; Qiu, Z.; Mazhar, U.; Wahiduzzaman, M.; Almazroui, M.; Nazrul Islam, M. Classification of aerosols over Saudi Arabia from 2004–2016. *Atmos. Environ.* **2020**, *241*, 117785. [CrossRef]
20. Mohammadpour, K.; Rashki, A.; Sciortino, M.; Kaskaoutis, D.G.; Boloorani, A.D. A statistical approach for identification of dust-AOD hotspots climatology and clustering of dust regimes over Southwest Asia and the Arabian Sea. *Atmos. Pollut. Res.* **2022**, *13*, 01395. [CrossRef]
21. Mohammadpour, K.; Sciortino, M.; Kaskaoutis, D.G. Classification of weather clusters over the Middle East associated with high atmospheric dust-AODs in West Iran. *Atmos. Res.* **2021**, *259*, 105682. [CrossRef]
22. Sogacheva, L.; Rodriguez, E.; Kolmonen, P.; Virtanen, T.H.; Saponaro, G.; de Leeuw, G.; Georgoulias, A.K.; Alexandri, G.; Kourtidis, K.; van der A, R.J. Spatial and seasonal variations of aerosols over China from two decades of multi-satellite observations. Part II: AOD time series for 1995–2017 combined from ATSR, ADV and MODIS C6.1 for AOD tendencies estimation. *Atmos. Chem. Phys.* **2018**, *18*, 16631–16652. [CrossRef]
23. Li, J.; Carlson, B.E.; Lacis, A.A. Application of spectral analysis techniques in the intercomparison of aerosol data: 1. An EOF approach to analyze the spatial-temporal variability of aerosol optical depth using multiple remote sensing data sets. *J. Geophys. Res. Atmos.* **2013**, *118*, 8640–8648. [CrossRef]
24. Li, J.; Carlson, B.E.; Lacis, A.A. Application of spectral analysis techniques in the intercomparison of aerosol data: Part III. Using combined PCA to compare spatiotemporal variability of MODIS, MISR, and OMI aerosol optical depth. *J. Geophys. Res. Atmos.* **2014**, *119*, 4017–4042. [CrossRef]
25. Ma, Q.; Zhang, Q.; Wang, Q.; Yuan, X.; Yuan, R.; Luo, C. A comparative study of EOF and NMF analysis on downward trend of AOD over China from 2011 to 2019. *Environ. Pollut.* **2021**, *288*, 117713. [CrossRef] [PubMed]
26. Abuelgasim, A.; Farahat, A. Effect of dust loadings, meteorological conditions, and local emissions on aerosol mixing and loading variability over highly urbanized semiarid countries: United Arab Emirates case study. *J. Atmos. Sol. Terr. Phys.* **2019**, *199*, 105215. [CrossRef]
27. Kim, J.; Lee, J.; Lee, H.C.; Higurashi, A.; Takemura, T.; Song, C.H. Consistency of the aerosol type classification from satellite remote sensing during the Atmospheric Brown Cloud—East Asia Regional Experiment campaign. *J. Geophys. Res.* **2007**, *112*, D22S33. [CrossRef]
28. Hsu, N.C.; Gautam, R.; Sayer, A.M.; Bettenhausen, C.; Li, C.; Jeong, M.J.; Tsay, S.-C.; Holben, B.N. Global and regional trends of aerosol optical depth over land and ocean using SeaWiFS measurements from 1997 to 2010. *Atmos. Chem. Phys.* **2012**, *12*, 8037–8053. [CrossRef]
29. De Meij, A.; Pozzer, A.; Lelieveld, J. Trend analysis in aerosol optical depths and pollutant emission estimates between 2000 and 2009. *Atmos. Environ.* **2012**, *51*, 75–85. [CrossRef]
30. Dubovik, O.; Holben, B.; Eck, T.F.; Smirnov, A.; Kaufman, Y.J.; King, M.D.; Tanré, D.; Slutsker, I. Variability of Absorption and Optical Properties of Key Aerosol Types Observed in Worldwide Locations. *J. Atmos. Sci.* **2002**, *59*, 590–608. [CrossRef]
31. Ridley, D.A.; Heald, C.L.; Kok, J.F.; Zhao, C. An observationally constrained estimate of global dust aerosol optical depth. *Atmos. Chem. Phys.* **2016**, *16*, 15097–15117. [CrossRef]
32. Smirnov, A.; Holben, B.; Dubovik, O.; O'Neill, N.T.; Eck, T.F.; Westphal, D.L.; Goroch, A.K.; Pietras, C.; Slutsker, I. Atmospheric Aerosol Optical Properties in the Persian Gulf Region. *J. Atmos. Sci.* **2002**, *59*, 620–634. [CrossRef]
33. Krishna Moorthy, K.; Babu, S.S.; Satheesh, S.K.; Lal, S.; Sarin, M.M.; Ramachandran, S. Climate implications of atmospheric aerosols and trace gases: Indian Scenario. In *Climate Sense*; Asrar, G., Ed.; Tudor Rose: Leicester, UK, 2009.
34. Al-Taani, A.A.; Nazzal, Y.; Howari, F.; Iqbal, J.; Bou-Orm, N.; Xavier, C.M.; Bărbulescu, A.; Sharma, M.; Dumitriu, C.S. Contamination assessment of heavy metals in agricultural soil, in the Liwa area (UAE). *Toxics* **2021**, *9*, 53. [CrossRef] [PubMed]
35. Bărbulescu, A.; Dumitriu, C.Ș. Assessing the water quality by statistical methods. *Water* **2021**, *13*, 1026. [CrossRef]
36. Bărbulescu, A.; Barbeș, L.; Dumitriu, C.S. Assessing the water pollution of Brahmaputra River using water quality indexes. *Toxics* **2021**, *9*, 297. [CrossRef] [PubMed]
37. Bărbulescu, A.; Dumitriu, C.S.; Ilie, I.; Barbes, S.B. Influence of Anomalies on the Models for Nitrogen Oxides. *Atmosphere* **2022**, *13*, 558. [CrossRef]
38. Nazzal, Y.H.; Bărbulescu, A.; Howari, F.; Al-Taani, A.A.; Iqbal, J.; Xavier, C.M.; Sharma, M.; Dumitriu, C.S. Assessment of metals concentrations in soils of Abu Dhabi Emirate using pollution indices and multivariate statistics. *Toxics* **2021**, *9*, 95. [CrossRef] [PubMed]
39. Aldababseh, A.; Temimi, M. Analysis of the long-term variability of poor visibility events in the UAE and the link with the climate dynamics. *Atmosphere* **2017**, *8*, 242. [CrossRef]
40. Nazzal, Y.; Bărbulescu, A.; Howari, F.M.; Yousef, A.; Al-Taani, A.A.; Al Aydaroos, F.; Naseem, M. New insight to dust storm from historical records, UAE. *Arab. J. Geosci.* **2019**, *12*, 396. [CrossRef]

41. Bărbulescu, A.; Nazzal, Y. Statistical analysis of the dust storms in the United Arab Emirates. *Atmos. Res.* **2020**, *231*, 104669. [CrossRef]
42. Lorenz, E.N. *Empirical Orthogonal Functions and Statistical Weather Prediction*; Science Report No. 1; MIT: Cambridge, MA, USA, 1956; 49p.
43. Fiore, A.M.; Jacob, D.J.; Mathur, R.; Martin, R.V. Application of empirical orthogonal functions to evaluate ozone simulations with regional and global models. *J. Geophys. Res. Atmos.* **2009**, *108*, 4431. [CrossRef]
44. Martin, R.V.; Jacob, D.J.; Logan, J.A.; Ziemke, J.M.; Washington, R. Detection of a lightning influence on tropical tropospheric ozone. *Geophys. Res. Lett.* **2000**, *27*, 1639–1842. [CrossRef]
45. Abdi, H.; Williams, L. Principal Component Analysis. *Wires Comput. Stat.* **2010**, *2*, 433–458. [CrossRef]
46. Zou, H.; Hastie, T.; Tibshirani, R. Sparse principal component analysis. *J. Comput. Graph. Stat.* **2006**, *15*, 265–286. [CrossRef]
47. Abid, A.; Zhang, M.J.; Bagaria, V.K.; Zou, J. Exploring patterns enriched in a dataset with contrastive principal component analysis. *Nat. Commun.* **2018**, *9*, 2134. [CrossRef] [PubMed]
48. Trendafilov, N.T. Stepwise estimation of common principal components. *Comput. Stat. Data Anal.* **2010**, *54*, 3446–3457. [CrossRef]
49. Kutzbach, J. Empirical eigenvectors in sea-level pressure, surface temperature, and precipitation complexes over North America. *J. Appl. Meteorol.* **1967**, *6*, 791–802. [CrossRef]
50. Cattell, R.B. The Scree test for the number of factors. *Multivar. Behav. Res.* **1966**, *1*, 245–276. [CrossRef]
51. Kaiser, H.F. The Application of Electronic Computers to Factor Analysis. *Educ. Psychol. Meas.* **1960**, *20*, 141–151. [CrossRef]
52. Jolliffe, I. *Principal Component Analysis*, 2nd ed.; Springer: Berlin/Heidelberg, Germany, 2002.
53. Rousseeuw, P.J. Silhouettes: A graphical aid to the interpretation and validation of cluster analysis. *Comput. Appl. Math.* **1987**, *20*, 53–65. [CrossRef]
54. Charrad, M.; Ghazzali, N.; Boiteau, V.; Niknafs, A. Package 'NbClust'. Available online: https://cran.r-project.org/web/packages/NbClust/NbClust.pdf (accessed on 25 September 2020).
55. Kwiatkowski, D.; Phillips, P.C.B.; Schmidt, P.; Shin, Y. Testing the null hypothesis of stationarity against the alternative of a unit root. *J. Econometr.* **1992**, *54*, 159–178. [CrossRef]
56. Yoon, J.; von Hoyningen-Huene, W.; Kokhanovsky, A.A.; Vountas, M.; Burrows, J.P. Trend analysis of aerosol optical thickness and Ångström exponent derived from the global AERONET spectral observations. *Atmos. Meas. Tech.* **2012**, *5*, 1271–1299. [CrossRef]
57. Karagulian, F.; Temimi, M.; Ghebreyesus, D.; Weston, M.; Kondapalli, N.K.; Valappil, V.K.; Aldababesh, A.; Lyapustin, A.; Chaouch, N.; Al Hammadi, F.; et al. Analysis of a severe dust storm and its impact on air quality conditions using WRF-Chem modeling, satellite imagery, and ground observations. *Air Qual. Atmos. Health* **2019**, *12*, 453–470. [CrossRef]
58. Nazzal, Y.; Orm, N.B.; Bărbulescu, A.; Howari, F.; Sharma, M.; Badawi, A.E.; Al-Taani, A.A.; Iqbal, J.; El Ktaibi, F.; Xavier, C.M.; et al. Study of atmospheric pollution and health risk assessment. A case study for the Sharjah and Ajman Emirates (UAE). *Atmosphere* **2021**, *12*, 1442. [CrossRef]

Article

Influence of Anomalies on the Models for Nitrogen Oxides and Ozone Series

Alina Bărbulescu [1], Cristian Stefan Dumitriu [2,*], Iulia Ilie [3] and Sebastian-Barbu Barbeş [4]

[1] Department of Civil Engineering, Transilvania University of Brașov, 5 Turnului Str., 900152 Brașov, Romania; alina.barbulescu@unitbv.ro

[2] SC Utilnavorep SA, 55, Aurel Vlaicu Av., 900055 Constanta, Romania

[3] Siemens Romania, 13A, Bd. Garii, 900055 Brașov, Romania; iulia.ilie1988@gmail.com

[4] Doctoral School of Civil Engineering, Technical University of Civil Engineering, 122-124 Lacul Tei Bvd., 020396 Bucharest, Romania; sebastian-barbu.barbes@phd.utcb.ro

* Correspondence: cris.dum.stef@gmail.com

Abstract: Nowadays, observing, recording, and modeling the dynamics of atmospheric pollutants represent actual study areas given the effects of pollution on the population and ecosystems. The existence of aberrant values may influence reports on air quality when they are based on average values over a period. This may also influence the quality of models, which are further used in forecasting. Therefore, correct data collection and analysis is necessary before modeling. This study aimed to detect aberrant values in a nitrogen oxide concentration series recorded in the interval 1 January–8 June 2016 in Timisoara, Romania, and retrieved from the official reports of the National Network for Monitoring the Air Quality, Romania. Four methods were utilized, including the interquartile range (IQR), isolation forest, local outlier factor (LOF) methods, and the generalized extreme studentized deviate (GESD) test. Autoregressive integrated moving average (ARIMA), Generalized Regression Neural Networks (GRNN), and hybrid ARIMA-GRNN models were built for the series before and after the removal of aberrant values. The results show that the first approach provided a good model (from a statistical viewpoint) for the series after the anomalies removal. The best model was obtained by the hybrid ARIMA-GRNN. For example, for the raw NO_2 series, the ARIMA model was not statistically validated, whereas, for the series without outliers, the ARIMA(1,1,1) was validated. The GRNN model for the raw series was able to learn the data well: $R^2 = 76.135\%$, the correlation between the actual and predicted values (r_{ap}) was 0.8778, the mean standard errors (MSE) = 0.177, the mean absolute error MAE = 0.2839, and the mean absolute percentage error MAPE = 9.9786. Still, on the test set, the results were worse: MSE = 1.5101, MAE = 0.8175, $r_{ap} = 0.4482$. For the series without outliers, the model was able to learn the data in the training set better than for the raw series ($R^2 = 0.996$), whereas, on the test set, the results were not very good ($R^2 = 0.473$). The performances of the hybrid ARIMA–GRNN on the initial series were not satisfactory on the test (the pattern of the computed values was almost linear) but were very good on the series without outliers (the correlation between the predicted values on the test set was very close to 1). The same was true for the models built for O_3.

Keywords: aberrant values; nitrogen oxides; ARIMA; GRNN; ARIMA–GRNN; isolation forest; LOF

Citation: Bărbulescu, A.; Dumitriu, C.S.; Ilie, I.; Barbeş, S.-B. Influence of Anomalies on the Models for Nitrogen Oxides and Ozone Series. *Atmosphere* **2022**, *13*, 558. https://doi.org/10.3390/atmos13040558

Academic Editor: Kenichi Tonokura

Received: 2 March 2022
Accepted: 28 March 2022
Published: 30 March 2022

Publisher's Note: MDPI stays neutral with regard to jurisdictional claims in published maps and institutional affiliations.

Copyright: © 2022 by the authors. Licensee MDPI, Basel, Switzerland. This article is an open access article distributed under the terms and conditions of the Creative Commons Attribution (CC BY) license (https://creativecommons.org/licenses/by/4.0/).

1. Introduction

Nowadays, ambient air pollution levels and trends have become a topic of interest worldwide because primary atmospheric pollutants (APPs) constitute a risk factor for the population and ecosystems [1–4]. Therefore, monitoring air quality, especially in urban or crowded areas, is essential for controlling pollution [5] and protecting human health.

Pollutants' dispersion into the atmosphere is a hazardous phenomenon, which is difficult to assess and sometimes unpredictable. Their diffusion depends on meteorological factors, such as the relative speed and wind direction, ambient temperature, atmospheric turbulence, and buoyant force [6,7]. The distinct mechanisms responsible for pollutant dispersion are molecular diffusion, turbulent diffusion, and transport due to wind. Generally, wind speed influences pollutants' distribution. High concentrations of pollutants reach the atmospheric layer and remain there if the wind speed is low and uniform. Atmospheric calm creates favorable conditions for the accumulation of pollutants in the source's vicinity [8].

Nitrogen oxides (NO_x) are gases containing various amounts of nitrogen and oxygen with high reactivity. NOx represents a family of seven chemical compounds (N_2O, NO, N_2O_2, N_2O_3, NO_2, N_2O_4, N_2O_5) [9] Nitrogen monoxide and dioxide (NO and NO_2) are the main NO_x found in the atmosphere, resulting from combustion processes (from electricity generation, industrial activities, and engine exhaust). They contribute to the apparition of acid rains and favor the accumulation of nitrates in the soil, leading to ecological disequilibrium [10]. Nitrogen oxides contribute to the greenhouse effect and smog formation, reducing the visibility in urban areas and the deterioration of water quality.

Nitrogen oxide (NO) is a colorless gas and a free radical. It is important that it is monitored s it is a precursor of tropospheric ozone, nitric acid, and particulate nitrate. Although NO does not directly affect acid deposition or the climate, nitric acid and ozone and particulate nitrate do. Natural NO reduces ozone in the upper stratosphere. NO emissions from jets that fly in the stratosphere also reduce stratospheric ozone. In urban zones, NO mixing ratios reach 0.1 ppmv in the early morning but may decrease to zero by midmorning due to the reaction with ozone. Outdoor levels of NO are not regulated in any country [11].

Nitrogen dioxide (NO_2) is a brown gas with a strong odor. NO_2 is an intermediary between NO emission and ozone (O_3) formation. It is also a precursor to nitric acid, a component of acid deposition. Natural NO_2, such as natural NO, reduces O_3 in the upper stratosphere. The primary source of NO_2 is NO oxidation. Minor sources are fossil fuel combustion and biomass burning. During combustion or burning, NO_2 emissions are about 5% to 15% of those of NO. In urban regions, NO_2 mixing ratios range from 0.1 to 0.25 ppmv. Outdoors, NO_2 is more relevant during the early morning than during midday or afternoon because sunlight breaks down most NO_2 past midmorning, which is usually the opposite to ozone [12].

NO's toxicity is four times lower than that of NO_2. Children are the most affected by exposure to nitrogen dioxide. NO_2 is very toxic for the population and animals [10,13]. Exposure to low concentrations of NO_2 affects lung tissue, and high pollutant concentrations may be fatal. The population exposed to low concentrations of nitrogen oxides may experience respiratory issues for a long time [2,4].

Therefore, outdoor levels of NO_2 are now regulated in many countries, including Romania [12,14,15]. Ozone is a relatively colorless gas at typical mixing ratios. O_3 exhibits an odor when its mixing ratio exceeds 0.02 ppmv. In urban smog, it is considered an air pollutant because of its harmful effects on humans, animals, plants, and materials. In the stratosphere, ozone's absorption of UV radiation provides a protective shield for terrestrial life. O_3 is not emitted. Its only source in the air is chemical reaction. O_3 is a pollutant produced in the atmosphere, and therefore it is not necessarily related to urban or industrial areas and may be seen in suburban or rural areas, in downwind zones from where the precursors are emitted. In urban air, ozone mixing ratios range from less than 0.01 ppmv at night to 0.5 ppmv (during the afternoon, downwind from the most polluted cities worldwide), with typical values of 0.15 ppmv during moderately polluted afternoons. It has a typical daily cycle characteristic of the positions with respect to the topography and the location where the precursors are emitted. Peak ozone mixing ratios are around 10 ppmv in the stratosphere [11].

In the last decade, special attention has been paid to mathematical modeling, the study of the pollutants diffusion from the atmosphere, developing new control systems, and reducing environmental pollution [16,17]. The diversity of actual models has imposed extraordinary rigor on their understanding and expanded their types for correct application depending on local or regional air pollution particularities. The transport and dispersion of pollutants in the atmosphere are complex phenomena that are not easy to translate into mathematical calculation systems, so many algorithms are accepted by simplifying hypotheses [18]. Under these conditions, the results of the estimates are more or less close to reality. Each model has its limits. The volume, type of input data, and mathematical complexity largely depend on the researchers' abilities because the data quality, accuracy, and discretization affect the integrity of the simulation results [19].

Modeling of the dissipation of NO_x from different sources has been achieved using different models, such as, for example, CALPUFF [20] (dispersion of traffic emissions in urban zones). Fallah-Shorshani et al. [21] used two air quality models to simulate local atmospheric dissipation of NO_x and its transformation to NO_2 using the Gaussian puff (CALPUFF) and street-canyon model (SIRANE). The SIRANE model is based on transformations involving NO, NO_2, and O_3 (in the Leighton cycle). Shekarrizfard et al. [22] reported CALMET-CALPUFF for the assessment of the effects of a regional transit policy on air quality and population exposure. Soulhac et al. [23] utilized the SIRANE dispersion model to assess the transfer of pollutants within and out of an urban canopy.

Stochastic models are statistical or semi-empirical techniques for estimating trends, periodicity, and the interrelationship between air quality and atmospheric measurements, and forecasting air pollution episodes. These models are instrumental in real-time forecasting or relatively short periods, where available information from measurements is relevant (immediate estimates) [24]. The most well-known model is the Box–Jenkins approach (for example, ARIMA and SARIMA).

Gocheva-Ilieva et al. [17] examined the concentrations of NO, NO_2, NO_x, and ground-level O_3 in a town in Bulgaria for one year using hourly data. The obtained SARIMA models demonstrated a very good fitting performance and short-term predictions for the next 72 h.

Kumar and Jain [25] used ARIMA, after a suitable variance stabilizing transformation of the concentration time series (O_3, CO, NO, and NO_2), to model data collected at a traffic station in Delhi (India). Zhu [26] compared the ARIMA and exponential smoothing models on 2014 concentrations of NO_2 and O_3 in the Yanqing county, Beijing, China. Munir and Mayfield [27] used auto-regressive integrated moving average with exogenous variables (ARIMAX) to model the distributions and temporal variability of NO_2 concentrations in Sheffield, UK, from August 2019 to September 2020. Using cross-validation ARIMAX, the authors found a strong correlation between the predicted values and the measured concentrations (the correlation coefficient was 0.84 and RMSE was 9.90). Hajmohammadi and Heydecker [28] developed a vector autoregressive moving average model to assess the air quality in London in 2017. The authors cross-validated the model using kriging to achieve spatial interpolation of NO, NO_2, and NO_x, respectively. Moreover, seasonal ARMA models of the air quality across London for 30 individual stations were validated. This study established that the VARMA model is appropriate for evaluating interventions, such as the Ultra-Low Emissions Zone.

Artificial neural networks (ANNs) have been widely used for modeling processes that present high variability and nonlinearities, such as those related to air pollution. Gardner and Dorling [29] employed a multilayer perceptron (MLP) artificial network to model NO and NO_2 concentrations in London and showed that the variation in emissions could be modeled using the time of day and day of the week as input variables.

Based on the literature findings, and\ given the superior performances of deterministic methods, Rahimi [30] utilized ANN to develop a model that provided accurate short-term (hourly) predictions of NO_x and NO_2 series in Tabriz, Iran. Dragomir et al. [31] presented an evaluation of the efficiency of artificial neural networks (ANNs) and the multiple linear regression (MLR) model for NO_2 prediction in 3 scenarios (by randomly eliminating (1) 25%, (2) 50%, or (3) 75% of the observed NO_2 data) in Brăila city, Romania, from 2009–2013. The analysis results demonstrated that the NO_2 values estimated using MLR and ANNs were similar to the measured NO_2 concentrations (the corresponding coefficients were (1) 0.580, 0.604; (2) 0.589, 0.565; and (3) 0.474, 0.483). The best outcomes were achieved for the ANN values in all scenarios.

Multilayer perceptron is a type of neural network used in the studies of Baawain and Al-Serihi [32], Jiang et al. [33], and Hrsut et al. [34] to model NO, NO_2, NO_X, O_3 [32], NO_2 [33], NO_x, and O_3 [34] in an industrial port, Shanghai, and a site in an urban residential area in Zagreb, Croatia, respectively. Moustris et al. [35] provided a 3-day forecast for the NO_2 and O_3 series in Athens using an MLP network. Agirre-Basurko et al. [36] compared the performances of MLP and linear regression approaches on O_3 and NO_2 series and Kukkonen et al. [37] on NO_2 series.

Another approach that has provided good results in predicting NO_x and NO_2 series is based on support vector regression and was utilized by Wang et al. [38] and Osowski and Garanty [39]. The last two authors also proposed a discrete wavelet decomposition for the data series.

Different scientists have searched for the best model for series forecasting. For example, Hajek and Olej [40] used SVR, TSFIS, and MLP for NO_2, NO_X, and O_3 prediction. Lin et al. [41] compared the ability of GRNN, SVR, MLP, and SARIMA to forecast NO_2 and NO_x concentrations. Singh et al. [42] utilized linear regression, MLP, GRNN, and RBF neural networks for NO_2 prediction in an urban area.

With the same idea, Liu et al. [43] presented a combined prediction model of the NO_2 concentration in Tianjin, China. The authors reported the results obtained using the discrete wavelet decomposition and neural network method. They concluded that when utilizing a series of pollutant concentrations with different frequencies, it is possible to describe the data characteristics better. A high-dimensional nonlinear learning algorithm was produced when the prediction model was built using an LSTM neural network, but the overall prediction accuracy was the highest. The best forecast of the NO_2 concentrations was obtained using the DWT-LSTM neural network method. Wang et al. [44] presented a hybrid approach consisting of the NOx emission prediction model based on CEEMDAN and AM-LSTM.

In a study examining population exposure to traffic-related NO_x air pollution, Shekarrizfard et al. [45] showed that improving the estimation of pollutant exposure is essential for estimating the effects of pollution.

Regardless of the chosen model type, it can only be used when the pollutant concentrations are known. Otherwise, an emissions inventory is helpful.

The National Inventory of Greenhouse Gas Emissions under the United Nations Framework Convention on Climate Change presents the levels of emissions/sequestration of greenhouse gases. They are structured according to the categories of activities and pollutants. The emissions represent aggregate annual values of the contribution of a particular type of source of a specific contaminant. The National Inventory of Air Pollutant Emissions reported to the Convention on Long-Range Transboundary Air Pollution Secretariat rearranges the data by national environmental principles. Finally, the conversion of data from national emission inventories is performed based on the national classification of economic activities, creating a relationship between environmental variables (emission level) and economic variables (value-added, turnover, etc.) according to the National Institute of Statistics methodology on account of air pollutant emissions (MAAPE-Air) [46].

In Romania, the National Air Quality Monitoring Network (NAQMN) [15] has 41 centers where data is collected from recording stations. After preliminary validation, data is transmitted for certification to the Air Quality Assessment Center of the National Agency for Environmental Protection. In Romania, Law no. 104/2011 [47] regulates the rules that ensure ambient air quality. Based on the air quality assessment, the number, type, and location of the fixed measurement points and assessed pollutants are determined. The agglomerations are classified into three classes (A, B, or C) based on the results of the national air quality assessment using measurements at fixed locations taken at the measuring stations of the Network of the National Air Quality Monitoring Authority, and the results obtained from the mathematical modeling of the dispersion of pollutants emitted into the air. The pollutants taken into account are sulfur dioxide, nitrogen dioxide, nitrogen oxides, particulate matter, lead, benzene, carbon monoxide, ozone, arsenic, cadmium, mercury, nickel, and benzo [15].

The specific air quality index, in short, "specific index", is a system used for coding the recorded concentrations for each of the monitored pollutants (SO_2, NO_2, O_3, PM2.5, and PM10) and is established for each of the automatic stations within the National Air Quality Monitoring Network as being the highest of the specific indices corresponding to the monitored pollutants. The general index and specific indices are represented by integers between 1 and 6, with each number corresponding to a color (1—good—turquoise, 2—acceptable—green, 3—moderate—yellow, 4—bad—red, 5—very bad—burgundy, 6—extremely bad—violet). The specific indices and the general index of the station are updated hourly [48]. For example, Figure 1 shows a recent map of the air quality in Romania.

Figure 1. Map of the air quality in Romania (updated 22 March 8:20:00) (retrieved from https://www.calitateaer.ro/public/home-page/?__locale=ro (accessed on 10 March 2022).

The critical concentration levels established by Romanian law [47] for NO_X/NO_2 is as follows: 400 µg/m^3—alert threshold; 200 µg/m^3 NO_2—hourly limit value for human health protection; 40 µg/m^3 NO_2—the annual limit value for the protection of human health; and 30 µg/m^3 NO_x—annual critical level for vegetation protection.

The results of studies have shown that the average number of days on which there is good air quality in big cities in Romania (Bucharest [49], Timisoara [50–52], Cluj-Napoca [53], Constanta, and the surrounding area [54,55], etc.) has decreased year by year.

Since NO_2 pollution in different European cities remains high (>40 µg/m^3 is the maximum accepted annual mean concentration) and given its harmful effects on population health [14,46], continuous monitoring is required.

Understanding the existence of anomalies existence is becoming an important topic in the investigation of air quality. Anomalies are values in a data series that are unusual or dissimilar from the remaining data. They may be irregular items resulting from unusual or unexpected events, indicating abnormal behavior [56,57]. The analysis of anomalies is necessary for the detection of the source of their occurrence [57]. Hawkins et al. [58] stated that the values of series collected in polluted areas can behave as anomalies (outliers).

Despite the importance of the detection of outliers in atmospheric sciences, only a few articles, especially in the last years, have investigated this aspect and proposed new approaches for the better selection of such values [56–60].

In the above context, this study aimed to identify the anomalies in a nitrogen oxide series in Timisoara, one of Romania's most prosperous industrial cities. The motivations for this study are as follows:

1. Only a few studies have been devoted to studying the existence of outliers in a pollutant series, with none of them using data collected in Romania.
2. Only a few articles have used hybrid approaches to model pollutant series, with most of them being based on atmospheric circulation models, not on the Box–Jenkins artificial neural network approach.
3. Very few studies have attempted to improve the quality of models after the removal of aberrant values from the time series.

Therefore, three models are proposed for a raw series including nitrogen oxides and ozone, and the series after the removal of outliers. Their performances are compared to determine the influence of the aberrant values on the models' quality.

2. Materials and Methods

2.1. Data

The geographical area of this study is Timiş county, located in the southwest Romania plain (Figure 2). The most important city in this county is Timișoara, situated at 45°44′ northern latitude and 21°13′ eastern longitude. It is one of the most prosperous economic and university cities. After 1990, transport, especially by cars, recorded an accelerated increase (reaching 1 car for every 2.66 inhabitants in 2017).

Figure 2. (**a**) Timișoara city (with the air monitoring stations, TM-1, TM-2, TM-4, and TM-5); (**b**) Map of Romania (http://www.destination360.com/europe/romania/map (accessed on 20 March 2022)).

Therefore, the pollution produced by this sector has proportionally increased.

The climate is moderate continental, with winds blowing from west and north-west, and an annual precipitation of 650 L/m^2. The atmospheric circulation favors the accumulation of pollutants emitted in industrial zones and car exhaust above the city.

Data (NO, NO_2, and NO_x and O_3 concentrations) recorded at the monitoring station TM2 (C. D. Loga Blvd.—45°45′16.88″ N; 21°14′05.91″ E, 92 m altitude) were downloaded daily from the NAQMN website [15] during the period 1 January–8 June 2016. They formed complete sets (Figure 3) without gaps. It is noted that the highest values were recorded for the NO_x series during the period March-April 2016 and for NO in the second half of May. The NO series exhibited the lowest variability. The existence of periods when the NO_x concentrations were much higher than the sum of the NO and NO_2 concentration is also noted, given that apart from NO and NO_2, NO_x incorporates other nitrogen oxide species that can accumulate in the atmosphere in periods of calm before participating in chemical reactions.

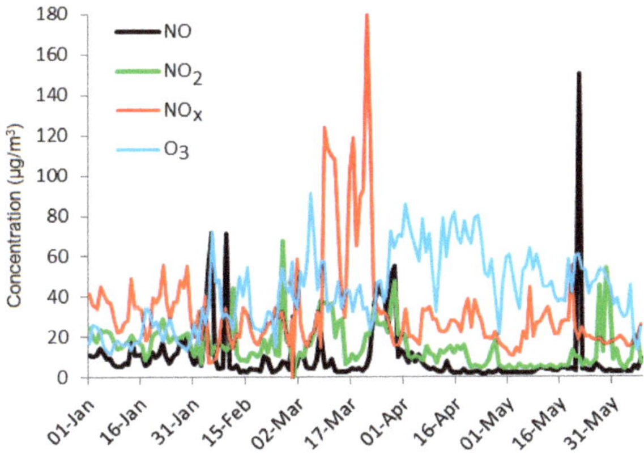

Figure 3. The pollutants series: NO, NO_2, NO_x, and O_3.

An example of the hourly air quality at the studied station during the period 1–21 March 2021 is presented in Figure 4a and the average annual concentration of NO_2 in Timisoara during the period 2000–2019 is presented in Figure 4b.

2.2. Methodology

2.2.1. Statistical Analysis

The hourly data were processed to build the average data series, which was studied. The statistical analysis consisted of normality, homoskedasticity, autocorrelation, and stationarity tests, using the Shapiro–Wilk and Fligner–Killeen test, Levene test, autocorrelation function, and KPSS test, respectively. The Pettitt test was used to address the existence of a change point (in mean) [3].

Anomaly (aberrant) detection is used in many domains, such as manufacturing error detection, attack detection in cybersecurity, stroke recognition in EEG measurement, etc.

Anomalies are observations that deviate significantly from the expected behavior and cannot be categorized as noise or measurement error, and thus cannot be easily discarded [61]. In the case of anomalies, the unexpected event might be the study object.

Fox et al. [61] define two types of anomalies: type I, affecting a single instance; and type II, where the anomalous behavior extends in time.

Anomaly detection can be studied in both the univariate and multivariate time domains, with the latter possibly implying multiple dimensions that display anomalies simultaneously or even waterfall effects. Here, we focused on the univariate case.

Most techniques used for anomaly detection in time series consider the time aspect, either in the vicinity or globally, using the entire data series to mark the anomalies. Four such methods were applied in this study [62]. One of the most popular, called the IQR method, considers values outside the interval (Q1 − 1.5 IQR, Q3 + 1.5 IQR) as anomalies (Q1 is the first quartile, Q3 is the third quartile, and IQR is the interquartile range). Sometimes, the term 1.5 is replaced by 3.

The second method employed in this study is isolation forest (IF) [63–65]. It relies on the concept of isolating unusual instances, as opposed to determining the properties of normal samples and then examining non-matching patterns. It achieves anomaly detection by building isolation trees (ITs), where anomalies are often represented as existing closer to the root of the IT, rather than higher at the leaves, where regular data points are found.

To build the trees, IF generates recursive partitioning of the dataset (Figure 5) by randomly selecting a dimension in the dataset, followed by a recursive split of the specific dimension anywhere between the minimum and maximum value of the remaining set.

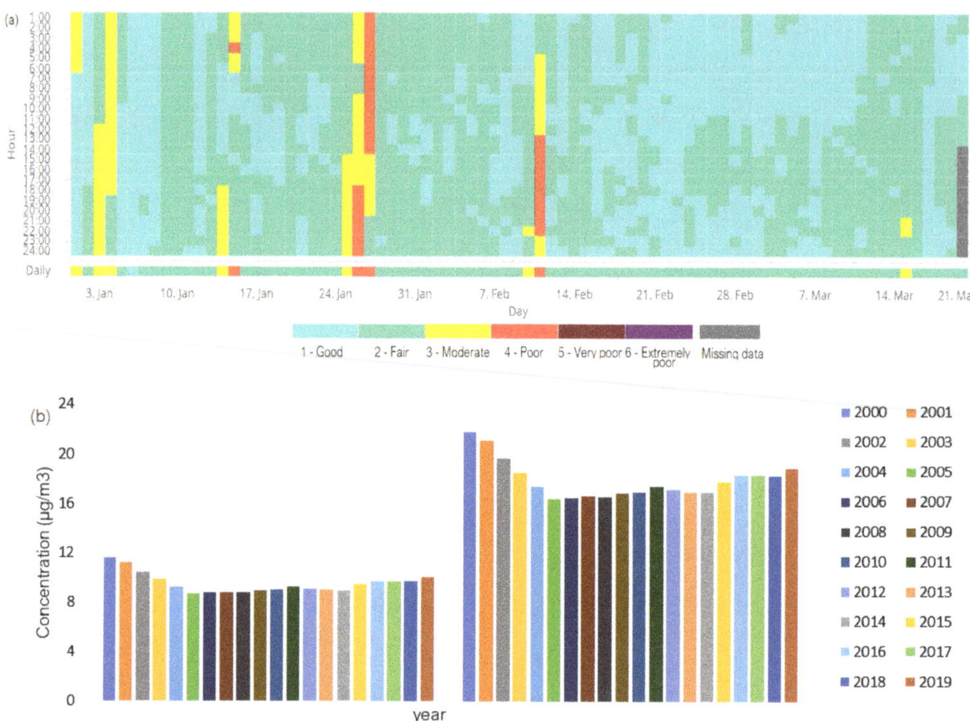

Figure 4. (**a**) Hourly air quality at the studied station during the period 1–21 March 2021. (**b**) Annual average concentration of NO_2.

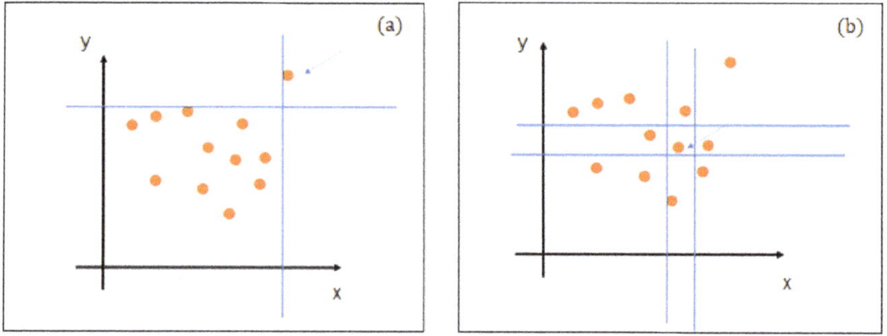

Figure 5. Recursive partitioning of the dataset. (**a**) shows much fewer splits needed to isolate an anomalous data point (indicated by arrow) compared to (**b**) where the data point indicated by arrow is normal.

A path length of a point x, $PL(x)$, is computed as the number of edges x that traverse an isolation tree from the root node until the traversal is terminated at an external node.

Computing the path length means to count the number of partitioning steps required to isolate a data point. The lower the path length or tree height value, the higher the probability of a specific instance being an anomaly.

The average path lengths for instances are then used to evaluate the probabilities of data points showing anomalous behavior.

The application of IF for anomaly detection has two main steps:

1. Building and training the isolation trees.
2. Assigning anomaly scores to data points based on PL by computing the tree height length as binary search trees.

   The anomaly score s of an instance x is defined as:

$$s(x, n) = 2^{-E(L(x))/c(n)}, \tag{1}$$

where $E(L(x))$ is the average of $L(x)$ from a collection of isolation trees, and $c(n)$ is the average of $L(n)$ given n instances.

3. Using the anomaly scores, the following decision is made:
   (a) If instances have an s value that is much smaller than 0.5, then they are considered normal instances;
   (b) If all the instances have s $\approx$ 0.5, then the entire sample does not have any distinct anomaly;
   (c) Instances with an s value larger than 0.5 are marked as anomalies [63].

While IQR and IF detect global outliers, LOF mainly identifies local outliers [42]. The decision regarding whether an outlier is local is made based on an evaluation of the associated probability, determined by the k-nearest neighbors (kNN) method [66].

To determine if a point p in a study set is an outlier, the following operations are performed in LOF [67] for p: (a) computation of the k-distance; (b) computation of the kNN; (c) calculation of the local reachability density; and (d) detection of the LOF score. Point p is classified as an outlier by comparing the score with a given threshold.

The last method utilized to detect both types of anomalies—local and global—in the data series is the generalized extreme studentized deviate test (GESD) [68]. Its stages are as follows.

- Analyze the existence of periodicity in the data series;
- Divide the series into non-overlapping intervals I_w;
- For each interval:
  - Determine the seasonal compound (if it exists);
  - Compute the median;
  - Extract the residual, as the difference between the values of the series, the median, and the seasonal component;
  - Run the ESD algorithm (with the median and mean absolute error in the computation of the test statistics) [69].
- Return the outliers obtained from the previous stage.

The advantage of this technique is that it can be used even if the timestamps are unknown.

The correlation between the four series and the series anomalies, respectively, is addressed by computing the correlation coefficients. In the case of low correlations, models were built only for the individual series.

2.2.2. Modeling

This work emphasizes how aberrant values (anomalies) influence the quality of models built using raw series and after their removal. ARIMA, GRNN, and hybrid ARIMA-GRNN models were built for the raw series and the series obtained after removing the aberrant values.

A time process (X_t, $t \in \mathbf{Z}$) is stationary if it satisfies the following conditions:

- $\forall t \in Z,\ M(X_t^2) < +\infty$;
- $\forall t \in Z,\ M(X_t) < +\infty$ and is invariant in time (M denotes the expectation);
- $\forall t, h \in Z,\ Cov(X_t,\ X_{t+h}) = \gamma(h)$ (i.e., the covariance of X_t and X_{t+h} depends only on the lag h).

Let us denote the d-the order difference of X_t by $\Delta^d X_t$, where B is the backshift operator.

A time process (X_t; $t \in \mathbf{Z}$) is called an autoregressive integrated moving average process ARIMA(p,d,q) if:

$$\Phi(B)\Delta^d X_t = \Theta(B)\varepsilon_t, \tag{2}$$

where Φ and Θ are respectively polynomials of p and q orders with roots higher than 1, respectively, and (ε_t, $t \in \mathbf{Z}$) is white noise [70].

Among two valid models, the best one is selected based on the Akaike criteria. The lower the AIC value, the better the model is [70].

An ARIMA(p, q) process is a particular case of ARIMA, with d = 0.

Generally, a stationary process can be approximated by an ARMA(p, q) model.

The generalized regression neural network belongs to the group probabilistic neural networks. It is composed of four layers (Figure 6) [71].

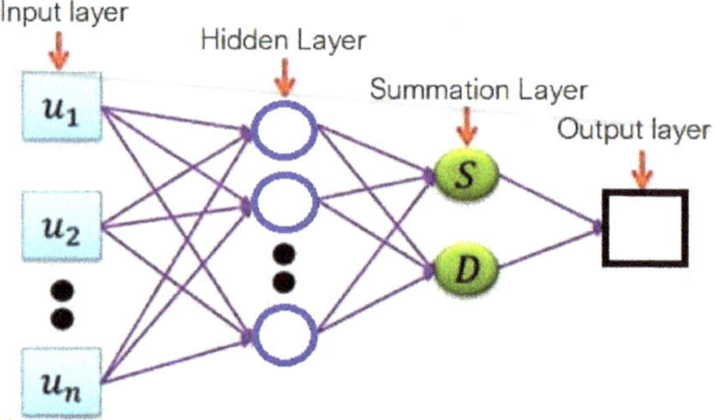

Figure 6. The structure of a GRNN.

The first one—input—contains the series values X = (x_1, ..., x_n). The second one—hidden—is composed of neurons that apply a kernel function to the distances between the training data and the prediction point. In this process, σ values are employed to compute the radius of influence. The best σ is determined when the network is trained to control the distributions of the kernel function. In this study, the Gaussian kernel was utilized, and the gradient algorithm was employed to estimate the best σ [71].

In this study, the interval 0.0001–10 was used to search for σ values in.

The number of neurons in the hidden layer after training is the same as the number of training samples involved in the modeling. The unnecessary neurons are removed based on the error minimization criterion during an optimization process [71,72].

The summation layer is composed of two neurons (D- and S-) that sum up the values collected from the previous layer. The only difference between them is that the D-summation neuron computes a weighted sum of the values resulting from the hidden layer [72].

The last layer (output) provides the ratios between the corresponding values from the D- and S- summation neurons.

To perform the modeling, the series was divided into a ratio training:test = 80:20, with the first part used for training, and the second part for testing. The number of iterations was fixed at 5000 (maximum) and 1000 (without improvement). The regressors were considered as lagged variables, with lags between 1 and 6. The algorithm was run with different regressors, and the best result was kept. The correlation between the actual and predicted values (r_{ap}), mean standard error (MSE), mean absolute error (MAE), mean absolute percentage error (MAPE), and R^2 were employed.

In the hybrid ARIMA–GRNN procedure, an ARIMA model was first built for the data series, and then the residual was modeled using GRNN. The same setting that was used in the GRNN algorithm for the data series was kept when running GRNN for the residuals in the ARIMA model.

The ability to capture nonlinearities, the use of nonparametric regression, and learning without backpropagation is recommended regarding GRNN to solve classification, regression, and forecast problems involving continuous variables [71,72]. These characteristics improve ARIMA's capabilities to model processes with phenomena with high linear dynamics.

Figure 7 shows a flowchart of the study.

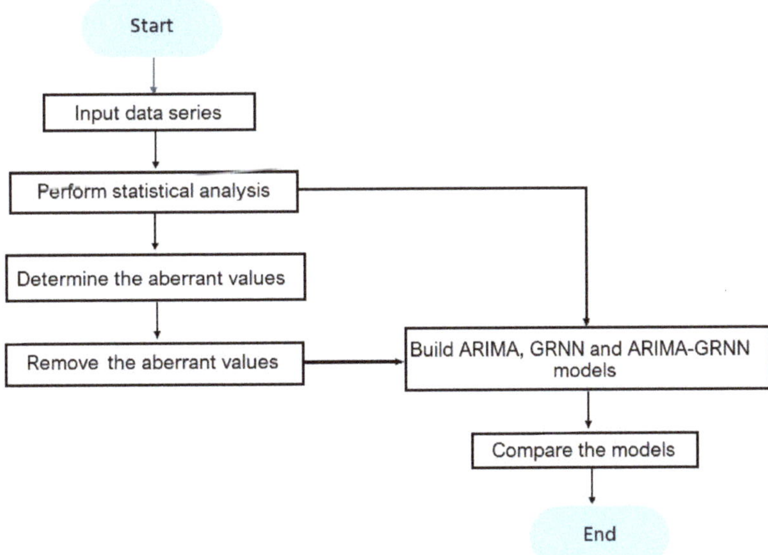

Figure 7. The flowchart of the study.

3. Results and Discussion

3.1. Results of the Statistical Analysis and the Anomaly Detection

The basic statistics of the average data series are presented in Table 1.

Table 1. Basic statistics of the pollutant series during the study period.

Statistics	NO_x	NO	NO_2	O_3
min ($\mu g/m^3$)	0.00	1.60	0.00	12.04
max ($\mu g/m^3$)	179.34	150.12	67.86	91.28
mean ($\mu g/m^3$)	32.63	9.87	15.67	42.72
stdev ($\mu g/m^3$)	24.81	16.27	10.53	18.71
cv	0.76	1.64	0.67	0.44
skew	3.00	5.28	1.78	0.33
kurt	10.82	37.00	4.64	−0.66

The NO and NO_x series display a very high range while the NO_2 and O_3 ranges are more than twofold lower compared to those of the first two series. The lowest average corresponds to NO. It is very small compared to the maximum, indicating that most series values are closer to the minimum than to the maximum. NOx showed low average values compared to the maximum for. All series had moderate standard deviations (stdev) and coefficient of variations, indicating a moderate dispersion of the data series around the average values. The series are right-skewed (skew >0), which is confirmed by the histograms shown in Figure 8. The kurtosis coefficient indicates leptokurtic distributions for all but the O_3 series (which is platykurtic).

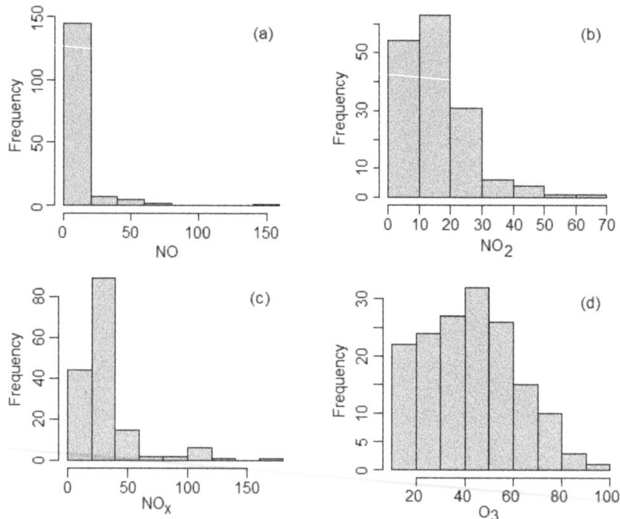

Figure 8. Histograms of the studied series: (**a**) NO, (**b**) NO_2, (**c**) NO_x, (**d**) O_3.

The normality and randomness hypotheses were rejected at the significance level of 5%. The homoscedasticity hypothesis was rejected for the NO_x series only (the *p*-value computed in the Levene test is 0.022). Figure 9 shows the presence of at least first-order autocorrelation for all the data series.

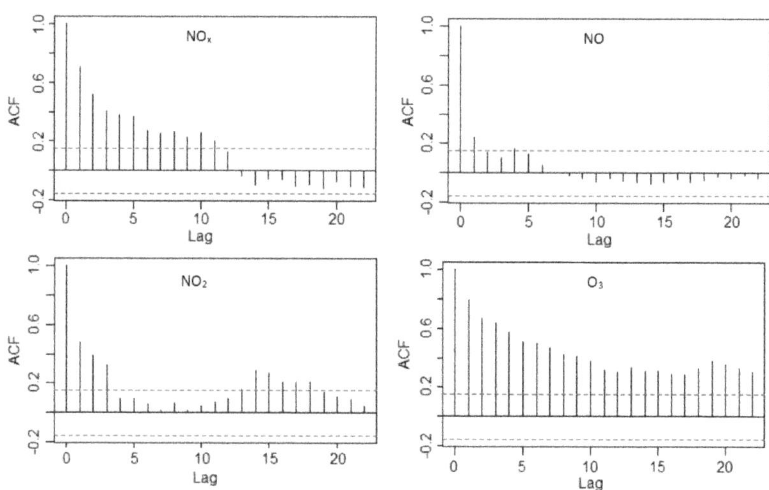

Figure 9. Charts of the autocorrelation functions (ACFs) for the data series. The blue lines represent the limits of the confidence intervals at a confidence level of 95%.

The KPSS test rejected hypothesis of level-stationarity for NO_2 and O_3, and trend-stationarity for NO_x and O_3.

After applying the change point test, the hypothesis that there is no change point could not be rejected for all the series. Two subseries were detected for each series. The change point and the subseries averages are presented as follows, where mean 1 is the average of the subseries containing the values before the change point, and mean 2 is the average of the subseries formed by the values after the change point:

- For NO: the change point is the 98th value, mean 1 = 12.611, mean 2 = 5.659;
- For NO_2: the change point is the 92nd value, mean 1 = 19.454, mean 2 = 10.544;
- For NO_X: the change point is the 87th value, mean 1 = 40.426, mean 2 = 23.348;
- For O_3: the change point is the 55th value, mean 1 = 25.554, mean 2 = 51.182.

So, the series presents high variability. The higher the variation is, the more difficult it is to find a good model.

The IQR method with a factor of 1.5 (and 3) detected the values situated outside the following intervals as outliers:

- $[-7.5305, 20.65750]$ and $[-18.101, 31.228]$ for NO;
- $[-10.1676, 39.0445]$ and $[-28.622, 57.499]$ for NO_2;
- $[-3.825, 57.975]$ and $[-27, 81.15]$ for NO_3;
- $[-14.195, 97.205]$ and $[-55.97, 148.98]$ for O_3.

This study was performed in the first case because the use of three reduces the domain of the anomalies. Therefore, based on this criterion, values recorded on the following days were outliers:

- 4, 5, and 9 February; 23–29 March; and 21 May for NO;
- 11 and 25 February; 7–11 and 23, 28, and 29 March; and 27, 29, and 30 May for NO_2;
- 1, 9–13, 16, 17, and 19–22 March; and 7 May for NO_x;
- 6, 7, 13, and 29 January; 5 February; 5 and 28 March; 1, 2, 6, 8, 12, 14, 15, 18, 21, and 22 April; and 7 June for O_3.

The NO, NO_2, and NO_x series, with the anomalies determined by IF, are presented in Figure 10. The aberrant values are mostly very high, especially for NO and NO_x.

IF provided more anomalies in comparison to IQR, but most of the aberrant values detected by the IQR method were also identified by IF. The aberrant values identified by IF included the values recorded on the following days:

- 1–10, 17, 18, and 22 January; 2, 3, 11, 25, 28, and 29 February; 7–11 and 23, 28, and 29 March; 27 April; 19 and 27–30 May; and 1, 3, and 6–8 June for NO;
- 11 and 25 February; 23 March; 27, 29, and 30 May; and 1, 6, and 9 June for NO_2;
- 1–5, 9, 13, 17, 22, and 29 January; 4–6, and 29 February; 1, 9–13, and 16–22 March; 7 and 19 May; and 4–8 June for NO_x;
- 1–7, 9, 13, 17, 18, 28, and 29 January; 1, 5–7, 13, 15, and 23 February; 5, 6, 22, and 28 March; 1, 2, 6, 15, 18, 21, 22, and 28 April; 4, 30, and 31 May; and 1–8 June for O_3.

Given the common origin of nitrogen oxides and the chemical reactions that occur when O_3 is present, as explained in the introduction, the correlations between the concentrations of the studied pollutants were investigated. Figure 11 presents (a) the correlations between the NO, NO_2, NO_x, and O_3 series and (b) the correlations between the series of anomalies detected by IF. While no significant correlations between the pollutant series were detected (the correlation coefficients range from −0.18 to 0.22), the highest correlations were identified between the O_3 anomalies and NO_x anomalies (NO_2 and NO anomalies, respectively), with a value of 0.51 (0.43 and 0.33, respectively). Still, these values do not show a strong correlation between the aberrant series.

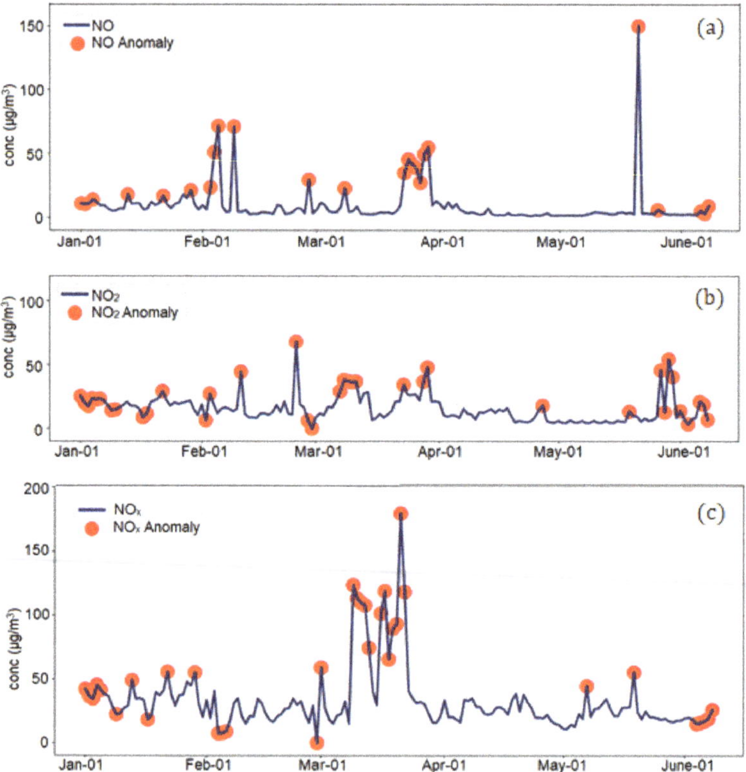

Figure 10. Series charts and the anomalies computed by the isolation forest for (**a**) NO, (**b**) NO_2, and (**c**) NO_x series.

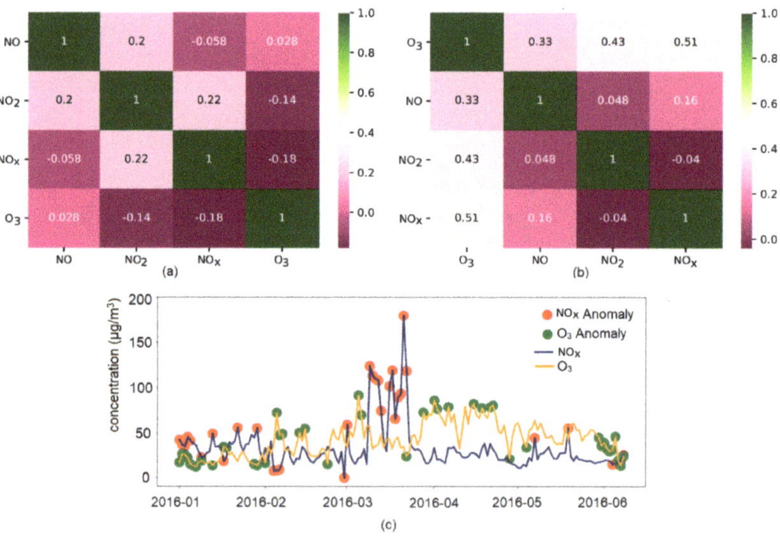

	NO	NO2	NOx	O3
NO	1	0.2	-0.058	0.028
NO2	0.2	1	0.22	-0.14
NOx	-0.058	0.22	1	-0.18
O3	0.028	-0.14	-0.18	1

	O3	NO	NO2	NOx
O3	1	0.33	0.43	0.51
NO	0.33	1	0.048	0.16
NO2	0.43	0.048	1	-0.04
NOx	0.51	0.16	-0.04	1

Figure 11. (**a**) Series correlations; (**b**) Correlations of the anomalies detected by IF; (**c**) NO_x and O_3 series and their anomalies.

Figure 11c depicts the NO_x and O_3 series, and their anomalies.

Figure 12 displays the series with the highlighted anomalies determined by LOF. Notice that the IF approach provided a higher number of anomalies than LOF. This result is due to the LOF algorithm only considering neighboring values rather than the entire series. Five common anomalies are provided by IF and LOF for NO, NO_x, and O_3, and seven for NO_2. The correlation between the series anomalies is close to zero. Figure 13 shows the anomalies detected by GESD. This algorithm did not find any anomalies in the O_3 series, 3 for NO_2 (25 February, 29 March, and 29 May), and 11 for NO_x (9–13 and 16–22 March). The outliers detected by this algorithm and IQR for NO are the same. Since no significant correlation between the data series was found, we did not search for a regression model, linking different variables. The next section contains the results of modeling the data series before and after the removal of the anomalies.

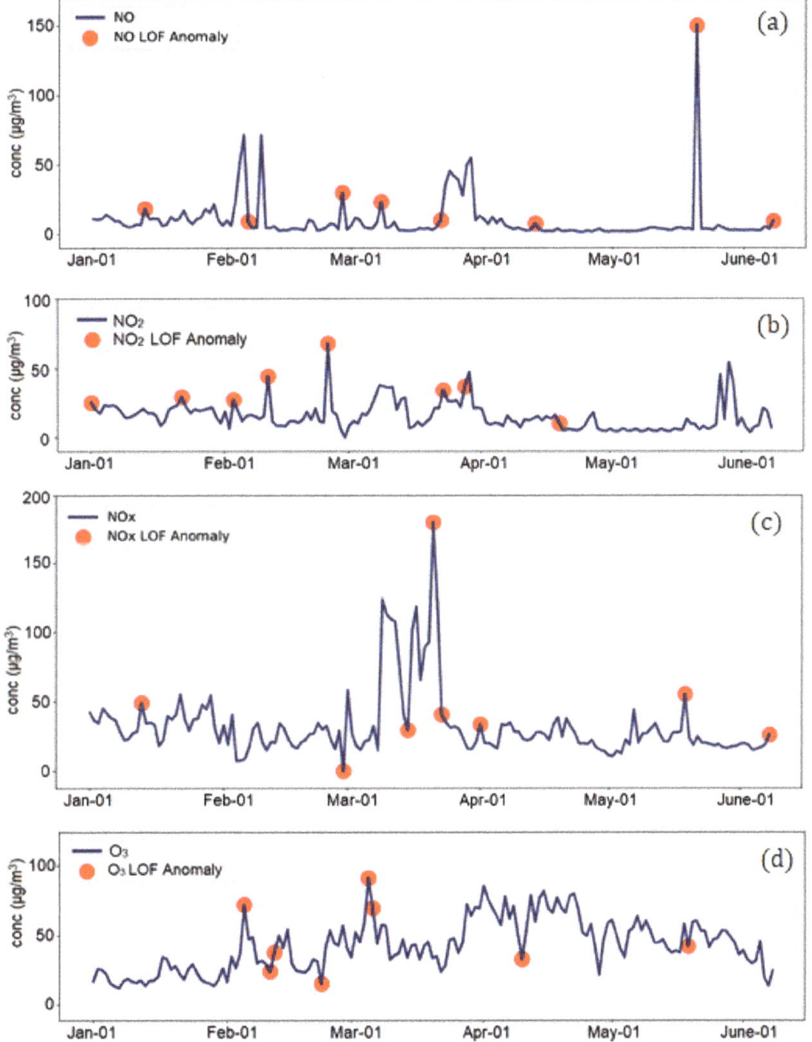

Figure 12. Series charts and anomalies computed by the LOF for (**a**) NO, (**b**) NO_2, (**c**) NO_x, (**d**) O_3.

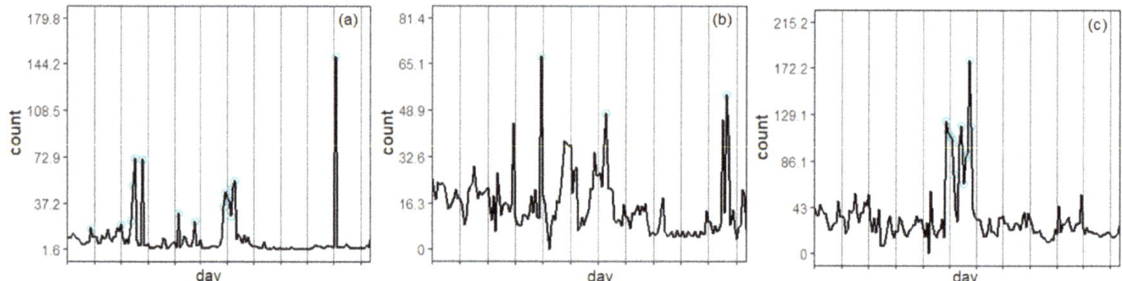

Figure 13. Series charts and the anomalies for (**a**) NO, (**b**) NO_2, and (**c**) NO_x, computed by GESD.

3.2. Models for the NO_2 Series

As presented in the previous section, the NO_2 series is not Gaussian. Since the normality of the series was achieved through a Box–Cox transformation with the parameter $\lambda = 0.130$, the series was firstly normalized and then stationarized by taking the first-order difference. Using the Akaike criterion and the capabilities of R software, the best ARIMA model for the transformed series (denoted NO_2BC) was the ARMA(1,1) type, with an autoregressive coefficient AR1 = 0.4728, moving average coefficient MA1 = -0.9069, and corresponding standard errors of the coefficients of 0.0973 and 0.0505. The values of the goodness of fit indicators for the model are a mean error (ME) = 0.0380, RMSE = 0.6488, MAE = 0.4543,—mean percentage error (MPE) = 0.268, and MAPE = 15.8283.

Figure 14a shows the NO_2BC series and the estimated one, whereas Figure 14b–d present the residual series, the residual autocorrelation function, and its histogram.

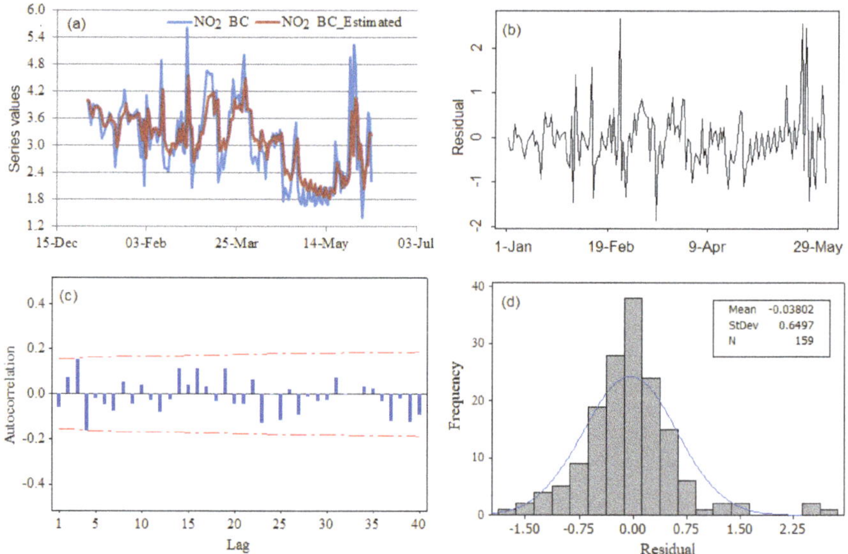

Figure 14. ARIMA model for the NO_2BC series. (**a**) NO_2BC series and the estimated one. (**b**) The residual series in the ARIMA model. (**c**) The residual autocorrelation function. (**d**) The histogram of the residual series.

Figure 14a shows good concordance between the recorded values (blue) and those estimated by the model (red). Figure 14c reveals no residual autocorrelation. The histogram (d) shows a mean value of the residuals of about zero and an almost symmetrical distribution of the residuals. The normality test of the residual series could not reject the normality hypothesis while the Levene test rejected the homoskedasticity one. Therefore, the residuals do not form white noise; so, the model could not be validated from a statistical viewpoint.

Figure 15 presents the chart of the GRNN model for the normalized NO_2 BC series after removing the exponential trend with the following equation:

$$(NO_2\ BC)_t = 5.8286 - 2.1721 \times \exp(0.00296t), \quad (3)$$

where $(NO_2\ BC)_t$ is the concentration of the value of the NO_2 BC series at the moment t.

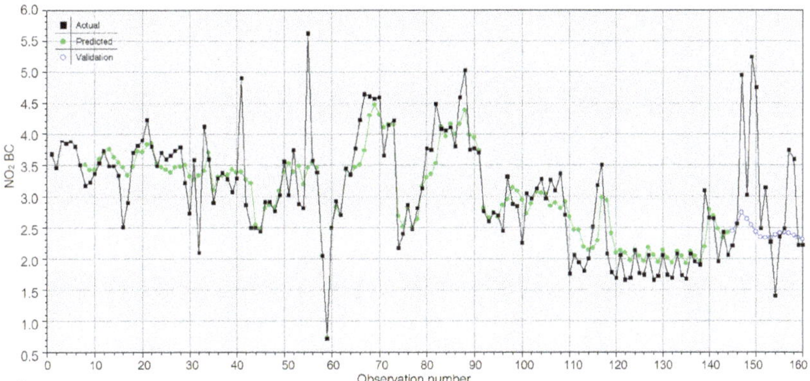

Figure 15. GRNN model for the NO_2BC series.

The model could learn the data well since the model's total variance on the training set is 76.135%, the correlation between the actual and predicted values is 0.8778, MSE = 0.177, MAE = 0.2839, and MAPE = 9.9786. Still, on the test set, the results are worse. For example, MSE = 1.5101, MAE = 0.8175, and $r_{ap} = 0.4482$.

Given that the ARIMA model could not be validated and the relative inability of GRNN to apply what was learnt in the training phase in the test, we searched for a hybrid model that could fit the data better and benefit from the ability of ARIMA to capture the linear behavior and the ability of GRNN to catch the nonlinear one. The raw series was considered to fit the ARIMA model, and then the residual series was subjected to GRNN modeling.

The best hybrid approach ARIMA-GRNN obtained for the NO_2 series is described as follows (Figure 16):

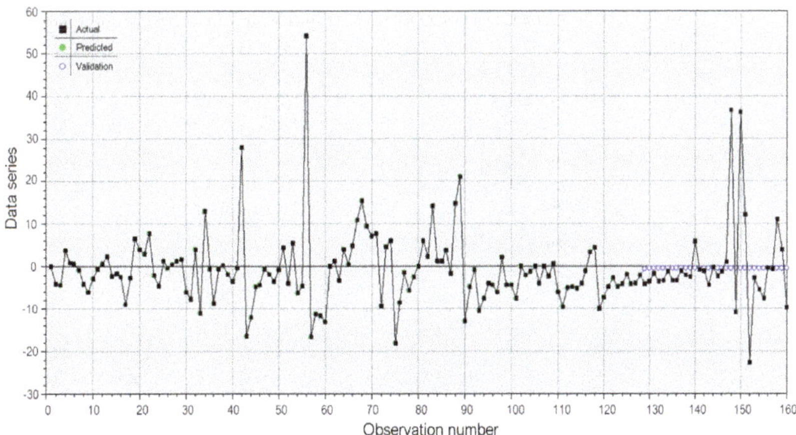

Figure 16. Hybrid ARIMA—GRNN model for the raw series.

- An ARIMA(2,1,1), with:
  - The autoregressive and moving average coefficients (and standard deviations) AR1 = 0.3584 (0.0834), AR2 = 0.1811 (0.0826), and MA1 = −0.9677 (0.0294);
  - MSE = 81.4417, MAE = 5.6679, the first-order residual autocorrelation = 0.97973;
  - AIC = 1161;
  - MAPE could not be computed (there is a value equal to 0);
- The GRNN model for the residual, with a lagged 1 variable as the regressor, and:
  - On the training set: R^2 = 99.635%, r_{ap} = 0.998178, MSE = 0.2562, MAE = 0.1112, MAPE = 27.4644.
  - On the test set: R^2 = 0.0635%, r_{ap} = 0.0578, MSE = 1222.97, MAE = 5.239, MAPE = 84.36.

Therefore, the GRNN model learnt the data well but could not use what it learnt for forecasting. Still, the new residuals are Gaussian.

Since the global anomalies were of interest, comparisons of the results provided by IQR, GESD, and IF were made to identify the values that were removed before the modeling. In the first stage, the common values provided by these methods were selected and removed from the data series. IQR was applied again to the new series in the second stage. Finally, the common values provided by IF remained after the first stage, and those from the second stage were removed. This procedure was chosen considering most anomalies detected.

The ARIMA model for the series without aberrant values (called NO_2New) was an ARIMA(1,1,1) type, with the following autoregressive and moving average coefficients (with the corresponding standard errors in brackets): AR1 = 0.4671 (0.0955) and MA1 = −0.9083 (0.0438), MSE = 15.95, MAE = 3.0694, MAPE = 30.76299, and AIC = 770.53. The residual variance in the ARIMA(1,1,1) model is 15.8890. The residuals' correlogram and their histogram (Figure 17) indicate that this series is not correlated and is Gaussian (confirmed by the Anderson–Darling test, where the p-value is 0.1269). The heteroskedasticity hypothesis was also rejected. Therefore, from a statistical viewpoint, the ARIMA(1,1,1) model is correct.

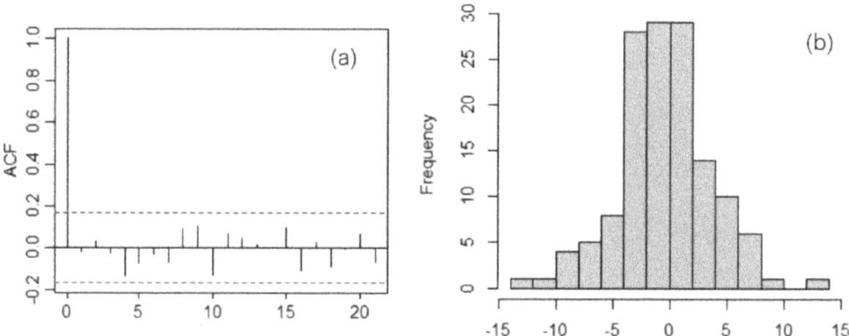

Figure 17. (**a**) Residual correlogram and (**b**) histogram in the ARIMA(1,1,1) model for the series after the removal of aberrant values.

The forecast for the next 48 moments based on the above model is shown in Figure 18 (the right-hand side), in blue, together with the confidence intervals at the confidence levels of 95% and 90% (different nuances of grey). The shape of the forecast series is not similar to that of the actual one. Its trend becomes almost linear after eight-time moments. Therefore, the model cannot be utilized in a future forecast, even if it was statistically validated.

The GRNN model for NO_2New is presented in Figure 19. The model learnt the data in the training set well (R^2 = 0.996). On the test set, MSE = 25.5047, MAE = 3.1555, and MAPE = 27.9311, but R^2 = 0.473 is not close to 1.

After comparing the GRNN performances on the initial series and that without aberrant values on the test set, the results of the last series are better. Still, the model should be improved because the blue dots—representing the computed values on the test set (validation in Figure 19) are not close enough to the recorded values, which were represented by the black line.

The hybrid ARIMA–GRNN model was built using the above ARIMA(1,1,1), whose residuals were modeled by GRNN (Figure 20).

The neural network learnt the data well. Indeed, on the left-hand side of Figure 20, the actual values and the computed ones (called predicted) are practically superposed on each other (the black and the green lines). It also performed well on the test set. On the right-hand side of Figure 20, the recorded values (black) and computed values (blue) are close. To confirm the model's goodness, Figure 21 displays the actual vs. predicted values in the residual modeling. The dots built by pairs of actual and predicted values of residuals are displayed along the diagonal (representing the ideal case of perfect superposition between the actual and computed values), indicating that the ARIMA-GRNN model performs very well. Therefore, the best model for the series without aberrant values is the ARIMA(1,1,1)–GRNN model.

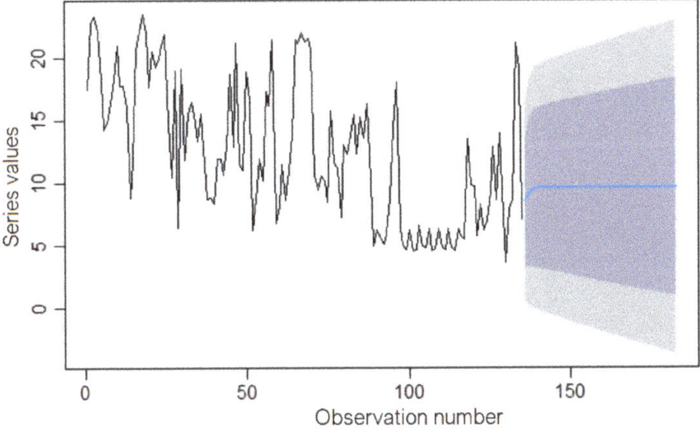

Figure 18. The forecast based on the ARIMA(1,1,1) model—the blue line—and the confidence intervals at 95% and 90%—different nuances of grey.

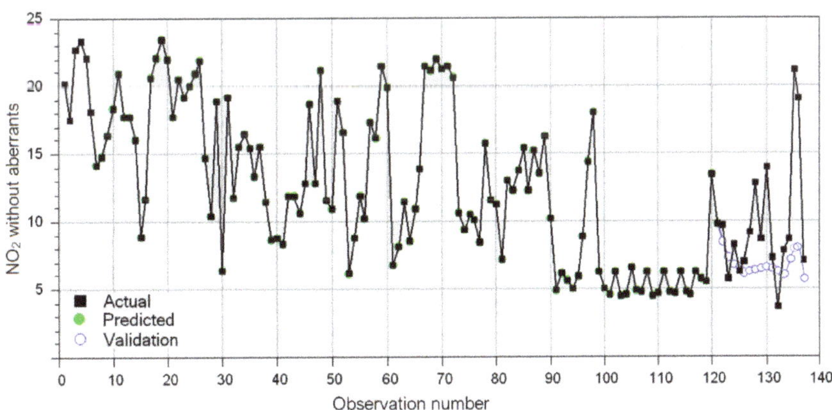

Figure 19. GRNN model for the NO_2 series after the removal of anomalies.

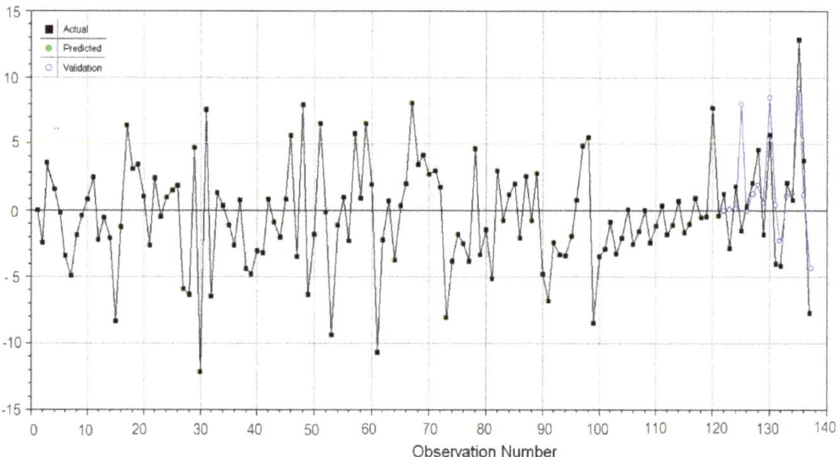

Figure 20. GRNN of the residual in the ARIMA(1,1,1) model for the series after the removal of anomalies.

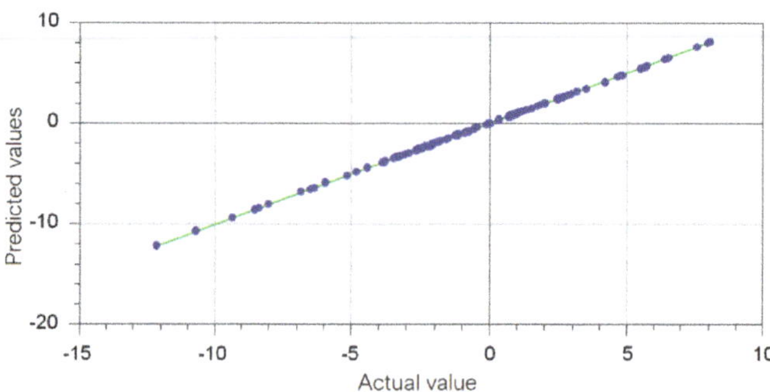

Figure 21. Actual vs. predicted values in the GRNN model of the residual from the ARIMA(1,1,1). after the removal of aberrant values.

Since similar results were obtained for the NO and NO_x series, the authors did not repeat the entire procedure.

3.3. Models for the O_3 Series

The same approach was followed to build models for the O_3 series. Given that high O_3 concentrations may negatively impact human health, a good forecast can provide information for early warning. The first approach provided an ARIMA(0,1,2) model for the raw data series. The series had to be stationarized before modeling (the degree of differentiation being 1). The moving average coefficients (with the standard errors in the brackets) are MA1 = −0.2971 (0.0789) and MA2 = −0.295(0.0884). The goodness of fit indicators are MSE = 69.72703, MAE = −5.392056, and MAPE = 21.79388. The MSE value is high due to the high variation in the errors. Despite their randomness, the residuals in the ARIMA(0,1,2) did not form white noise because they are not Gaussian (the p-value in the Anderson–Darling test is 0.0055 < 0.005) or homoskedastic. Figure 22 displays the residuals in the ARIMA(0,1,2) model for O_3, their histogram, and the correlogram. The residuals chart in Figure 22 confirms the existence of high residual values. Since the model could not be validated, its improvement was necessary.

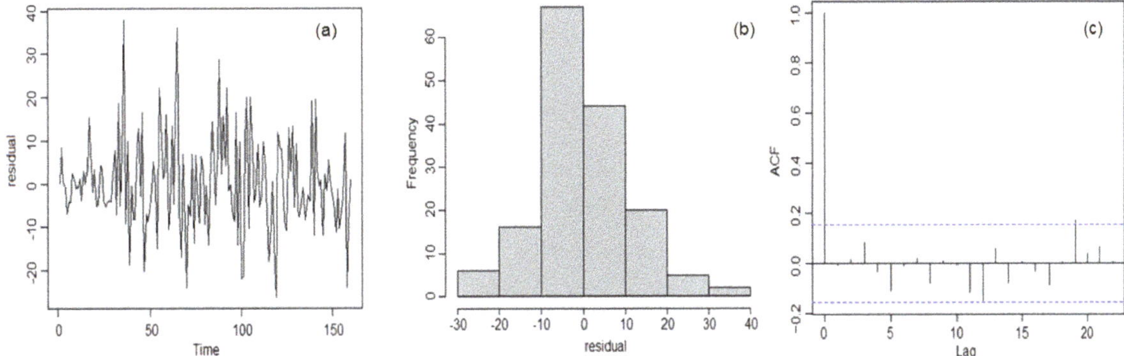

Figure 22. (**a**) The residual, (**b**) the histogram, and (**c**) the correlogram of the residual in the ARIMA(0,1,2) model for O_3.

The neural-network approach provided a GRNN model (Figure 23) that learnt the data well but did not perform well on the test set. For example, on the training set, the correlation between the actual and predicted values is 0.8634 while on the test set, it is only 0.5282. On the test set, the computed values (represented by blue circles) do not have the same pattern as the recorded data (the black line).

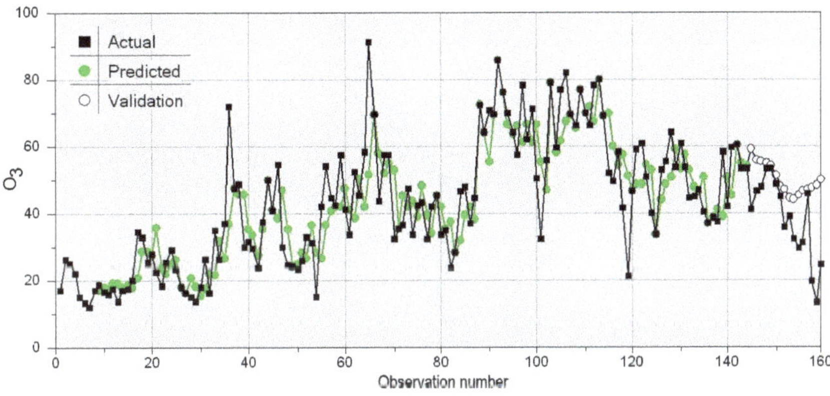

Figure 23. The GRNN model for the O_3 series.

The hybrid ARIMA-GRNN provided R^2 = 99.681%, correlation between actual and computed values of 0.9984, MSE = 0.3965, MAE = 0.0606, and MAPE = 38.64744 on the training set. Still, the hybrid model did not perform well on the test set, since R^2 = 5.898%, and the correlation between the actual and computed values = 0.333, so it cannot be used for prediction.

After removing the aberrant values from the O_3 series, and performing the Mann–Kendall test [73], the hypothesis that there is no monotonic trend was rejected. Using the nonparametric method of Sen [74], it was found that the series presents an increasing trend, with a slope of 0.310673. The KPSS test revealed nonstationarity in the level of this series. It was found that the best model was ARIMA(0,1,0) with a drift of 0.310673 (the same as the slope). The goodness of fit indicators showed very low residual values (RMSE = 0.00022, MAE = 0.00233, MAPE = 0.000844), with no residual correlation. Given the model's quality, it is not necessary to improve it.

From this model, it was found that the O_3 series had an increasing trend over the study period, which must be observed in the future, since the O_3 concentration may reach a level that is dangerous for the population.

4. Conclusions

The detection of aberrant values in time series has been a problem of interest for a long time, given that their presence may influence the modeling results. Moreover, forecasting based on derived models may be significantly biased by the existence of aberrant values. Therefore, this study investigated the influence of the presence of anomalies on a series of nitrogen oxide concentrations.

Given that some methodologies are used to search for different kinds of anomalies (local or global), first, the results provided by LOF, IQR, IF, and GESD were compared. Since the focus was placed on global aberrant values, their selection was made before using the last three algorithms for modeling.

Three models were built for each NO_2 raw series and after the removal of anomalies: –ARIMA, GRNN, and a hybrid GRNN-ARIMA.

In the case of the NO_2 series, the building of three models was necessary to improve the initial model, even in the absence of anomalies. This was motivated by the following reasons. An ARIMA model, for example, is not necessarily the best choice, given that the residual must be white noise (a fact that is not always true). A GRNN model is not appropriate because the R^2 value or the correlation between the actual and predicted values is not very high on both the training and test sets. The selection of the regressors in the artificial intelligence-based approaches is not obvious. Their selection and number are essential for determining the best model. Even in the absence of outliers, improvement of the model is necessary to obtain a good forecast in the next stage. From this point of view, the best model is one that provides the best forecast.

It was shown that the removal of anomalies resulted in better models than when they were present. The ARIMA model for the raw data series could not be statistically validated whereas, for the series without anomalies, it was correct from a statistical viewpoint. The hybrid approach was also better than the ARIMA and GRNN on both NO_2 series.

The hybrid approach provided the best model for the O_3 raw series. After the removal of aberrant values, the ARMA(0,1,0) with drift provided the best model for the series evolution. Given that the model was statistically validated and the residual was extremely low, it was unnecessary to search for another model. It was proved that the O_3 series presents a significant increasing trend (at a significance level of 5%). Given that high ozone concentrations are harmful to the population's health, keeping the ozone level under observation is necessary.

As a future work in the same research direction, dynamical system approaches, such as phase space reconstruction, will be introduced to analyze the dynamics of atmospheric pollutants.

Author Contributions: Conceptualization, A.B. and I.I.; methodology, A.B. and I.I.; software, C.S.D.; validation, A.B., I.I. and C.S.D.; formal analysis, S.-B.B.; investigation, I.I.; resources, A.B.; data curation, S.-B.B.; writing—original draft preparation, A.B.; writing—review and editing, C.S.D.; visualization, A.B. and I.I.; supervision, A.B.; project administration, A.B.; funding acquisition, A.B. All authors have read and agreed to the published version of the manuscript.

Funding: This research received no external funding.

Institutional Review Board Statement: Not applicable.

Informed Consent Statement: Not applicable.

Data Availability Statement: Data are available for free download at https://www.calitateaer.ro/public/home-page/?locale=en (accessed on 15 November 2021).

Conflicts of Interest: The authors declare no conflict of interest.

References

1. Azid, A.; Juahir, H.; Toriman, M.E.; Kamarudin, M.K.A.; Saudi, A.S.M.; Hasnam, C.N.C.; Aziz, N.A.A.; Azaman, F.; Latif, M.T.; Zainuddin, S.F.M.; et al. Prediction of the level of air pollution using principal component analysis and artificial neural network techniques: A case study in Malaysia. *Water Air Soil Pollut.* **2014**, *225*, 2063. [CrossRef]
2. Manisalidis, I.; Stavropoulou, E.; Stavropoulos, A.; Bezirtzoglou, E. Environmental and health impacts of air pollution: A review. *Front. Public Health* **2020**, *8*, 14. [CrossRef] [PubMed]
3. Bărbulescu, A. *Applications in Environmental Sciences*; Studies on Time Series; Springer: Cham, Switzerland, 2016.
4. Ghorani-Azam, A.; Riahi-Zanjani, B.; Balali-Mood, M. Effects of air pollution on human health and practical measures for prevention in Iran. *J. Res. Med. Sci.* **2016**, *21*, 65. [CrossRef] [PubMed]
5. Al-Taani, A.; Nazzal, Y.; Howari, F.; Iqbal, J.; Bou-Orm, N.; Xavier, C.M.; Bărbulescu, A.; Sharma, M.; Dumitriu, C.S. Contamination assessment of heavy metals in agricultural soil, in the Liwa area (UAE). *Toxics* **2021**, *9*, 53. [CrossRef]
6. Arnaudo, E.; Farasin, A.; Rossi, C. A comparative analysis for air quality estimation from traffic and meteorological data. *Appl. Sci.* **2020**, *10*, 4587. [CrossRef]
7. Dominick, D.; Latif, M.T.; Juahir, H.; Aris, A.Z.; Zain, S.M. An assessment of influence of meteorological factors on PM_{10} and NO_2 at selected stations in Malaysia. *Sustain. Environ. Res.* **2012**, *22*, 305–315.
8. Leelőssy, Á.; Molnár, F.; Izsák, F.; Havasi, Á.; Lagzi, I. Mészáros R. Dispersion modeling of air pollutants in the atmosphere: A review. *Cent. Eur. J. Geosci.* **2014**, *6*, 257–278. [CrossRef]
9. EPA. Nitrogen Oxides (NOx), Why and How They Are Controlled. Technical Bulletin. 1999. Available online: https://www3.epa.gov/ttn/catc/dir1/fnoxdoc.pdf (accessed on 25 March 2021).
10. Thurston, G.D. Outdoor Air Pollution: Sources, Atmospheric Transport, and Human Health Effects. In *International Encyclopedia of Public Health*; Heggenhougen, H.K., Ed.; Academic Press: Cambridge, MA, USA, 2008; pp. 700–712. [CrossRef]
11. Millán, M.M.; Sanz, J.; Salvador, R.; Mantilla, E. Atmospheric dynamics and ozone cycles related to nitrogen deposition in the western Mediterranean. *Environ. Poll.* **2002**, *118*, 167–186. [CrossRef]
12. Leonardo da Vinci Programme, Pilot Project no RO/02/B/F/PP–141004. Training Module for Environmental Pollution Control. pp. 14–15, 18–19. Available online: http://leonardo.unibuc.ro/products/textbook.html (accessed on 2 March 2022).
13. Addison, C.C. *Nitrogen Oxides, AccessScience*; McGraw-Hill Education: New York, NY, USA, 2018. [CrossRef]
14. EEA. Assessing the Risks to Health from Air Pollution. 2021. Available online: https://www.eea.europa.eu/publications/assessing-the-risks-to-health (accessed on 15 April 2021).
15. NAWMN2021. Available online: https://www.calitateaer.ro/public/description-page/general-info-page/?locale=en (accessed on 15 April 2021).
16. Hajek, P.; Olej, V. Predicting common air quality index—The case of Czech microregions. *Aerosol Air Qual. Res.* **2015**, *15*, 544–555. [CrossRef]
17. Gocheva-Ilieva, S.G.; Ivanov, A.V.; Voynikova, D.S.; Boyadzhiev, D.T. Time series analysis and forecasting for air pollution in small urban area: An SARIMA and factor analysis approach. *Stoch. Env. Res. Risk. Assess.* **2014**, *28*, 1045–1060. [CrossRef]
18. Burden, F.R.; Forstner, U.; McKelvie, I.D.; Guenther, A. Time-Series Analysis. In *Environmental Monitoring Handbook*; McGraw-Hill Professional: New York, NY, USA, 2002; Available online: https://www.accessengineeringlibrary.com/content/book/9780071351768/back-matter/appendix1 (accessed on 15 February 2021).
19. Bontempi, G.; Ben Taieb, S.; Le Borgne, Y.A. Machine learning strategies for time series forecasting. In *Business Intelligence*; Aufaure, M.A., Zimányi, E., Eds.; Springer: Berlin/Heidelberg, Germany, 2013; Volume 138, pp. 59–73. [CrossRef]
20. CALPUFF Modeling System. Available online: www.src.com (accessed on 10 April 2021).
21. Fallah-Shorshani, M.; Shekarrizfard, M.; Hatzopoulou, M. Evaluation of regional and local atmospheric dispersion models for the analysis of traffic-related air pollution in urban areas. *Atmos. Environ.* **2017**, *167*, 270–282. [CrossRef]
22. Shekarrizfard, M.; Faghih-Imani, A.; Tétreault, L.F.; Yasmin, S.; Reynaud, F.; Morency, P.; Plante, C.; Drouin, L.; Smargiassi, A.; Eluru, N.; et al. Regional assessment of exposure to traffic-related air pollution: Impacts of individual mobility and transit investment scenarios. Sustain. *Cities Soc.* **2017**, *29*, 68–76. [CrossRef]
23. Soulhac, L.; Nguyen, C.V.; Volta, P.; Salizzoni, P. The model SIRANE for atmospheric urban pollutant dispersion. PART III: Validation against NO_2 yearly concentration measurements in a large urban agglomeration. *Atmos. Environ.* **2017**, *167*, 377–388. [CrossRef]
24. Bai, L.; Wang, J.; Ma, X.; Lu, H. Air pollution forecasts: An overview. *Int. J. Environ. Res. Public Health* **2018**, *15*, 780. [CrossRef]
25. Kumar, U.; Jain, V.K. ARIMA Forecasting of Ambient Air Pollutants (O_3, NO, NO_2 and CO). *Stoch. Environ. Res. Risk Assess.* **2010**, *4*, 751–760. [CrossRef]
26. Zhu, J. Comparison of ARIMA model and exponential smoothing model on 2014 air quality index in Yanqing county, Beijing, China. *Appl. Comput. Math.* **2015**, *4*, 456. [CrossRef]
27. Munir, S.; Mayfield, M. Application of density plots and time series modelling to the analysis of nitrogen dioxides measured by low-cost and reference sensors in urban areas. *Nitrogen* **2021**, *2*, 167–195. [CrossRef]
28. Hajmohammadi, H.; Heydecker, B. Multivariate time series modelling for urban air quality. *Urban Clim.* **2021**, *37*, 100834. [CrossRef]
29. Gardner, M.W.; Dorling, S.R. Neural network modeling and prediction of hourly NO_x and NO_2 concentrations in urban air in London. *Atmos. Environ.* **1999**, *33*, 709–719. [CrossRef]

30. Rahimi, A. Short-term prediction of NO_2 and NO_x concentrations using multilayer perceptron neural network: A case study of Tabriz, Iran. *Ecol Process* **2017**, *6*, 4. [CrossRef]
31. Dragomir, C.M.; Voiculescu, M.; Constantin, D.-E.; Georgescu, L.P. Prediction of the NO_2 concentration data in an urban area using multiple regression and neuronal networks. *AIP Conf. Proc.* **2015**, *1694*, 040003. [CrossRef]
32. Baawain, M.S.; Al-Serihi, A.S. Systematic Approach for the Prediction of Ground-Level Air Pollution (around an Industrial Port) Using an Artificial Neural Network. *Aerosol Air Qual. Res.* **2014**, *14*, 124–134. [CrossRef]
33. Jiang, D.; Zhang, Y.; Hu, X.; Zeng, Y.; Tan, J.; Shao, D. Progress in Developing an ANN Model for Air Pollution Index Forecast. *Atmos. Environ.* **2004**, *38*, 7055–7064. [CrossRef]
34. Hrust, L.; Klaić, Z.B.; Križan, J.; Antonić, O.; Hercog, P. Neural Network Forecasting of Air Pollutants Hourly Concentrations Using Optimised Temporal Averages of Meteorological Variables and Pollutant Concentrations. *Atmos. Environ.* **2009**, *43*, 5588–5596. [CrossRef]
35. Moustris, K.P.; Ziomas, I.C.; Paliatsos, A.G. 3- Day-ahead Forecasting of Regional Pollution Index for the Pollutants NO_2, CO, SO_2, and O_3 Using Artificial Neural Networks in Athens, Greece. *Water Air Soil Pollut.* **2010**, *209*, 29–43. [CrossRef]
36. Agirre-Basurko, E.; Ibarra-Berastegi, G.; Madariaga, I. Regression and Multilayer Perceptron-based Models to Forecast Hourly O_3 and NO_2 Levels in the Bilbao Area. *Environ. Modell. Softw.* **2006**, *21*, 430–446. [CrossRef]
37. Kukkonen, J.; Partanen, L.; Karppinen, A.; Ruuskanen, J.; Junninen, H.; Kolehmainen, M.; Niska, H.; Dorling, S.; Chatterton, T.; Foxall, R.; et al. Extensive Evaluation of Neural Network Models for the Prediction of NO_2 and PM_{10} Concentrations, Compared with a Deterministic Modelling System and Measurements in Central Helsinki. *Atmos. Environ.* **2003**, *37*, 4539–4550. [CrossRef]
38. Wang, W.; Men, C.; Lu, W. Online Prediction Model Based on Support Vector Machine. *Neurocomputing* **2008**, *71*, 550–558. [CrossRef]
39. Osowski, S.; Garanty, K. Forecasting of the Daily Meteorological Pollution using Wavelets and Support Vector Machine. *Eng. Appl. Artif. Intell.* **2007**, *20*, 745–755. [CrossRef]
40. Hajek, P.; Olej, V. Ozone Prediction on the Basis of Neural Networks, Support Vector Regression and Methods with Uncertainty. *Ecol. Inf.* **2012**, *12*, 31–42. [CrossRef]
41. Lin, K.P.; Pai, P.F.; Yang, S.L. Forecasting Concentrations of Air Pollutants by Logarithm Support Vector Regression with Immune Algorithms. *Appl. Math. Comput.* **2011**, *217*, 5318–5327. [CrossRef]
42. Singh, K.P.; Gupta, S.; Kumar, A.; Shukla, S.P. Linear and Nonlinear Modeling Approaches for Urban Air Quality Prediction. *Sci. Total Environ.* **2012**, *426*, 244–255. [CrossRef] [PubMed]
43. Liu, B.; Zhang, L.; Wang, Q.; Chen, J. A novel method for regional NO_2 concentration Prediction using discrete Wavelet transform and an LSTM network. *Comput. Intel. Neurosc.* **2021**, *2021*, 6631614. [CrossRef] [PubMed]
44. Wang, X.; Liu, W.; Wang, Y.; Yang, G. A hybrid NOx emission prediction model based on CEEMDAN and AM-LSTM. *Fuel* **2022**, *310C*, 122486. [CrossRef]
45. Shekarrizfard, M.; Faghih-Imani, A.; Hatzopoulou, M. An examination of population exposure to traffic-related air pollution: Comparing spatially and temporally resolved estimates against long-term average exposures at the home location. *Environ. Res.* **2016**, *147*, 435–444. [CrossRef]
46. ECA 2018. Available online: https://op.europa.eu/webpub/eca/special-reports/air-quality-23-2018/en/ (accessed on 15 April 2021).
47. Law 24/15 June 2011 on Ambient Air Quality. Available online: https://www.calitateaer.ro/export/sites/default/.galleries/Legislation/national/Lege-nr.-104_2011-calitatea-aerului-inconjurator.pdf_2063068895.pdf (accessed on 15 March 2022). (In Romanian).
48. Quality Indices. Available online: https://www.calitateaer.ro/public/monitoring-page/quality-indices-page/?__locale=ro (accessed on 22 March 2022).
49. Iorga, G. Air pollution monitoring: A case study from Romania. In *Air Quality—Measurement and Modeling*; Sallis, P., Ed.; InTech: London, UK, 2016. [CrossRef]
50. Bărbulescu, A.; Barbeş, L. Mathematical modeling of sulfur dioxide concentration in the western part of Romania. *J. Environ. Manag.* **2017**, *204*, 825–830. [CrossRef]
51. Bărbulescu, A.; Barbeş, L. Modeling the carbon monoxide dissipation in Timisoara, Romania. *J. Environ. Manag.* **2017**, *204*, 831–838. [CrossRef]
52. Bărbulescu, A.; Barbeş, L. Statistical assessment and modeling of benzene level in atmosphere in Timiş County, Romania. *Int. J. Environ. Sci. Tech.* **2022**, *19*, 817–828. [CrossRef]
53. Levei, L.; Hoaghia, M.A.; Roman, M.; Marmureanu, L.; Moisa, C.; Levei, E.A.; Ozunu, A.; Cadar, O. Temporal trend of PM_{10} and associated human health risk over the past decade in Cluj-Napoca city, Romania. *Appl. Sci.* **2020**, *10*, 5331. [CrossRef]
54. Bărbulescu, A.; Barbeş, L.; Nazzal, Y. New model for inorganic pollutants dissipation on the northern part of the Romanian Black Sea coast. *Rom. J. Phys.* **2018**, *63*, 806.
55. Bărbulescu, A.; Barbeş, L. Models for pollutants' correlation in the Romanian littoral. *Rom. Rep. Phys.* **2014**, *66*, 1189–1199.
56. Torres, J.M.; Nieto, P.J.G.; Alejano, L.; Reyes, A.N. Detection of outliers in gas emissions from urban areas using functional data analysis. *J. Hazard. Mater.* **2011**, *186*, 144–149. [CrossRef]
57. Shaadan, N.; Jemain, A.A.; Latif, M.T.; Deni, S.M. Anomaly detection and assessment of PM10 functional data at several locations in the Klang Valley, Malaysia. *Atmos. Poll. Res.* **2015**, *6*, 365–375. [CrossRef]

58. Hakins, S.J.; Gibbs, P.E.; Pope, N.D.; Burt, G.R.; Chesman, B.S.; Bray, S.; Proud, S.V.; Spence, S.K.; Southward, A.J.; Southward, G.A.; et al. Recovery of polluted ecosystems: The case for long-term studies. *Marine Environ. Resear* **2002**, *54*, 215–222. [CrossRef]
59. Martínez, J.; Saavedra, Á.; García-Nieto, P.J.; Piñeiro, J.I.; Iglesias, C.; Taboada, J.; Sancho, J.; Pastor, J. Air quality parameters outliers detection using functional data analysis in the Langreo urban area (Northern Spain). *Appl. Math Comput.* **2014**, *241*, 1–10. [CrossRef]
60. van Zoest, V.M.; Stein, A.; Hoek, G. Outlier Detection in Urban Air Quality Sensor Networks. *Water Air Soil Pollut.* **2018**, *229*, 111. [CrossRef]
61. Fox, A.J. Outliers in Time Series. *J. Royal Stat. Soc. Ser. B* **1972**, *34*, 350–363. [CrossRef]
62. Blázquez-García, A.; Conde, A.; Mori, U.; Lozano, J.A. A Review on outlier/Anomaly Detection in Time Series Data. *ACM Comput. Surv.* **2021**, *54*, 1–33. [CrossRef]
63. Liu, F.T.; Ting, K.M.; Zhou, Z.-H. Isolation forest. In Proceedings of the 2008 Eighth IEEE International Conference on Data Mining, Pisa, Italy, 15–19 December 2008; pp. 413–422. [CrossRef]
64. Liu, F.T.; Ting, K.M.; Zhou, Z.-H. Isolation-based anomaly detection. *ACM T. Knowl. Discov. D.* **2012**, *6*, 3. [CrossRef]
65. Cheng, Z.; Zou, C.; Dong, J. Outlier detection using isolation forest and local outlier factor. In Proceedings of the RACS '19: Proceedings of the Conference on Research in Adaptive and Convergent Systems, Chongqing, China, 24–27 September 2019; pp. 161–168. [CrossRef]
66. Souiden, I.; Brahmi, Z.; Toumi, H. A Survey on Outlier Detection in the Context of Stream Mining: Review of Existing Approaches and Recommadations. In *Intelligent Systems Design and Applications. ISDA 2016. Advances in Intelligent Systems and Computing*; Madureira, A., Abraham, A., Gamboa, D., Novais, P., Eds.; Springer: Cham, Switzerland, 2017; Volume 557, pp. 372–383.
67. Alghushairy, O.; Alsini, R.; Soule, T.; Ma, X. A Review of Local Outlier Factor Algorithms for Outlier Detection in Big Data Streams. *Big Data Cogn. Comput.* **2021**, *5*, 1. [CrossRef]
68. Vallis, O.; Hochenbaum, J.; Kejariwal, A. A Novel Technique for Long-Term Anomaly Detection in the Cloud. In Proceedings of the 6th USENIX Workshop on Hot Topics in Cloud Computing, Philadelphia, PA, USA, 17–18 June 2014; Available online: https://www.usenix.org/system/files/conference/hotcloud14/hotcloud14-vallis.pdf (accessed on 4 December 2021).
69. Rosner, B. Percentage Points for a Generalized ESD Many-Outlier Procedure. *Technometrics* **1983**, *25*, 165–172. [CrossRef]
70. Brockwell, P.J.; Davis, R.A. *Introduction to Time Series and Forecasting*; Springer: New York, NY, USA, 2002.
71. Specht, D.F. A General Regression Neural Network. *IEEE Trans. Neural Netw.* **1991**, *2*, 568–576. [CrossRef] [PubMed]
72. Zaknich, A. *Neural Networks for Intelligent Signal Processing*; World Scientific: Hackensack, NJ, USA, 2003.
73. Hipel, K.W.; McLeod, A.I. *Time Series Modelling of Water Resources and Environmental Systems*; Elsevier Science: New York, NY, USA, 1994.
74. Sen, P.K. Estimates of the regression coefficient based on Kendall's tau. *J. Am. Stat. Assoc.* **1968**, *63*, 1379–1389. [CrossRef]

 MDPI

Article

Characteristics of Particulate Matter at Different Pollution Levels in Chengdu, Southwest of China

Yi Huang [1,2,*,†], Li Wang [1,†], Xin Cheng [2], Jinjin Wang [2], Ting Li [2], Min He [2], Huibin Shi [2], Meng Zhang [2], Scott S. Hughes [3] and Shijun Ni [2]

1 State Key Laboratory of Geohazard Prevention and Geoenvironment Protection, College of Ecology and Environment, Chengdu University of Technology, Chengdu 610059, China; Wangli_0420@126.com
2 Department of Geochemistry, Chengdu University of Technology, Chengdu 610059, China; chengxin17@cdut.edu.cn (X.C.); jinjin.wangjj@gmail.com (J.W.); lt18382264420@126.com (T.L.); hm151___7@126.com (M.H.); shihuibin958@163.com (H.S.); zm01292021@163.com (M.Z.); nsj@cdut.edu.cn (S.N.)
3 Department of Geosciences, Idaho State University, Pocatello, ID 83209, USA; hughscot@isu.edu
* Correspondence: huangyi@cdut.edu.cn
† These authors contributed equally to this work and should be considered dual-first authors.

Abstract: Air pollution is becoming increasingly serious along with social and economic development in the southwest of China. The distribution characteristics of particle matter (PM) were studied in Chengdu from 2016 to 2017, and the changes of PM bearing water-soluble ions and heavy metals and the distribution of secondary ions were analyzed during the haze episode. The results showed that at different pollution levels, heavy metals were more likely to be enriched in fine particles and may be used as a tracer of primary pollution sources. The water-soluble ions in $PM_{2.5}$ were mainly Sulfate-Nitrate-Ammonium (SNA) accounting for 43.02%, 24.23%, 23.50%, respectively. SO_4^{2-}, NO_3^-, NH_4^+ in PM_{10} accounted for 34.56%, 27.43%, 19.18%, respectively. It was mainly SO_4^{2-} in PM at Clean levels ($PM_{2.5}$ = 0~75 μg/m^3, PM_{10} = 0~150 μg/m^3), and mainly NH_4^+ and NO_3^- at Light-Medium levels ($PM_{2.5}$ = 75~150 μg/m^3, PM_{10} = 150~350 μg/m^3). At Heavy levels ($PM_{2.5}$ = 150~250 μg/m^3, PM_{10} = 350~420 μg/m^3), it is mainly SO_4^{2-} in $PM_{2.5}$, and mainly NH_4^+ and NO_3^- in PM_{10}. The contribution of mobile sources to the formation of haze in the study area was significant. SNA had significant contributions to the PM during the haze episode, and more attention should be paid to them in order to improve air quality.

Keywords: particulate matter; heavy metals; Sulfate-Nitrate-Ammonium; pollution levels; mobile sources

Citation: Huang, Y.; Wang, L.; Cheng, X.; Wang, J.; Li, T.; He, M.; Shi, H.; Zhang, M.; Hughes, S.S.; Ni, S. Characteristics of Particulate Matter at Different Pollution Levels in Chengdu, Southwest of China. *Atmosphere* **2021**, *12*, 990. https://doi.org/10.3390/atmos12080990

Academic Editor: Alina Barbulescu

Received: 27 June 2021
Accepted: 26 July 2021
Published: 31 July 2021

Publisher's Note: MDPI stays neutral with regard to jurisdictional claims in published maps and institutional affiliations.

Copyright: © 2021 by the authors. Licensee MDPI, Basel, Switzerland. This article is an open access article distributed under the terms and conditions of the Creative Commons Attribution (CC BY) license (https://creativecommons.org/licenses/by/4.0/).

1. Introduction

PM_{10} refers to particles with an aerodynamic equivalent diameter less than or equal to 10 μm in ambient air, and $PM_{2.5}$ refers to particles with an aerodynamic equivalent diameter less than or equal to 2.5 μm in ambient air. The composition of atmospheric particles is complex, including heavy metals, water-soluble ions, carbonaceous components, and so on from multiple sources [1,2]. In addition, with small particle sizes and large surface area, atmospheric particulates have adverse effects on the atmospheric environment and public health. In recent years, there have been many haze pollution incidents occurring in a number of locations across China, which has caused more and more public concerns and attention paid to air pollution [3–5].

The Sichuan Basin is in southwestern China. The topography of hills and basins, coupled with the climate conditions of high humidity and low wind speed, leads to atmospheric pollution easily in this area [6,7]. It is the fourth highest haze area following the Beijing-Tianjin-Hebei area, Yangtze River Delta, and Pearl River Delta. Its pollution characteristics are of high particle concentration and low visibility [8–10]. The special terrain and the humid climate of Chengdu are not conducive to the diffusion of particulate matter and are prone to secondary pollutant (SNA) conversion and generation [6,10,11].

In recent years, although the implementation of pollution prevention and control measures has improved the air environment in Chengdu, the region still has a problem with air pollution [12]. The research topics in this region mainly include the analysis of particulate pollution characteristics [13,14], the impact of meteorological conditions on particulate pollution [15], the lag effect of particulate pollution on related diseases [16], and source apportionment [17]. It is reported that the heavy metals in the atmospheric particulate matter (PM) in Chengdu are mainly arsenic (As), lead (Pb), copper (Cu), nickel (Ni), zinc (Zn), iron (Fe), and manganese (Mn) [18,19]. Among them, arsenic (As) mainly comes from industrial smelters. Pb, Cu, Ni, and Zn mainly come from the exhaust of motor vehicles and the wear of tires and brake pads whereas Fe and Mn are mainly from dust generated during vehicle driving [20,21]. Sulfate-Nitrate-Ammonium (SNA) are water-soluble ions that greatly contribute to PM concentration [22,23]. Air pollution in Chengdu has obvious seasonal distribution characteristics, which are closely related to the meteorological factors of the city [24]. It is reported that the mass concentration of SNA is the highest in winter and the lowest in summer [25]. The high temperature in summer and autumn is conducive to the conversion of sulfur dioxide (SO_2) to sulfate (SO_4^{2-}) while adverse to the stable existence of ammonium nitrate (NH_4NO_3). Although the low temperature in winter inhibits the conversion of gaseous precursors, it is beneficial to the stable existence of NH_4NO_3 [26,27].

However, it is known from the previous studies that the concentration of particulate matter increases with the increase of pollution, but the mechanism of particle concentration and composition change is different under different pollution levels [28]. A recent study showed that the rapid increase of $PM_{2.5}$ at light pollution level in Beijing was caused by regional transportation, while the rise from heavy to severe was mainly caused by an increase in the proportion of secondary inorganic components [29]. The air pollution in cities in southern China has been easily overlooked. Up to now, there is no detailed report on the various characteristics of atmospheric particulate matter at various pollution levels in Chengdu, southwest China according to our investigation.

The purpose of this study is to find out how heavy metals and water-soluble ions in PM in Chengdu, China during the haze periods are distributed and changed at different pollution levels. Therefore, we investigated $PM_{2.5}$ and PM_{10} in Chengdu in southwest China. The changes in heavy metal elements and water-soluble ions corresponding to the pollution level and their contribution to particulate matter are discussed. The effects of SO_4^{2-}, NO_3^-, and NH_4^+ on the particulate matter were emphatically explored. The secondary production of sulfate and nitrate will be shown to be important in high pollution level scenarios, and the same with the heavy metal analysis.

2. Materials and Methods

2.1. Study Site and Sample Collection

Chengdu is located in the western part of the Sichuan basin, surrounded by the western part of the Longquan Mountains and the eastern part of the Qionglai Mountains. The sampling site was located in Shilidian, Chenghua District, Chengdu (104°08′ E, 30°40′ N), the capital of Sichuan Province in the western part of the Sichuan Basin. Chengdu is densely populated, about 1000 people/km^2 [30,31], with the annual temperature 15.2~16.6 °C, the annual precipitation 873 mm~1265 mm, the annual sunshine 23–30%, the average annual wind speed 1.3 m/s, and the average annual relative humidity 80% [32]. Shilidian is surrounded by major cities in Sichuan, including Deyang and Mianyang (Figure 1).

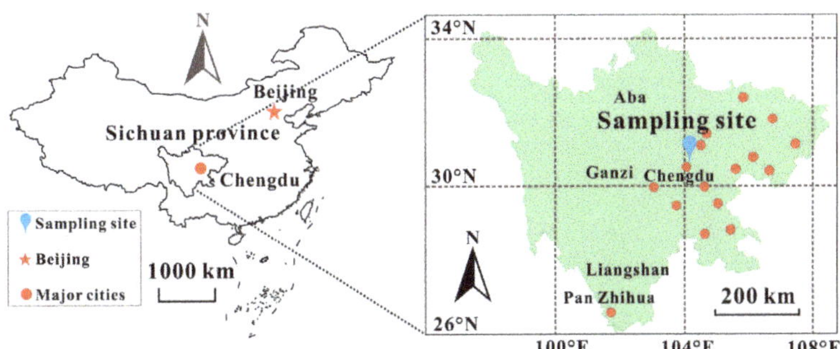

Figure 1. Sichuan region of China and sampling site in Chengdu.

From March 2016 to January 2017, $PM_{2.5}$ and PM_{10} were collected at the Chengdu University of Technology in the urban area of Chengdu with no chemical enterprises and tall buildings. The sampling period was 24 h and we collected 72 $PM_{2.5}$ samples and 72 PM_{10} samples at the same time, with a total of 144 samples. The sampling instrument was a TH-150C medium flow atmospheric sampler (Wuhan Tianhong, Wuhan, China), with a calibrated flow rate of 100 L/min. Two kinds of filter membranes made of quartz and Teflon, respectively, (Whatman, Buckinghamshire, UK) were chosen. The Teflon filters are used for the heavy metal analysis because Teflon filters have low heavy metal background content, and the quartz filters are used for the water-soluble ion analysis. After the samples were collected, the sampling membranes were placed in clearly marked sample boxes immediately. At the same time, the meteorological data at the Shilidian meteorological monitoring station were recorded, including temperature, air pressure, wind speed, relative humidity, etc. Samples were collected under stable weather conditions, with weak wind at speeds less than 1.5 m/s, thus the contribution from pollutants transported long distances are likely small. The samples in this study mainly represent the local atmospheric conditions in Chengdu.

2.2. Mass Concentration Analysis

Before sampling, the Teflon filter membrane (Whatman, Φ90 mm, Buckinghamshire, UK) is equilibrated for at least 24 h at a temperature of 20 ± 5 °C and a relative humidity of 50 ± 5%. Quartz filter membranes (Whatman, Φ90 mm) were wrapped in aluminum foil, baked in a muffle furnace (SX-8-13, Beijing) at 500 °C for 4 h to remove the background organic matter, and then placed in the same environment as the Teflon filters for at least 24 h. After the filter membrane, use a one-hundred thousandth balance (Sartorius, Göttingen, Germany, CPA225D) was used to weigh each filter 3 times to ensure that the difference between any two weighing values did not exceed 0.04 mg. After the filter membranes were weighed, they were all wrapped in aluminum foil, and put in a sealed bag, and stored at −4 °C until analyzed. A pretreated blank filter membrane was used as a background. Before sample collection, the cutting head of the sampler filter membrane grid, sealing gasket, and other places that may be in contact with the filter membrane were wiped two to three times with high-grade pure absolute ethanol to prevent impurities from entering the filter membrane during the sampling process. Refer to "Ambient Air PM_{10} and $PM_{2.5}$ Measurement-Gravimetric Method" (HJ618-2011) for details on the method used to calculate the mass concentrations of $PM_{2.5}$ and PM_{10}.

2.3. Heavy Metals Analysis

The concentrations of heavy metals of the samples were then analyzed. Before the experiment, all Teflon vials were thoroughly cleaned with 20% hot nitric acid solution (70 °C) and deionized water to avoid contamination. Subsequently, 1/2 of a Teflon filter was dissolved with 1 mL of nitric acid (HNO_3) and 1 mL hydrofluoric acid (HF) in a closed-cap Teflon vial for 48 h at 180 °C. After that, the mixed solution was steamed to near dry, and then re-dissolved twice with 1 mL HNO_3 (120 °C). After the last re-dissolution, HNO_3 (1 mL), Rh solution (1 mL of 1000 ng/mL), and 5 mL deionized water were added and kept in Teflon vials for 6 h (100 °C). At this point, the sample pre-treatment was completed. The concentrations of the heavy metals were analyzed by inductively coupled plasma-mass spectrometry (ICP-MS, Perkin Elmer Corp., Norwalk, USA). The reference material GSS-4 was used to ensure the analytical accuracy with recovery between 94.3% and 103.6%. In addition, for 10% of the samples analysis was repeated and reagent blanks were also used to check the quality of the analysis. A total of eight metal elements were measured, including arsenic (As), cadmium (Cd), chromium (Cr), copper (Cu), nickel (Ni), lead (Pb), vanadium (V), and zinc (Zn). Their detection limits are: As (0.30 ng/m^3), Cd (0.01 ng/m^3), Cr (0.10 ng/m^3), Cu (0.04 ng/m^3), Ni (0.04 ng/m^3), Pb (0.03 ng/m^3), V (0.08 ng/m^3), and Zn (0.10 ng/m^3).

2.4. Water-Soluble Ions Determination

The main steps for the determination of water-soluble ions in the sample are as follows: putting 1/4 of the quartz filter into a 50 mL PET bottle with 20 mL of ultra-pure water and sonicated (25 °C, power 50%, Kunshan Ultrasound Instrument Co., Ltd., Kunshan, China, KQ-700DB) for 0.5 h. The bottle was transferred to a water bath shaker (Changzhou Putian Instrument Manufacturing Co., Ltd., Putian, China, SHA-CA) at room temperature and kept shaking for 30 min. The extract was then filtered through a 0.22 μm filter membrane. Anions of fluoride (F^-), chloride (Cl^-), nitrate (NO_3^-), sulfate (SO_4^{2-}), and cations of sodium (Na^+), ammonium (NH_4^+), potassium (K^+), calcium (Ca^{2+}), and magnesium (Mg^{2+}) were determined by ion chromatography (Metrohm 792). The anion column used was a Metrosep A Supp 5—150/4.0; the cation column used in ion chromatography is Metrosep C4-150. The flow rate was 0.7 mL/min. The sampling time for each run was 20 min. The anion eluent was sodium carbonate/sodium bicarbonate, fully dissolve the two in ultrapure water, and dilute them in a 100 mL volumetric flask, as a stock solution. The stock solution diluted 100 times is used as the eluent for anion determination. The cation eluent was 7.25 mM HNO_3 and 0.02 M methanesulfonic acid. The ultrapure water, reagent solutions, and samples used in the test were filtered through a 0.45 μm filter membrane. Their detection limits are F^- (0.010 μg/m^3), Cl^- (0.012 μg/m^3), NO_3^- (0.027 μg/m^3), SO_4^{2-} (0.030 μg/m^3) and Na^+ (0.019 μg/m^3), NH_4^+ (0.020 μg/m^3), K^+ (0.025 μg/m^3), Ca^{2+} (0.037 μg/m^3) and Mg^{2+} (0.020 μg/m^3).

2.5. SOR and NOR Analysis

The concentrations of sulfate, nitrate, and ammonium are related to the concentration of gaseous precursors: sulfur dioxide (SO_2), nitrogen oxides (NO_x), and ammonia (NH_3), and their conversion rates to particles generated in the atmosphere. Here SOR (sulfur oxidation rate) and NOR (nitrogen oxidation rate) are used to describe the formation of secondary aerosol species. The measured values of SO_2 and NO_2 come from the Chengdu Shilidian permanent monitoring site. Based on Ma et al. [29], the calculation formulas of SOR and NOR are:

$$SOR = nSO_4^{2-}/(nSO_4^{2-} + nSO_2)$$

$$NOR = nNO_3^-/(nNO_3^- + nNO_2)$$

where n is the molar concentration of the species. When SOR > 0.1, it indicates that there is a process of SO_2 oxidation to SO_4^{2-} in the particles. When NOR > 0.001, it is said that there is a process of oxidation of NO_2 to NO_3^- in the particulate matter. The higher value of SOR or NOR, the higher the oxidation rate of the pollutant [33].

3. Results and Discussion

3.1. PM Mass Concentration

Thirty samples, eight samples, twenty-one samples, and thirteen samples were analyzed in spring, summer, autumn, and winter, respectively. The concentrations of $PM_{2.5}$ and PM_{10} showed obvious seasonal distribution characteristics. The changes of $PM_{2.5}$ concentration with seasons (spring to winter) were: 98.62 $\mu g/m^3$, 66.75 $\mu g/m^3$, 84.02 $\mu g/m^3$, and 159.74 $\mu g/m^3$. PM_{10} concentration changes with seasons (spring to winter) were: 169.87 $\mu g/m^3$, 107.22 $\mu g/m^3$, 167.16 $\mu g/m^3$, and 260.30 $\mu g/m^3$. The concentrations of $PM_{2.5}$ and PM_{10} were the highest in winter and the lowest in summer. An inversion easily forms in winter, which prevents particles from diffusing. While the movement of atmospheric molecules and the atmospheric oxidation capacity is enhanced because of the high temperature in summer, which is conducive to the diffusion of atmospheric particles.

According to the "Ambient Air Quality Index (AQI) Technical Regulations (Trial)" (Ministry of Environmental Protection of China, 2012), the PM concentration is divided into four levels (Clean: $PM_{2.5}$ = 0~75 $\mu g/m^3$, PM_{10} = 0~150 $\mu g/m^3$; Light-Medium: $PM_{2.5}$ = 75~150 $\mu g/m^3$, PM_{10} = 150~350 $\mu g/m^3$; Heavy: $PM_{2.5}$ = 150~250 $\mu g/m^3$, PM_{10} = 350~420 $\mu g/m^3$; Severe: $PM_{2.5}$ > 250 $\mu g/m^3$, PM_{10} > 420 $\mu g/m^3$). The particulate matter concentration exceeding the clean level ($PM_{2.5}$ = 0~75 $\mu g/m^3$, PM_{10} = 0~150 $\mu g/m^3$) is defined as a haze incident. Haze incidents during the sampling period mainly occurred from March to May 2016 and November 2016 to January 2017 (Figure 2), so PM in these periods was analyzed. Figure 3a showed that there were 32 samples at the clean levels, 55 samples at the Light-Medium levels, and 12 samples at the severe levels. It is worth noting that the pollution level based on $PM_{2.5}$ did reach the Severe levels on 3 January 2017.

The concentrations of $PM_{2.5}$ that increase with the change in pollution levels were on average 48.80 $\mu g/m^3$, 109.84 $\mu g/m^3$, and 186.21 $\mu g/m^3$ for Clean, Light-medium, and Heavy levels, respectively. The concentrations of PM_{10} increase with the change in pollution levels were 98.49 $\mu g/m^3$, 226.53 $\mu g/m^3$, and 383.21 $\mu g/m^3$ for Clean, Light-medium, and Heavy levels, respectively (Figure 3b). From the perspective of the increase in particle concentration, the growth rate of PM_{10} is faster than the growth rate of $PM_{2.5}$, indicating that coarse particles ($PM_{2.5-10}$) have a certain contribution to the growth of PM_{10}. $PM_{2.5}/PM_{10}$ from Clean to Heavy pollution decreases first and then increases slightly, indicating $PM_{2.5}$ contributed the most to PM_{10} at Clean levels and the least to PM_{10} at Severe levels.

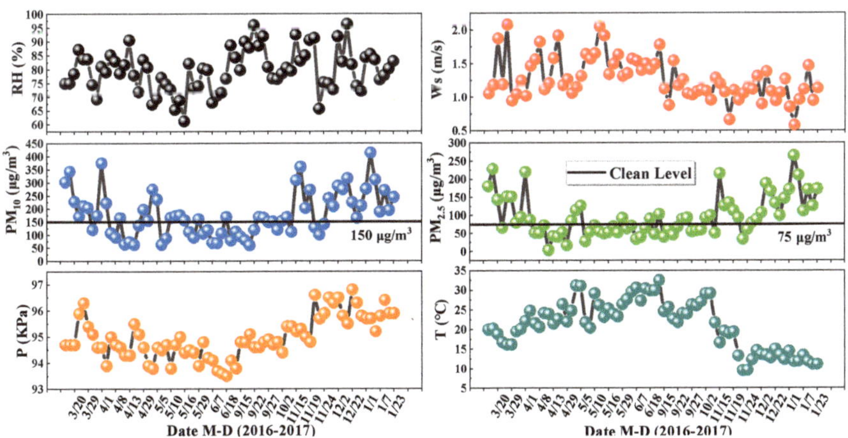

Figure 2. Time series of changes in PM mass concentration and related meteorological conditions.

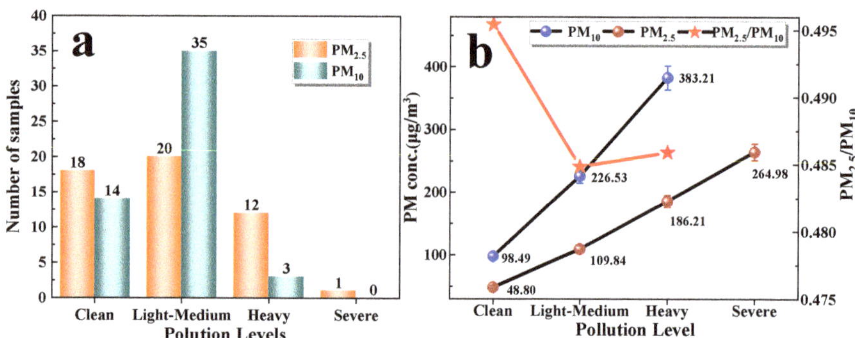

Figure 3. The number of samples that $PM_{2.5}$ and PM_{10} were at different pollution levels (**a**); the concentrations of $PM_{2.5}$ and PM_{10} at different pollution levels (**b**).

There is a correlation between the concentration of PM and related climatic conditions (Table 1). The concentration of PM is significantly negatively correlated with wind speed, temperature, and ozone, and significantly positively correlated with relative humidity, atmospheric pressure, CO, NO_2, and SO_2. The correlation for $PM_{2.5}$ and PM_{10} with temperature, CO, and NO_2 are similar. $PM_{2.5}$ has a stronger correlation with relative humidity, atmospheric pressure, ozone, and SO_2, while PM_{10} has a stronger correlation with wind speed. This shows that the influence of meteorological conditions on fine particles is greater.

Table 1. Correlation coefficient of PM with meteorological parameters and gas-phase species.

	Ws (m/s)	RH (%)	P (kPa)	T (°C)	CO ($\mu g \cdot m^{-3}$)	NO_2 ($\mu g \cdot m^{-3}$)	O_3 ($\mu g \cdot m^{-3}$)	SO_2 ($\mu g \cdot m^{-3}$)
$PM_{2.5}$	−0.75	0.98	0.93	−0.99	0.99	0.98	−0.99	0.96
PM_{10}	−0.98	0.87	0.77	−0.99	0.97	0.98	−0.91	0.93

3.2. Heavy Metals Characteristics and the Potential Use

The content of heavy metals in PM at different pollution levels is shown in Figure 4. The content of heavy metals varies greatly at different levels of pollution (average values are shown in Tables S1 and S2). At each pollution level, the heavy metal content in PM_{10} was significantly higher than that of $PM_{2.5}$. With the increase of pollution level, the total amount of heavy metals in the particles gradually increased, but the degree of increase gradually decreased. In $PM_{2.5}$ and PM_{10}, the order of heavy metal content at each pollution level was Zn > Cu> Pb > Cr > As > Ni > V > Cd. It is reported that Zn, Cu, Cr, Pb mainly come from exhaust emissions of motor vehicles or the wear of brake pads and tires [34–36], and Pb, As, Ni come from coal and petroleum combustion [13,37]. Cd is related to industrial processes [13,38], and V may come from mining or soil fertilizer use [39]. Lead, zinc, and copper account for a relatively high proportion, which is related to automobile exhaust. Urban traffic jams are becoming more and more serious, leading to frequent braking and start-up of vehicles, which aggravates the emission of heavy metals in the exhaust gas. Beijing is the city with the largest number of cars in China, and car exhaust has been studied in Beijing as a factor [40,41]. Chengdu is the second-largest city in the country for car ownership, so the contribution of car exhaust to Chengdu's atmospheric particulate matter is also significant [42].

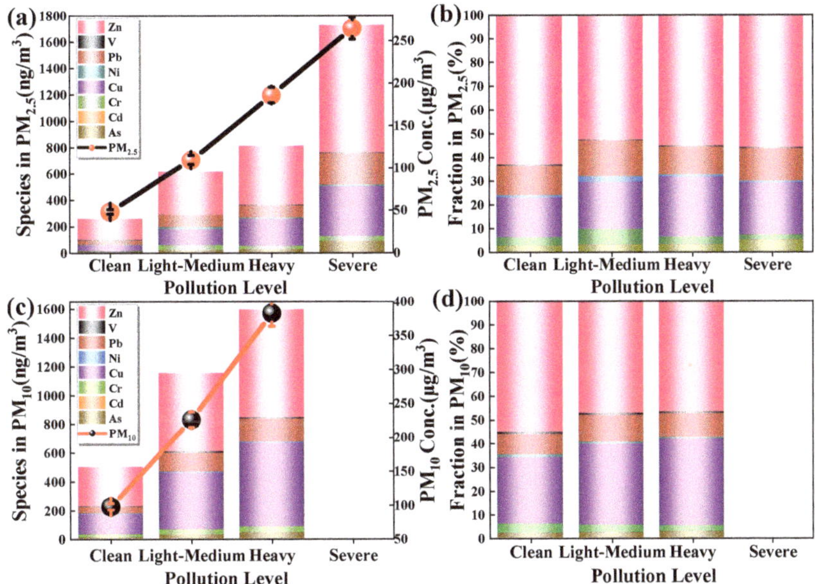

Figure 4. (**a**,**b**) are the content and percentage of heavy metals in $PM_{2.5}$ at different pollution levels; (**c**,**d**) are the content and percentage of heavy metals in PM_{10} at different pollution levels.

The relative percentage content of all heavy metals is almost constant at each pollution level. The content of heavy metals (HM) per particle at different pollution levels is shown in Figure S1 (Supplementary Information). It can be seen that the heavy metals per particle changes with the increase of particle concentration. The heavy metals content per particle in $PM_{2.5}$ is always higher than that for PM_{10}.

At the Light-Medium level, the ratio of heavy metals in $PM_{2.5}$ to heavy metals in PM_{10} is the largest, indicating that heavy metals are mainly concentrated in fine particles at this pollution level. At the Heavy levels, the content of heavy metals in PM_{10} and $PM_{2.5}$ is the smallest, and the contribution of heavy metals in $PM_{2.5}$ to that in PM_{10} is the smallest, indicating that heavy metals enriched in coarser particles may be discharged into the atmosphere at this pollution level. At the severe level, the heavy metal content increased sharply. For example, on 3rd January 2017, it was found that the wind speed was the lowest during the study period (0.5 m/s). The wind speed on the previous day (2 January) was relatively higher (0.9 m/s) and from the northwest. It is speculated that the heavy metal content on January 3rd sharply increased due to the metal sources carried by the wind from the northwest of Chengdu.

3.3. *Ions in PM*

3.3.1. Ions Characteristics at Different Pollution Levels

The ions in PM have significant differences at different pollution levels (Figure 5). From Clean to the subsequent pollution levels, the ion content in the particles increased gradually (see Tables S3 and S4 for the average values). The order of ion content in $PM_{2.5}$ was $SO_4^{2-} > NO_3^- > NH_4^+ > Cl^- > K^+ > Na^+ > Ca^{2+} > F^- > Mg^{2+}$. Among them, SO_4^{2-}, NO_3^-, and NH_4^+ accounted for 43.02%, 24.23% and 23.50% of the total ion content, respectively. The order of ion content, in PM_{10} was $SO_4^{2-} > NO_3^- > NH_4^+ > Ca^{2+} > K^+ > Cl^- > Na^+ > Mg^{2+} > F^-$, while SO_4^{2-}, NO_3^-, NH_4^+ accounted for 34.56%, 27.43%, 19.18% of the total ion content, respectively. The results showed that the secondary ions (SO_4^{2-}, NO_3^-, NH_4^+) were the main ions in Chengdu atmospheric particles.

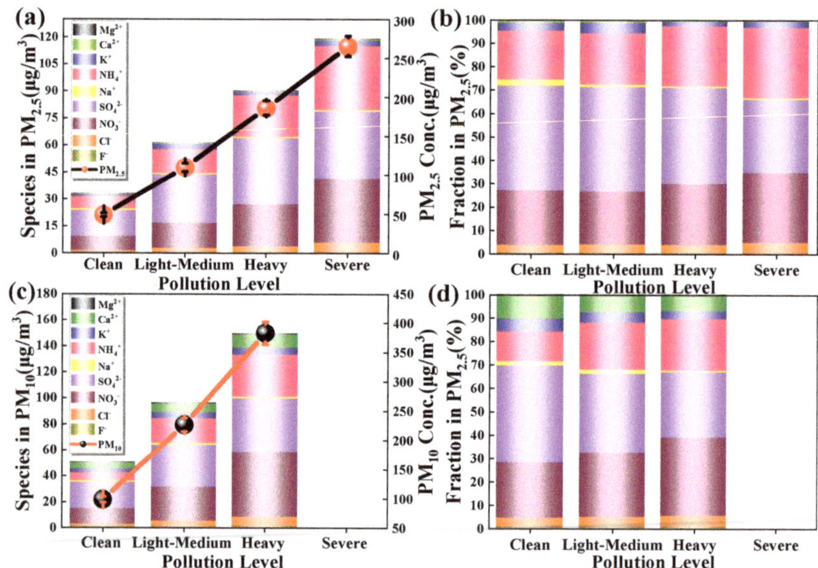

Figure 5. (**a**,**b**) are the content and percentage of water-soluble ions in $PM_{2.5}$ at different pollution levels, respectively; (**c**,**d**) are the content and percentage of water-soluble ions in PM_{10} at different pollution levels.

However, unlike the heavy metal percentage distribution, the relative percentage of each ion varies significantly at different pollution levels. In PM_{10}, the percentage of Ca^{2+} was significantly higher than that of $PM_{2.5}$, indicating that Ca^{2+} is more likely to be enriched in coarse particles. Coarse particles often are dust, and $CaCO_3$ is a major component of dust, which is the same as previous studies in Chengdu [13,27]. From Clean to the subsequent pollution levels, the relative percentages of NO_3^-, NH_4^+ gradually increased, but the relative percentages of Ca^{2+}, K^+, Na^+, and SO_4^{2-} gradually decreased, and the relative percentages of Mg^{2+}, Cl^-, and F^- were basically stable. This result demonstrated atmospheric polluting processes in Chengdu were mainly caused by particles with ions such as NO_3^- and NH_4^+ during the research period.

It is reported that with the control of SO_2 pollution in China, the sulfate content in PM has been significantly reduced [43,44]. At the same time, NO_3^- and SO_4^{2-} will interact, and NO_x will catalyze the conversion of SO_2 to SO_4^{2-} [45]. The oxidation of a large amount of SO_2 will not only produce SO_4^{2-} but also promote the formation of NO_3^- on water particles [46]. Therefore, SO_2, as the precursor of sulfate, is oxidized, as NOx is converted to NO_3^-, and the conversion of SO_2 should be slow and reduced.

Figure S2 (Supplementary Information) shows the content of ions per particle at the different pollution levels. The content of ions in particles is obviously different at different pollution levels. From Clean to Heavy or Severe, the content of ions in $PM_{2.5}$ and PM_{10} decreased gradually. The ion content per particle in $PM_{2.5}$ is always greater than that for PM_{10} at each pollution level, but the ratio of ion content per particle in $PM_{2.5}$ to PM_{10} decreases gradually from Clean to Heavy. The results indicate that ions may mainly enrich fine particles, but the proportion of ions in coarser particles gradually increases as the particle concentration increases.

3.3.2. Characteristics of Sulfate-Nitrate-Ammonium (SNA)

Figure 6 shows the changes of parameters related to SNA (SO_4^{2-}, NO_3^-, NH_4^+) at different pollution levels. From Clean to the subsequent pollution levels, SOR is always greater than NOR, but the degree the two increases with the pollution levels are different. NOR in $PM_{2.5}$ increased from 0.11 (Clean) to 0.22 (Severe), and SOR increased from 0.43

(Clean) to 0.61 (Severe). NOR in PM_{10} increased from 0.16 (Clean) to 0.29 (Heavy), and the SOR increased from 0.52 (Clean) to 0.60 (Heavy). Early studies have shown that, when SOR is greater than 0.1, there is a photochemical reaction of SO_2 in the atmosphere [47]. This result indicates that SO_2 is more susceptible to secondary conversion than NO_2. The sulfate and nitrate in this study were largely formed through secondary reactions.

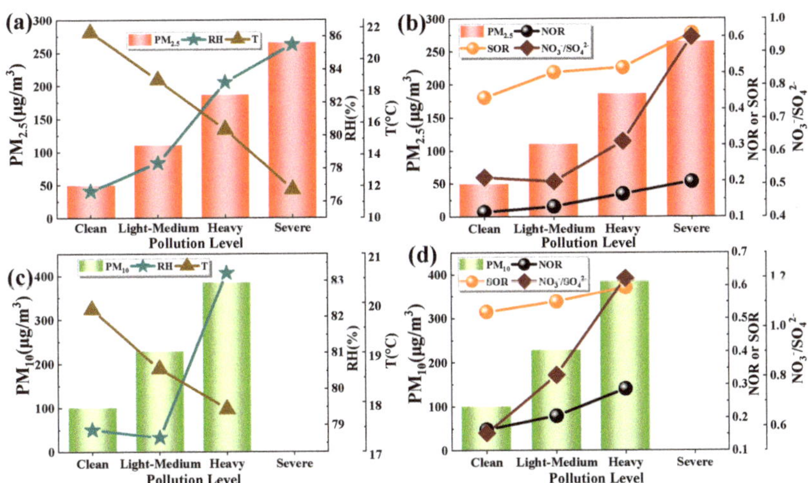

Figure 6. Changes of (**a**) RH, T, $PM_{2.5}$; (**b**)SOR, NOR, NO_3^-/SO_4^{2-}, and $PM_{2.5}$; (**c**) RH, T, PM_{10}; (**d**) SOR, NOR, NO_3^-/SO_4^{2-} and PM_{10} at different pollution levels (RH: relative humidity; SOR/NOR: sulfur/nitrogen oxidation rate; T: temperature).

The formation of SNA is closely related to meteorological conditions (relative humidity and temperature) [48–50]. When the relative humidity is low, the main reaction is a gas-phase reaction, and when the relative humidity is high, the main reaction is a heterogeneous reaction on particles [51,52]. According to Pandis and Seinfeld [53], the liquid-phase oxidation of SO_2 may be an important way to generate SO_4^{2-}, while NO_3^- is mainly generated by gas-phase oxidation of NO_x. So, the effect of humidity on SO_4^{2-} is more significant. With the increase in pollution levels, the relative humidity increased from 76% to 83%. As the air approached saturation, the particle concentration increased, and the temperature decreased ($PM_{2.5}$: 21.7–11.8 °C; PM_{10}: 19.8–17.8 °C). It can be seen that nitrate and sulfate in this study tended to form through heterogeneous reactions with the change of pollution level. Sulfate and nitrate are important hygroscopic ions, which can promote the hygroscopic growth of atmospheric particles and have a great impact on visibility and temperature [3,54,55]. The NO_3^-/SO_4^{2-} ratio has large differences at different pollution levels, which gradually increase with the increase of pollution levels, and the aerosol ions will be easier to absorb moisture [56]. NO_3^- represents mobile source, and SO_4^{2-} represents fixed source. The NO_3^-/SO_4^{2-} ratio is often used to indicate whether particulate matter is dominated by mobile source or fixed source. The NO_3^-/SO_4^{2-} ratio increased from 0.52 (Clean) to 0.95 (Severe) in $PM_{2.5}$, and increased from 0.57 (Clean) to 1.20 (Heavy) in PM_{10}. The results show that the contribution of pollution caused by mobile sources to the increase of PM is gradually increasing. In the fine particles, it is a mainly fixed pollution source at different pollution levels. While in the coarse particles, it is a mainly fixed pollution source at Clean and Light-Medium levels, and mainly mobile sources at the Heavy.

Figure 7 shows the correlation between SNA and PM at different pollution levels. The relative contribution of SNA to the increase of $PM_{2.5}$ and PM_{10} at each pollution level is different. At Clean levels, the contribution of SO_4^{2-} to the increase of $PM_{2.5}$ is 11.5%, which is much larger than the contribution of other ions, and the relative contribution of SNA to PM_{10} is very small. At Light-Medium levels, the contributions of SO_4^{2-}, NH_4^+, NO_3^- to PM differ ($PM_{2.5}$: 7.18%, 10.9%, 9.93%; PM_{10}: 10.4%, 14.0%, 14.0%). Sulfates contributed more to $PM_{2.5}$ at the Clean levels because of the trend to form ammonium sulfate during the formation of SNA in the inhomogeneous phase, which impeded the formation of ammonium nitrate. This result is reflected in more contribution of nitrate to PM at the Light-Medium levels, compared to the Clean levels. At Heavy levels, the contributions of SO_4^{2-}, NH_4^+, NO_3^- to PM are significantly different ($PM_{2.5}$: 24.0%, 3.76%, 11.7%; PM_{10}: 24.0%, 40.8%, 59.7%). This result shows sulfate is more likely to be enriched in fine particles at each level. Nitrate and ammonium salts are easily concentrated in fine particles at Clean and Light-medium pollution levels, while they are easily concentrated in coarse particles at Heavy pollution levels (NO_3^-: 59.7%; NH_4^+: 40.8%). The secondary conversion of SO_2 is mainly liquid-phase reaction, which is closely related to relative humidity. The relative humidity of the Heavy level is the largest (83%), so the contribution of sulfate to particles is also the largest at this level. The secondary reaction of NO_2 is a mainly gas-phase reaction. The atmospheric temperature is lower than other levels at Heavy levels, which is not conducive to the secondary generation of NO_2. However, it has been reported that it is conducive to the stable existence of NH_4NO_3. Therefore, the contribution of NO_3^- to PM is relatively large at Heavy levels [57].

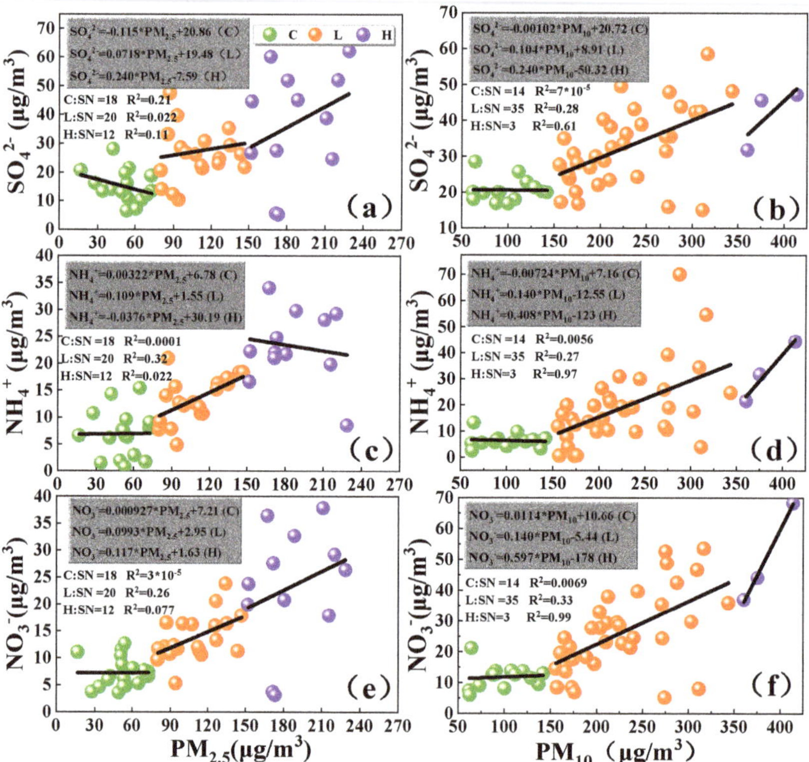

Figure 7. Linear regression of SO_4^{2-} with (**a**) $PM_{2.5}$ and (**b**) PM_{10}, NH_4^+ with (**c**) $PM_{2.5}$ and (**d**) PM_{10}, NO_3^- with (**e**) $PM_{2.5}$ and (**f**) PM_{10} at different pollution levels ($p < 0.05$; C: Clean; L: Light-Medium; H: Heavy; SN means sample number).

4. Conclusions

The present study analyzed the distribution and changes of heavy metals and water-soluble ions in $PM_{2.5}$ and PM_{10} during the haze periods from March 2016 to January 2017 in Chengdu, China at different pollution levels. It revealed the concentration of PM was closely related to meteorological conditions and the effect on fine particles is more significant. Heavy metals were more easily enriched in fine particles at different pollution levels, and the relative percentage content was basically stable. However, the relative percentage of water-soluble ions varied with the pollution level, and the relative percentage of NO_3^- and NH_4^+ increased gradually. The water-soluble ions in the particles during the study were mainly SO_4^{2-}, NO_3^- and NH_4^+ and mainly from secondary reactions. Furthermore, the contribution of SNA to the increase of PM was variable at different pollution levels. It was mainly SO_4^{2-} in PM at Clean levels, and mainly NH_4^+ and NO_3^- at Light-Medium levels. At Heavy levels, it is mainly SO_4^{2-} in $PM_{2.5}$, and mainly NH_4^+ and NO_3^- in PM_{10}. Mobile sources are contributing more to the occurrence of haze in Chengdu, which should have more attention paid to it. The results of this research not only enrich the air pollution research in Chengdu, China, but also provide a reference for the urban air pollution research with the same background. The deficiency lies in the lack amount of PM_{10} samples under Heavy and Severe pollution levels. The next step will be to study the source analysis of PM quantitatively.

Supplementary Materials: The following are available online at https://www.mdpi.com/article/10.3390/atmos12080990/s1, Figure S1: Changes of heavy metals per particle at different pollution levels, Figure S2. Change in ions per particle at different pollution levels, Table S1: Average mass concentrations of heavy metals in $PM_{2.5}$ at different pollution levels (ng/m^3), Table S2: Average mass concentrations of heavy metals in PM_{10} at different pollution levels (ng/m^3), Table S3: Average mass concentration of ions in $PM_{2.5}$ at different pollution levels (μg/m^3), Table S4: Average mass concentration of ions in PM_{10} at different pollution levels (μg/m^3).

Author Contributions: Y.H.: Conceptualization, Resources, Funding acquisition; L.W.: Data curation, Writing—Original draft preparation; X.C.: Methodology, Validation; J.W.: Methodology, Supervision; T.L. and M.H.: Investigation; H.S. and M.Z.: Visualization; S.S.H.: Writing—Reviewing and Editing; S.N.: Project administration. All authors have read and agreed to the published version of the manuscript.

Funding: This research was funded by the National Natural Science Foundation of China, grant number 41977289 and the Major Science and Technology Projects of Sichuan Province, grant number 21CXTD0015.

Institutional Review Board Statement: Not applicable.

Acknowledgments: This study was supported by the National Natural Scientific Foundation of China (41977289), and the Science and Technology Department of Sichuan Province (2021JDTD0013).

Conflicts of Interest: The authors declare that they have no known competing financial interests or personal relationships that could have appeared to influence the work reported in this paper.

References

1. Begam, G.R.; Vachaspati, C.V.; Ahammed, Y.N.; Kumar, K.R.; Reddy, R.R.; Sharma, S.K.; Saxena, M.; Mandal, T.K. Seasonal characteristics of water-soluble inorganic ions and carbonaceous aerosols in total suspended particulate matter at a rural semi-arid site, Kadapa (India). *Environ. Sci. Pollut. Res.* **2017**, *24*, 1719–1734. [CrossRef] [PubMed]
2. Wang, H.B.; Qiao, B.Q.; Zhang, L.M.; Yang, F.M.; Jiang, X. Characteristics and sources of trace elements in $PM_{2.5}$ in two megacities in Sichuan Basin of southwest China. *Environ. Pollut.* **2018**, *242*, 1577–1586. [CrossRef] [PubMed]
3. Bai, Z.P.; Ji, Y.; Pi, Y.Q.; Yang, K.X.; Wang, L.; Zhang, Y.Q.; Zhai, Y.D.; Yan, Z.G.; Han, X.D. Hygroscopic analysis of individual Beijing haze aerosol particles by environmental scanning electron microscopy. *Atmos. Environ.* **2018**, *172*, 149–156. [CrossRef]
4. Chen, J.Y.; Jiang, X.Y.; Shi, C.L.; Liu, R.C.; Lu, R.; Zhang, L. Association between gaseous pollutants and emergency ambulance dispatches for asthma in Chengdu, China: A time-stratified case-crossover study. *Environ. Health Prev. Med.* **2019**, *24*, 20. [CrossRef]
5. Zhu, X.J.; Qiu, H.; Wang, L.Y.; Duan, Z.Q.; Yu, H.Y.; Deng, R.; Zhang, Y.Y.; Zhou, L. Risks of hospital admissions from a spectrum of causes associated with particulate matter pollution. *Sci. Total Environ.* **2019**, *656*, 90–100. [CrossRef]

6. Chen, Y.; Xie, S.D. Long-term trends and characteristics of visibility in two megacities in southwest China: Chengdu and Chongqing. *J. Air Waste Manag. Assoc.* **2013**, *63*, 1058–1069. [CrossRef]
7. Ning, G.C.; Wang, S.G.; Ma, M.J.; Ni, C.J.; Shang, Z.W.; Wang, J.X.; Li, J.X. Characteristics of air pollution in different zones of Sichuan Basin, China. *Sci. Total Environ.* **2018**, *612*, 975–984. [CrossRef] [PubMed]
8. Wang, J.; Zhou, M.; Liu, B.S.; Wu, J.H.; Peng, X.; Zhang, Y.F.; Han, S.Q.; Feng, Y.C.; Zhu, T. Characterization and source apportionment of size-segregated atmospheric particulate matter collected at ground level and from the urban canopy in Tianjin. *Environ. Pollut.* **2016**, *219*, 982–992. [CrossRef]
9. Fu, X.X.; Wang, X.M.; Hu, Q.H.; Li, G.H.; Ding, X.; Zhang, Y.L.; He, Q.F.; Liu, T.Y.; Zhang, Z.; Yu, Q.Q.; et al. Changes in visibility with $PM_{2.5}$ composition and relative humidity at a background site in the Pearl River Delta region. *J. Environ. Sci.* **2016**, *40*, 10–19. [CrossRef]
10. Zhang, W.; Jiang, Y.; Luo, B. Characteristics of water-soluble ions in $PM_{2.5}$ in winter in west Sichuan plain. *Chin. J. Environ. Sci. Technol.* **2018**, *41*, 111–116.
11. Lin, Y.; Ye, Z.X.; Yang, H.J.; Zhang, J.; Zhu, Y.M. Analysis of pollution characteristics and sources of atmospheric PM_1 in downtown Chengdu. *China Environ. Sci.* **2017**, *37*, 3220–3226.
12. Li, J.Q.; Zhang, J.K.; Dong, G.M.; Deng, J.L.; Liu, Z.R.; Wang, Y.S. Characteristics of water-soluble inorganic ions in atmospheric $PM_{2.5}$ in Chengdu in the later period of "Action Plan for Air Pollution Prevention and Control". *Environ. Sci.* **2021**, 1–10. (In Chinese) [CrossRef]
13. Wang, Q.Y.; Cao, J.J.; Shen, Z.X.; Tao, J.; Xiao, S.; Luo, L.; He, Q.Y.; Tang, X.Y. Chemical characteristics of $PM_{2.5}$ during dust storms and air pollution events in Chengdu, China. *Particuology* **2013**, *11*, 70–77. [CrossRef]
14. Tao, J.; Zhang, L.M.; Engling, G.; Zhang, R.J.; Yang, Y.H.; Cao, J.J.; Zhu, C.S.; Wang, Q.Y.; Luo, L. Chemical composition of $PM_{2.5}$ in an urban environment in Chengdu, China: Importance of springtime dust storms and biomass burning. *Atmos. Res.* **2013**, *122*, 270–283. [CrossRef]
15. Xiao, K.; Wang, Y.K.; Wu, G.; Fu, B.; Zhu, Y.Y. Spatiotemporal Characteristics of Air Pollutants (PM_{10}, $PM_{2.5}$, SO_2, NO_2, O_3, and CO) in the Inland Basin City of Chengdu, Southwest China. *Atmosphere* **2018**, *9*, 74. [CrossRef]
16. Zeng, W.; Zhang, Y.C.; Wang, L.; Wei, Y.L.; Lu, R.; Xia, J.J.; Chai, B.; Liang, X. Ambient fine particulate pollution and daily morbidity of stroke in Chengdu, China. *PLoS ONE* **2018**, *13*, e0206836. [CrossRef]
17. Wang, H.B.; Tian, M.; Chen, Y.; Shi, G.M. Seasonal characteristics, formation mechanisms and geographical origins of $PM_{2.5}$ in two megacities in Sichuan Basin, China. *Atmos. Chem. Phys.* **2018**, *18*, 865–881. [CrossRef]
18. Chen, Y.; Xie, S.D.; Luo, B. Analysis of composition and pollution characteristics of fine particulate matter in Chengdu, 2012–2013. *Chin. J. Environ. Sci.* **2016**, *36*, 1021–1031.
19. Tao, J.; Gao, J.; Zhang, L.B.; Zhang, R.S.; Che, H.Z.; Zhang, Z.Z.; Lin, Z.J.; Jing, J.; Cao, J.J.; Hsu, S.C. $PM_{2.5}$ pollution in a megacity of southwest China: Source apportionment and implication. *Atmos. Chem. Phys.* **2014**, *14*, 8679–8699. [CrossRef]
20. Li, L.L.; Tan, Q.W.; Zhang, Y.H.; Feng, M.; Qu, Y.; An, J.L.; Liu, X.G. Characteristics and source apportionment of $PM_{2.5}$ during persistent extreme haze events in Chengdu, southwest China. *Environ. Pollut.* **2017**, *230*, 718–729. [CrossRef] [PubMed]
21. Shi, G.L.; Tian, Y.Z.; Ma, T.; Song, D.L.; Zhou, L.D.; Han, B.; Feng, Y.C.; Russell, A.G. Size distribution, directional source contributions and pollution status of PM from Chengdu, China during a long-term sampling campaign. *J. Environ. Sci.* **2017**, *56*, 1–11. [CrossRef]
22. Zhou, J.B.; Xing, Z.Y.; Deng, J.J.; Du, K. Characterizing and sourcing ambient $PM_{2.5}$ over key emission regions in China I: Water-soluble ions and carbonaceous fractions. *Atmos. Environ.* **2016**, *135*, 20–30. [CrossRef]
23. Song, M.D.; Liu, X.G.; Tan, Q.W.; Feng, M.; Qu, Y.; An, J.L.; Zhang, Y.H. Characteristics and formation mechanism of persistent extreme haze pollution events in Chengdu, southwestern China. *Environ. Pollut.* **2019**, *251*, 1–12. [CrossRef]
24. Zeng, S.L.; Zhang, Y. The Effect of Meteorological Elements on Continuing Heavy Air Pollution: A Case Study in the Chengdu Area during the 2014 Spring Festival. *Atmosphere* **2017**, *8*, 71. [CrossRef]
25. Feng, Y.P.; Zhang, J.K.; Huang, X.J.; Liu, Q.; Zhang, W.; Zhang, J.Q. Pollution characteristics of water-soluble inorganic ions in chengdu in summer and winter. *Huan Jing Ke Xue = Huanjing Kexue/[Bian Ji, Zhongguo Ke Xue Yuan Huan Jing Ke Xue Wei Yuan Hui "Huan Jing Ke Xue" Bian Ji Wei Yuan Hui.]* **2020**, *41*, 3012–3020. [CrossRef]
26. Jiang, Y.; He, G.Y.; Luo, B.; Chen, J.W.; Wang, B.; Du, Y.S.; Du, M. Characteristics of inorganic water-soluble ion pollution in atmospheric particulate matter in Chengdu plain. *Chin. J. Environ. Sci.* **2016**, *37*, 2863–2870.
27. Tao, Y.L.; Li, Q.K.; Zhang, J.S.; Li, Q.; Li, X.D. Seasonal variation characteristics of particle size distribution and water-soluble ion composition in Chengdu. *Chin. J. Environ. Sci.* **2017**, *38*, 4034–4043.
28. Song, X.Y.; Li, J.J.; Shao, L.Y.; Zheng, Q.M.; Zhang, D.Z. Inorganic ion chemistry of local particulate matter in a populated city of North China at light, medium, and severe pollution levels. *Sci. Total Environ.* **2019**, *650*, 566–574. [CrossRef]
29. Ma, Q.X.; Wu, Y.F.; Zhang, D.Z.; Wang, X.J.; Xia, Y.J.; Liu, X.Y.; Tian, P.; Han, Z.W.; Xia, X.G.; Wang, Y.; et al. Roles of regional transport and heterogeneous reactions in the $PM_{2.5}$ increase during winter haze episodes in Beijing. *Sci. Total Environ.* **2017**, *599–600*, 246–253. [CrossRef] [PubMed]
30. Cui, Y.; Yin, F.; Deng, Y.; Volinn, E.; Chen, F.; Ji, K.; Zeng, J.; Zhao, X.; Li, X.S. Heat or Cold: Which One Exerts Greater Deleterious Effects on Health in a Basin Climate City? Impact of Ambient Temperature on Mortality in Chengdu, China. *Int. J. Environ. Res. Public Health* **2016**, *13*, 1225. [CrossRef] [PubMed]
31. Chengdu Yearbook. Available online: http://www.chengduyearbook.com/ (accessed on 10 October 2017).

32. Chengdu Bureau of Statistics. Available online: http://cdstats.chengdu.gov.cn/ (accessed on 10 October 2017).
33. Kang, C.-M.; Lee, H.S.; Kang, B.-W.; Lee, S.-K.; Sunwoo, Y. Chemical characteristics of acidic gas pollutants and $PM_{2.5}$ species during hazy episodes in Seoul, South Korea. *Atmos. Environ.* **2004**, *38*, 4749–4760. [CrossRef]
34. Mukhtar, A.; Limbeck, A. Recent developments in assessment of bio-accessible trace metal fractions in airborne particulate matter: A review. *Anal. Chim. Acta* **2013**, *774*, 11–25. [CrossRef] [PubMed]
35. Gao, P.; Lei, T.T.; Jia, L.M.; Song, Y.; Lin, N.; Du, Y.Q.; Feng, Y.J.; Zhang, Z.H.; Cui, F.Y. Exposure and health risk assessment of $PM_{2.5}$-bound trace metals during winter in university campus in Northeast China. *Sci. Total Environ.* **2017**, *576*, 628–636. [CrossRef] [PubMed]
36. Jin, Y.L.; O'Connor, D.; Ok, Y.S.; Tsang, D.C.W.; Liu, A.; Hou, D.Y. Assessment of sources of heavy metals in soil and dust at children's playgrounds in Beijing using GIS and multivariate statistical analysis. *Environ. Int.* **2019**, *124*, 320–328. [CrossRef]
37. Tian, H.Z.; Zhao, D.; He, M.C.; Wang, Y.; Cheng, K. Temporal and spatial variation characteristics of atmospheric emissions of Cd, Cr, and Pb from coal in China. *Atmos. Environ.* **2012**, *50*, 157–163. [CrossRef]
38. Liu, G.; Li, J.H.; Wu, D.; Xu, H. Chemical composition and source apportionment of the ambient $PM_{2.5}$ in Hangzhou, China. *Particuology* **2015**, *18*, 135–143. [CrossRef]
39. Hu, R.; Zhou, X.L.; Wang, Y.N.; Fang, Y.M. Survey of atmospheric heavy metal deposition in Suqian using moss contamination. *Hum. Ecol. Risk Assess. Int. J.* **2019**, *26*, 1–15. [CrossRef]
40. Yang, Y.R.; Liu, X.G.; Qu, Y.; Wang, J.L.; An, J.L.; Zhang, Y.H.G.; Zhang, F. Formation mechanism of continuous extreme haze episodes in the megacity Beijing, China, in January 2013. *Atmos. Res.* **2015**, *155*, 192–203. [CrossRef]
41. Wang, Y.H.; Liu, Z.R.; Zhang, J.K.; Hu, B.; Ji, D.S.; Yu, Y.C.; Wang, Y.S. Aerosol physicochemical properties and implications for visibility during an intense haze episode during winter in Beijing. *Atmos. Chem. Phys.* **2015**, *15*, 3205–3215. [CrossRef]
42. Wu, M.; Wu, D.; Xia, J.R.; Zhao, T.L.; Yang, Q.J. Analysis of pollution characteristics and sources of $PM_{2.5}$ chemical components in winter in Chengdu. *Chin. J. Environ. Sci.* **2019**, *40*, 76–85.
43. Yang, Y.R.; Liu, X.G.; Qu, Y.; An, J.L.; Jiang, R.; Zhang, Y.H.; Sun, Y.L.; Wu, Z.J.; Zhang, F. Characteristics and formation mechanism of continuous hazes in China: A case study during the autumn of 2014 in the North China Plain. *Atmos. Chem. Phys.* **2015**, *15*, 8165–8178. [CrossRef]
44. Zhang, X.Y.; Wang, J.Z.; Wang, Y.Q.; Liu, H.L.; Sun, J.Y.; Zhang, Y.M. Changes in chemical components of aerosol particles in different haze regions in China from 2006 to 2013 and contribution of meteorological factors. *Atmos. Chem. Phys.* **2015**, *15*, 12935–12952. [CrossRef]
45. He, H.; Wang, Y.S.; Ma, Q.X.; Ma, J.Z.; Chu, B.W.; Ji, D.S.; Tang, G.Q.; Liu, C.; Zhang, H.X.; Hao, J.M. Mineral dust and NO_x promote the conversion of SO_2 to sulfate in heavy pollution days. *Sci. Rep.* **2014**, *4*, 4172. [CrossRef]
46. Wang, Q.Z.; Zhuang, G.S.; Huang, K.; Liu, T.N.; Lin, Y.F.; Deng, C.R.; Fu, Q.Y.; Fu, J.S.; Chen, J.K.; Zhang, W.J.; et al. Evolution of particulate sulfate and nitrate along the Asian dust pathway: Secondary transformation and primary pollutants via long-range transport. *Atmos. Res.* **2016**, *169*, 86–95. [CrossRef]
47. Ohta, S.; Okita, T. A chemical characterization of atmospheric aerosol in Sapporo. *Atmos. Environ. Part A Gen. Top.* **1990**, *24*, 815–822. [CrossRef]
48. Li, X.D.; Yang, Z.; Fu, P.Q.; Yu, J.; Lang, Y.C.; Liu, D.; Ono, K.; Kawamura, K. High abundances of dicarboxylic acids, oxocarboxylic acids, and α-dicarbonyls in fine aerosols ($PM_{2.5}$) in Chengdu, China during wintertime haze pollution. *Environ. Sci. Pollut. Res.* **2015**, *22*, 12902. [CrossRef]
49. Wang, G.H.; Zhang, R.Y.; Gomez, M.E.; Yang, L.X.; Zamora, M.L.; Hu, M.; Lin, Y.; Peng, J.F.; Guo, S.; Meng, J.J.; et al. Persistent sulfate formation from London Fog to Chinese haze. *Proc. Natl. Acad. Sci. USA* **2016**, *113*, 13630–13635. [CrossRef] [PubMed]
50. Li, Y.J.; Sun, Y.L.; Zhang, Q.; Li, X.; Li, M.; Zhou, Z.; Chan, C.K. Real-time chemical characterization of atmospheric particulate matter in China: A review. *Atmos. Environ.* **2017**, *158*, 270–304. [CrossRef]
51. He, K.B.; Yang, F.M.; Duan, F.K. *Atmospheric Particulate Matter and Regional Combined Pollution*; Science Press: Beijing, China, 2011; 450p.
52. Sun, Y.L.; Wang, Z.F.; Fu, P.Q.; Jiang, Q.; Yang, T.; Li, J.; Ge, X.L. The impact of relative humidity on aerosol composition and evolution processes during wintertime in Beijing, China. *Atmos. Environ.* **2013**, *77*, 927–934. [CrossRef]
53. Pandis, S.N.; Seinfeld, J.H. Sensitivity analysis of a chemical mechanism for aqueous-phase atmospheric chemistry. *J. Geophys. Res. Atmos.* **1989**, *94*, 1105–1126. [CrossRef]
54. Wu, Y.X.; Yin, Y.; Gu, X.S.; Tan, H.B.; Wang, Y. Observation on moisture absorption of atmospheric aerosols in northern suburbs of Nanjing. *Chin. J. Environ. Sci.* **2014**, *34*, 1938–1949.
55. Denjean, C.; Caquineau, S.; Desboeufs, K.; Laurent, K.; Maille, M.; Rosado, M.Q.; Vallejo, P.; Mayol-Bracero, O.L.; Formenti, P. Long-range transport across the Atlantic in summertime does not enhance the hygroscopicity of African mineral dust. *Geophys. Res. Lett.* **2015**, *42*, 7835–7843. [CrossRef]
56. Sun, J.X.; Liu, L.; Xu, L.; Wang, Y.Y.; Wu, Z.J.; Hu, M.; Shi, Z.B.; Li, Y.J.; Zhang, X.Y.; Chen, J.M.; et al. Key Role of Nitrate in Phase Transitions of Urban Particles: Implications of Important Reactive Surfaces for Secondary Aerosol Formation. *J. Geophys. Res. Atmos.* **2018**, *123*, 1234–1243. [CrossRef]
57. Guo, S.; Hu, M.; Wang, Z.B.; Slanina, J.; Zhao, Y.L. Size-resolved aerosol water-soluble ionic compositions in the summer of Beijing: Implication of regional secondary formation. *Atmos. Chem. Phys. Discuss.* **2009**, *9*, 23955–23986.

Article

Study of Atmospheric Pollution and Health Risk Assessment: A Case Study for the Sharjah and Ajman Emirates (UAE)

Yousef Nazzal [1], Nadine Bou Orm [1], Alina Barbulescu [2,*], Fares Howari [1], Manish Sharma [1], Alaa E. Badawi [1], Ahmed A. Al-Taani [1], Jibran Iqbal [1], Farid El Ktaibi [1], Cijo M. Xavier [1] and Cristian Stefan Dumitriu [3,*]

1 College of Natural and Health Sciences, Zayed University, Abu Dhabi P.O. Box 144534, United Arab Emirates; yousef.nazzal@zu.ac.ae (Y.N.); Nadine.BouOrm@zu.ac.ae (N.B.O.); fares.howari@zu.ac.ae (F.H.); manish.sharma@zu.ac.ae (M.S.); Alaa.Badawi@zu.ac.ae (A.E.B.); Ahmed.Al-Taani@zu.ac.ae (A.A.A.-T.); Jibran.Iqbal@zu.ac.ae (J.I.); farid.elktaibi@zu.ac.ae (F.E.K.); cijo.xavier@zu.ac.ae (C.M.X.)
2 Faculty of Civil Engineering, Transilvania University of Brasov, 2 Turnului Street, 500025 Brasov, Romania
3 SC Utilnavorep SA, 900055 Constanta, Romania
* Correspondence: alina.barbulescu@unitbv.ro (A.B.); cris.dum.stef@gmail.com (C.S.D.)

Abstract: Dust is a significant pollution source in the United Arab Emirates (UAE) that impacts population health. Therefore, the present study aims to determine the concentration of heavy metals (Cd, Pb, Cr, Cu, Ni, and Zn) in the air in the Sharjah and Ajman emirates' urban areas and assesses the health risk. Three indicators were used for this purpose: the average daily dose (ADD), the hazard quotient (HQ), and the health index (HI). Data were collected during the period April–August 2020. Moreover, the observation sites were clustered based on the pollutants' concentration, given that the greater the heavy metal concentration is, the greater is the risk for the population health. The most abundant heavy metal found in the atmosphere was Zn, with a mean concentration of 160.30 mg/kg, the concentrations of other metals being in the following order: Ni > Cr > Cu > Pb > Cd. The mean concentrations of Cd, Pb, and Cr were within the range of background values, while those of Cu, Ni, and Zn were higher than the background values, indicating anthropogenic pollution. For adults, the mean ADD values of heavy metals decreased from Zn to Cd (Zn > Ni > Cr > Cu > Pb > Cd). The HQ (HI) suggested an acceptable (negligible) level of non-carcinogenic harmful health risk to residents' health. The sites were grouped in three clusters, one of them containing a single location, where the highest concentrations of heavy metals were found.

Keywords: heavy metals; pollution; concentration; indicators; health risk assessment

Citation: Nazzal, Y.; Orm, N.B.; Barbulescu, A.; Howari, F.; Sharma, M.; Badawi, A.E.; A. Al-Taani, A.; Iqbal, J.; Ktaibi, F.E.; Xavier, C.M.; et al. Study of Atmospheric Pollution and Health Risk Assessment: A Case Study for the Sharjah and Ajman Emirates (UAE). *Atmosphere* **2021**, *12*, 1442. https://doi.org/10.3390/atmos12111442

Academic Editor: Daniele Contini

Received: 12 September 2021
Accepted: 27 October 2021
Published: 1 November 2021

Publisher's Note: MDPI stays neutral with regard to jurisdictional claims in published maps and institutional affiliations.

Copyright: © 2021 by the authors. Licensee MDPI, Basel, Switzerland. This article is an open access article distributed under the terms and conditions of the Creative Commons Attribution (CC BY) license (https://creativecommons.org/licenses/by/4.0/).

1. Introduction

Heavy metals are the most common and hazardous chemicals in the environment due to their toxicity, persistence, and bioaccumulation [1,2]. Even at low concentrations, heavy metals (lead (Pb), chromium (Cr), nickel (Ni), arsenic (As), mercury (Hg), cadmium (Cd), cobalt (Co), zinc (Zn), and copper (Cu)) are known for their high toxicity [3]. These pollutants originate from anthropogenic and natural processes [4].

Anthropogenic processes that lead to the release of heavy metals and other pollutants include industrial, agricultural, mining, and metallurgical activities. Automobile exhaust, smelting, insecticides, and fossil burning are activities that contribute significantly to environmental pollution with heavy metals, e.g., lead, arsenic, copper, zinc, nickel, vanadium, mercury, selenium, and tin [4]. On the other hand, sources of natural emissions of these metals include sea-salt sprays, volcanic eruptions, forest fires, and wind-borne soil particles.

Rock-weathering is another source of heavy metals released into the atmosphere [5]. Several studies demonstrated that high levels of heavy metals result from natural emissions and vehicles' exhaust in the traffic [6,7].

A significant ecological and public health concern is associated with the environmental contamination and heavy metals' ultimate toxic effect [8–15]. Although many heavy metals are essential micronutrients necessary for various biochemical and physiological processes and functions [8], excessive exposure to these agents results in a wide range of adverse health effects and diseases [16]. Each metal has a distinctive toxicological profile and action mechanism. These toxicological effects depend on exposed individuals' age, gender, genetics, and nutritional status. Limiting access to arsenic, cadmium, chromium, lead, and mercury is a health priority given their systemic toxicity and carcinogenic effect on the population [17].

The rapid economic and industrial development in the United Arab Emirates (UAE) has markedly impacted the country's air quality, where gases and dust are being emitted into the air in exceedingly high concentrations, rendering air pollution a critical public health problem [18–21]. Recent studies have demonstrated that road dust is a significant source of air pollution with heavy metals [21–23] and is a leading factor affecting human health [21,24,25]. Indeed, in the UAE, results of ecological risk assessments showed that Cd and Hg in road dust constitute a high public health risk [12,18]. The primary sources of heavy metal in road dust are soil materials, vehicle exhaust emissions, atmospheric deposition, and industrial and commercial activities [26–28]. The vehicles' emissions—including a complex mixture of metals from tires, brakes, parts wear and tear, and suspended road dust—are perhaps the most important source of air pollution with heavy metals [21,26,29–32] in urban areas. Long-term inhalation, ingestion, and dermal contact of these factors are associated with a wide range of acute or chronic health adverse effects [24,26] by their accumulation in the vital organs, such as the brain, liver, bones, and kidneys [33,34].

Copper is a nutrient for humans, but exposure to high concentrations can produce diseases, as Taylor et al. [35] presented in their reviews on the literature about the effects of Cu on human health. Pb is regarded as a mutagen and probable carcinogen, producing renal tumors and disturbing the reproductive and nervous systems [36]. Exposure to increased concentration of Zn has toxic effects, rarely resulting in intoxication and interferring with Cu uptake [37]. The health effects produced by Ni can be cardiovascular diseases, contact dermatitis, respiratory diseases (respiratory tract cancer, lung fibrosis, and asthma) [38,39]. Inhalation and ingestion of contaminated food and water are the main ways of introducing Ni to the organism [40]. Cadmium is a toxic metal for the population and animals, deposited in the environment by agricultural and industrial pollution [41]. Its accumulation in the human body through inhalation and ingestion provokes different types of cancer. The primary way chromium (especially in the form of Cr(III) and Cr(VI)) enters the organism is through inhalation [42], affecting the respiratory tract by producing rhinitis, pharyngitis, bronchitis.

Therefore, the present study was performed to determine the levels of heavy metals in the road dust from urban areas in the Sharjah and Ajman emirates (UAE) and to evaluate these agents' impact on public health. Clustering the observation sites (based on the studied metals' concentrations in the atmospheric dust and health indicators) was performed to determine the most polluted zones and those with the highest risk for the population.

2. Materials and Methods

2.1. Study Area

Sharjah is the third emirate in the UAE, in terms of population number. Sharjah city, the capital of this emirate, is situated at 25°21′27″ N latitude and 55°23′27″ E longitude. Ajman is the fifth largest emirate in the UAE, and its capital, with the same name, is located at 25°24′49″ N latitude and 55°26′44″ E longitude (Figure 1).

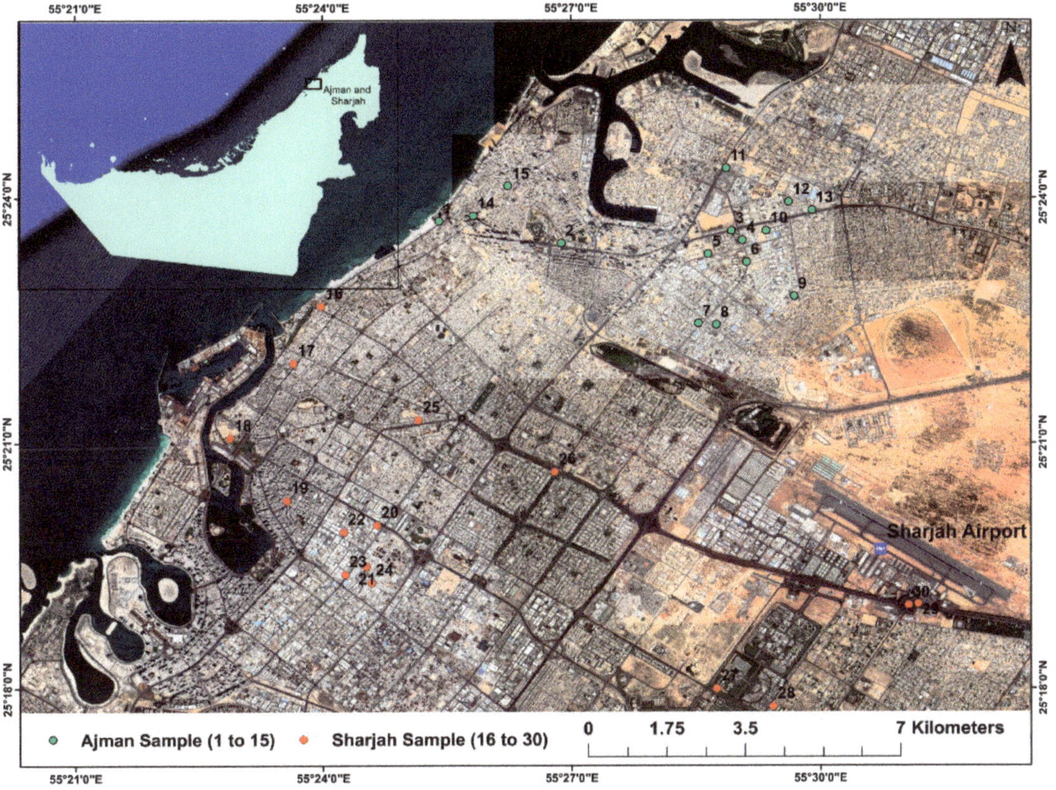

Figure 1. Study area location and sampling map.

The articles [21,25] present an extensive analysis of the climate in the region. Still, we summarize here some aspects related to the climate in the Sharjah and Ajman emirates. The study area belongs to a hot desert with warm winters and scorching and humid summers. Rainfall is generally light and erratic and occurs almost entirely from November to April. About two-thirds of annual precipitations fall in February and March [43].

The chart from Figure 2 presents the average temperatures and precipitation evolution. Figure 3 shows the cloudy, sunny, and precipitation days, precipitation amounts, maximum temperatures, and wind speed recorded at the Sharjah International Airport. Two sampling sites are situated nearby (29 and 30).

The wind rose for Sharjah International Airport (Figure 4) shows that most often throughout the year the wind blows from west to east or east to west, with speeds between 12 and 19 km/h.

Ajman has a similar climate as Sharjah.

Land use/Land cover (LULC) is the placement of activities and physical structures within a specific geographical area. It is a crucial metric for determining how human activities interact with the natural world [44]. The local, regional, and global environments are under tremendous stress due to changing land-use practices. The degradation of air quality is one of the most important environmental effects of urbanization.

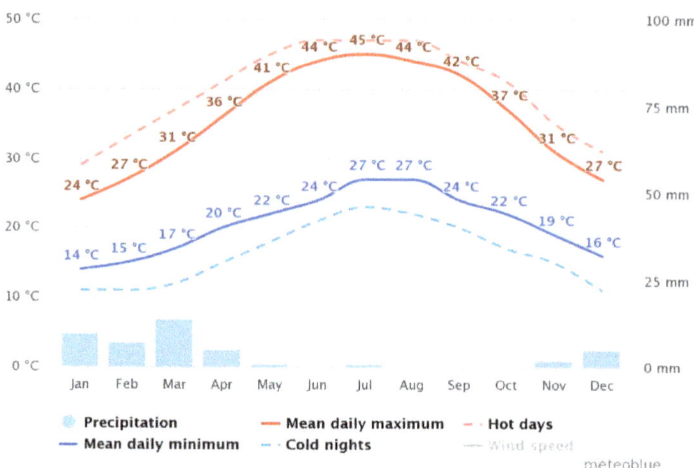

Figure 2. Average temperature and precipitation in Sharjah (International Airport).

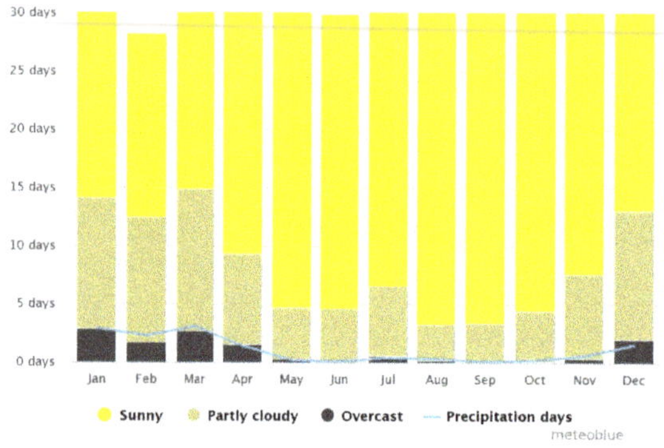

(a)

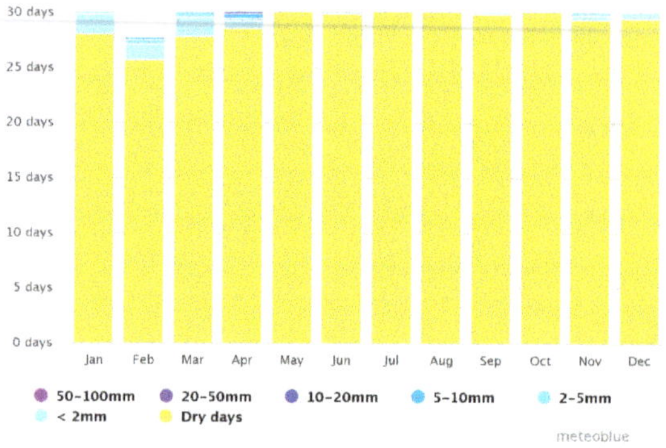

(b)

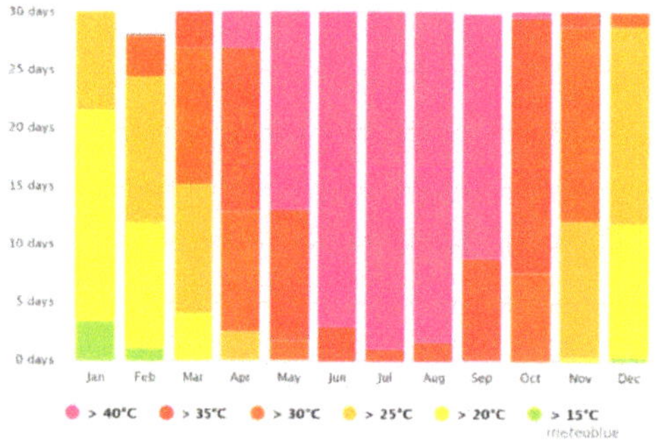

(c)

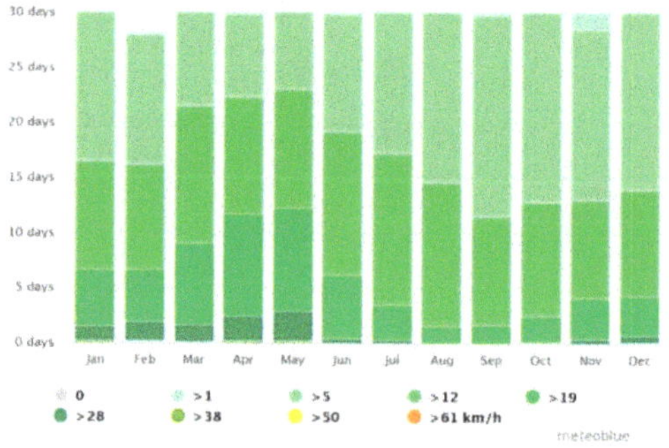

(d)

Figure 3. (**a**) Cloudy, sunny, and precipitation days; (**b**) precipitation amounts; (**c**) maximum temperatures; (**d**) wind speed.

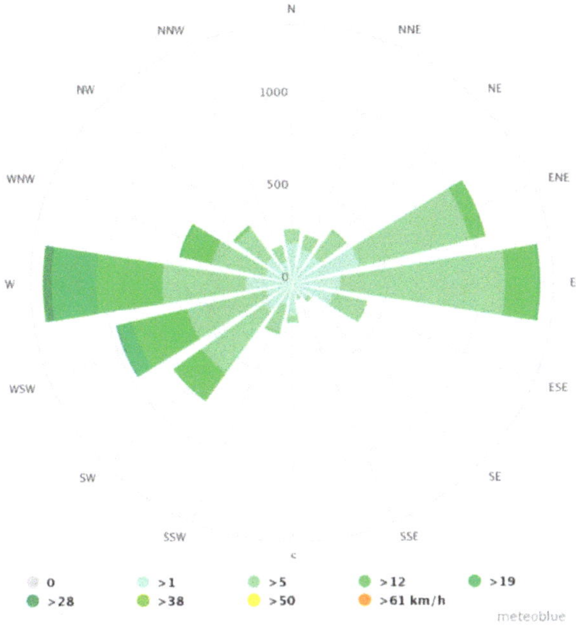

Figure 4. Wind rose for Sharjah International Airport.

Environmental and social factors, such as land use, community design, transportation networks, have been shown to have a significant impact on public health [45]. Many variables could cause particulate pollution, such as dust from construction, domestic garbage, and vehicle exhaust, but most pollution can be associated with land-use changes. Understanding the response mechanisms of urban particle pollution is crucial for pollution prevention and environmental protection [46].

To better understand the study area, we used recently released Landsat 8 satellite images for LULC mapping and monitoring in the region (Figure 5).

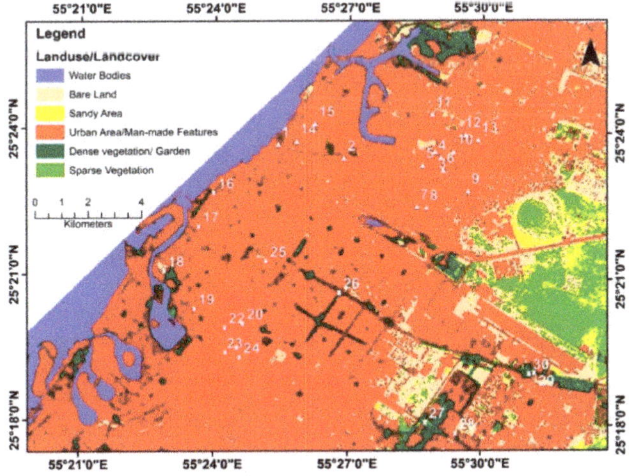

Figure 5. Landuse/Landcover (LULC) map of the study area.

Results of the land-cover analysis (Figures 5 and 6) show that 66% of the study area (187.61 km^2) mostly includes urban area/human-made features, which includes industrial sites, petrol pumps, hotels, tourist areas, residential and commercial buildings, airport, etc.

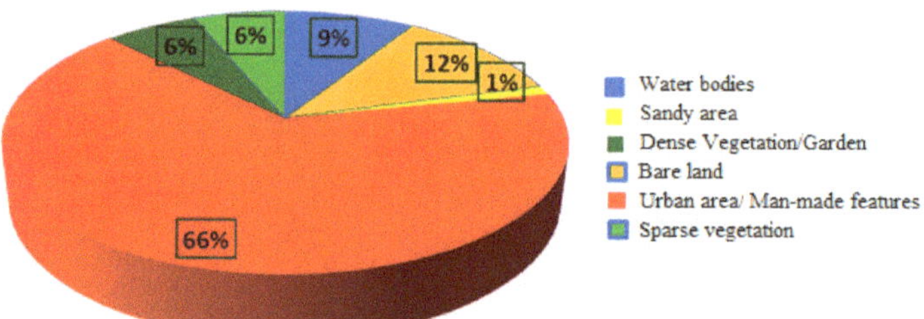

Figure 6. Pie chart showing the LULC percentage distribution of the studied region (from the LANDSAT 8).

Other land uses do not directly emit air pollutants but attract vehicular sources such as bus terminals, shopping centers, warehouses, etc.

The major categories of the land use and the associated surfaces in the study area are:

- Sparse vegetation: date palms, *Prosopis juliflora*, etc. (18.07 km^2);
- Water bodies: water in the terrestrial area and nearby sea (25.35 km^2);
- Dense vegetation/Garden: human-made garden areas and concentrated vegetation (17.77 km^2);
- Urban area/Human-made features: industrial areas, petrol pumps, hotels, tourist areas, residential and commercial buildings, airports, etc. (187.61 km^2);
- Sandy area (3.37 km^2)
- Bare land (33.52 km^2).

2.2. *Instruments and Methods*

2.2.1. Samples Collection

Dust samples were collected from thirty different Sharjah and Ajman emirates locations for five months (April–August 2020) using large dust traps placed at the height of 4 m above the ground level. Collected samples (150 at each site) were safely packed and moved to a desiccator before transporting to the laboratory. Samples were air-dried for 48 h to avoid moisture in a well-protected area. Then, each sample was sieved using a mechanical sieve shaker (Retsch, AS 200) with a 63µm filter to remove any large particles. A six-stage Anderson cascade impactor (Tecora, Italy) with an intake flow rate of 28.3 L/min was used to segregate dust particles.

Dust with a diameter lower than 10 µm was collected on the glass disks in the cascade impactor. The size ranges were 10 µm, 9.0 µm, 7.0 µm, 5.8 µm, 4.7 µm, and 3.3 µm. A cellulose nitrate filter with 100 mm diameter and 3 µm pore size was used as a backup filter.

2.2.2. Reagents, Standards, and Laboratory Ware

All experiments were performed using analytical reagent (AR) grade chemicals. The reference standard, check standard, and reagents were purchased from Sigma Aldrich. A 1:1 acid mixture was prepared using conc. nitric acid (69% *v*/*v*) and hydrochloric acid (37% *v*/*v*). Ultra-pure water with chemical resistivity of 18.2 MΩ.cm was obtained from a Merck Millipore (Massachusetts, USA) water purification system in the lab. For the sample oxidation, 30% hydrogen peroxide was used. Class-A grade glassware was utilized throughout the analysis. All glassware and plasticware were washed 5–6 times with ultrapure water followed by 10% nitric acid to remove contaminations and then air-dried. The Mars-6 system (CEM, North Carolina, USA) was employed to digest the samples. ICP-OES analysis was performed using a Perkin Elmer (Ohio, USA) Avio 200 system.

2.2.3. Samples Analysis

Sample digestion was performed by following USEPA 3050B [47] procedure. A total of 0.2 g of each sample was accurately weighed and transferred to Teflon vessels for microwave-assisted digestion. Afterwards, 10 mL of 1:1 HCl: HNO_3 were added to the digestion vessel, mixed the slurry well, and digested it using the microwave digestion system at 95 °C for 5 min. The slurry was cooled and then added to 5 mL conc. HNO_3. It was then heated and refluxed at 95 °C for 5 min, cooled, followed by the careful addition of 10% H_2O_2 for oxidation. The solutions were carefully transferred to 100 mL volumetric flasks, made up to mark with water, and filtered using Whatman 41 filters. The filtered solutions were moved to the ICP-OES system and analyzed for heavy metals. Replicate analyses were carried out on each sample.

Strict quality control and quality assurance procedures were followed to prepare and analyze samples, laboratory blanks, check standards, and standard spiked samples. Laboratory blanks were prepared using the same reagents used for the digestion without adding dust samples. The laboratory blank values for each metal were much lower than those of metals' concentrations in the target samples. Method detection (*MDL*) was calculated using the equation:

$$MDL = Mean + 2\,9 \times SD \tag{1}$$

where *Mean* is the average concentration and *SD* is the standard deviation of blanks [48]. The *MDL* values ranged between 0.02 µg/kg (Cd) and 25.2 µg/kg (K). The metals' recovery percentages (spiked and standard) were between 95% and 105%. The precision of repeated analysis was determined (for every metal) by computing the coefficient of variation, which was less than 3%.

2.3. *Heath Risk Assessment*

In this study, the impact of the pollution on the population exposed to dust metals has been assessed by computing the *ADD* (mg/kg/day) of pollutants via ingestion (ADD_{ing}), dermal contact (ADD_{derm}), and inhalation (ADD_{inh}). The utilized formulas are (2)–(4) [24,47].

$$ADD_{ing} = \frac{c \times R_{ing} \times CF \times EF \times ED}{BW \times AT}, \tag{2}$$

$$ADD_{derm} = \frac{c \times SA \times CF \times SL \times ABS \times EF \times ED}{BW \times AT}, \tag{3}$$

$$ADD_{inh} = \frac{c \times R_{inh} \times EF \times ED}{PEF \times BW \times AT}, \tag{4}$$

where the notations' meanings are given in Table 1.

Table 1. Exposure factors for dose models (adult).

Factor	Definition	Unit	Value	Reference
c	Concentration of the contaminant in dusts	mg/kg	-	This study
R_{ing}	Ingestion rate of soil	mg/day	100	[49]
AT	Average time	days	365 × ED	
BW	Average body weight	kg	55.9	Environmental site [50]
CF	Conversion factor	kg/mg	1×10^{-6}	
EF	Exposure frequency	days/year	35	
ED	Exposure duration	year	24	[50]

Table 1. *Cont.*

Factor	Definition	Unit	Value	Reference
SA	Surface area of the skin that contacts the dust	cm^2	5000	
R_{inh}	Inhalation rate	m^3/day	20	
SL	Skin adherence factor for dust	mg/cm^2	1	[50]
ABS	Dermal absorption factor (chemical specific)	-	0.001	
PEF	Particle emission factor	m^3/kg	1.32×10^9	

The model used in this study to calculate people's exposure to dust metals is based on those developed by the Environmental Protection Agency of the United States [24].

The reference dose (*RfD*) estimates the maximum acceptable risk on a population group (in this case, adults) through daily exposure during a lifetime. An unfavorable health effect during a lifetime can be signaled using the threshold of *RfD* value. No adverse health effect is concluded if the *ADD* value is lower than the reference dose. If the *ADD* value is higher than the *RfD*, the exposure pathway will likely cause harmful human health effects [24]. The reference dose (*RfD*) values of heavy metals for the ingestion, dermal contact, and inhalation are presented in Table 2 [50].

Table 2. Values of RfD for the six studied heavy metals [50].

Metal	Ingestion	Dermal	Inhalation
Cd	0.0010	0.00005	0.0030
Pb	0.0035	0.00053	0.0035
Cr	0.0050	0.00025	0.000029
Cu	0.0370	0.0011	0.0400
Ni	0.0200	0.0010	0.0210
Zn	0.300	0.0600	0.3200

After computing *ADD*, the hazard quotient (*HQ*), related to non-carcinogenic toxic risk, was calculated by dividing the daily dose by a specific reference dose (*RfD*).

$$HQ = \frac{ADD}{RfD} \tag{5}$$

The last index determined in this study is the hazard index (*HI*), representing the cumulative non-carcinogenic risk. It is estimated by summing up the hazard quotients for ingestion (HQ_{ing}), dermal (HQ_{derm}), and inhalation(HQ_{inh}):

$$HI = HQ_{ing} + HQ_{derm} + HQ_{inh} \tag{6}$$

2.4. Sites Classification

The last objective of this study was to classify the sites based on the metals concentrations in the samples and on the indexes computed in the previous section. To this aim, the k-means algorithm was utilized after choosing the optimal number of clusters by the elbow method [51,52]. A comparison of the clusters' contents was finally provided to determine the concordance between the pollution level and the health risk in the zones contained by the groups.

3. Results and Discussion

3.1. Analysis of the Heavy Metals' Concentrations

The average concentrations in the samples at the observation sites are presented in Table 3.

Table 3. Average concentration values of the metals in the samples.

Site no	Location	Latitude	Longitude	Pb (ppm)	Copper (ppm)	Zn (ppm)	Ni (ppm)	Cr (ppm)	Cd (ppm)
1	Sheraton hotel tourist area	25°23′43″	55°25′24″	6.06	34.84	89.80	142.34	61.49	0.02
2	Alnuaimiay tourist area	25°23′27″	55°26′53″	11.57	67.41	115.11	173.49	89.45	0.02
3	Ajman industrial areas and petrol stations	25°23′36″	55°28′56″	15.19	66.76	190.50	167.21	82.39	0.01
4		25°23′29″	55°29′04″	34.28	65.71	470.49	165.65	80.78	0.02
5		25°23′19″	55°28′39″	16.22	61.75	132.38	156.81	66.81	0.01
6		25°23′13″	55°29′07″	37.77	57.37	377.30	148.97	64.71	0.02
7		25°22′28″	55°28′32″	32.31	53.58	150.32	146.17	63.54	0.02
8		25°22′27″	55°28′45″	44.84	47.67	185.83	136.30	61.97	0.02
9		25°22′48″	55°29′41″	40.21	42.14	316.49	134.68	61.71	0.02
10		25°23′36″	55°29′21″	21.45	41.64	115.19	134.66	58.90	0.01
11	Ajman residential and commercial area	25°24′22″	55°28′52″	13.99	40.33	170.67	134.37	58.47	0.01
12		25°23′57″	55°29′37″	14.92	40.24	133.33	129.59	55.35	0.01
13	Adnoc Ajman	25°23′51″	55°29′54″	9.49	39.92	83.48	115.79	49.99	0.01
14	Ajman commercial area	25°23′47″	55°25′49″	16.47	37.53	101.15	114.93	49.84	0.01
15		25°24′09″	55°26′14″	11.06	35.41	106.18	108.56	49.67	0.01
16	Sharjah residential and commercial areas	25°22′41″	55°23′59″	4.54	35.16	121.45	98.72	47.61	0.01
17		25°21′59″	55°23′39″	18.49	32.99	229.41	97.02	45.50	0.01
18	Sharjah-bus station	25°21′4″	55°22′53″	20.46	31.11	152.75	96.55	44.94	0.01
19	Sharjah commerial area	25°20′18″	55°23′34″	11.06	29.22	124.60	93.76	44.85	0.01
20	Sharjah industrial area	25°19′06″	55°24′39″	52.74	28.73	192.01	90.92	41.25	0.01
21		25°19′30″	55°24′31″	24.01	27.90	127.34	89.87	39.87	0.01
22		25°19′55″	55°24′15″	20.59	27.32	105.58	84.01	38.11	0.01
23		25°19′24″	55°24′16″	15.89	25.26	106.31	83.19	37.69	0.01
24		25°19′18″	55°24′35″	4.08	25.25	55.95	79.86	35.29	0.01
25	Sharjah airport highway	25°21′17″	55°25′9″	16.15	24.53	126.11	79.42	34.81	0.01
26		25°20′39″	55°26′48″	7.05	20.69	66.94	78.66	34.19	0.01
27	Sharjah University	25°18′0″	55°28′45″	18.11	20.44	106.82	70.03	34.11	0.02
28		25°17′47″	55°29′26″	16.96	17.92	275.41	69.88	33.80	0.02
29	Sharjah airport	25°19′2″	55°31′12″	22.29	16.43	151.21	62.22	30.02	0.02
30		25°19′1″	55°31′5″	24.92	15.13	129.01	61.76	26.42	0.02

The most abundant metal measured was Zn, with a mean concentration of 160.304 mg/kg. The average concentrations of the other studied metals were, in decreasing order, Ni > Cr > Cu > Pb > Cd. The mean concentrations of Cd, Pb, and Cr were within the range of background values. The mean concentrations of Cu, Ni, and Zn were higher than the background values, indicating anthropogenic pollution.

Based on the experimental data, the maps reflecting the concentration of the metals are presented in Figure 7.

The minimum, mean, and maximum levels of heavy metals (Cd, Pb, Cr, Cu, Ni, and Zn) in the dust samples collected from the studied areas in Sharjah and Ajman are presented in Table 4.

Table 4. Extreme values of the heavy metals concentrations in the 30 samples.

Metal	Heavy Metals Concentrations in Samples (mg/kg)				Background Values of the World (mg/Kg)
	Mean	Min	Max	Std. Dev.	
Cd	0.013	0.005	0.018	0.003	0.35
Pb	20.105	4.075	52.737	12.000	35
Cr	50.783	26.416	89.445	16.100	70
Cu	37.011	15.125	67.411	15.200	30
Ni	111.513	61.762	173.486	35.600	50
Zn	160.304	55.953	470.493	92.100	90

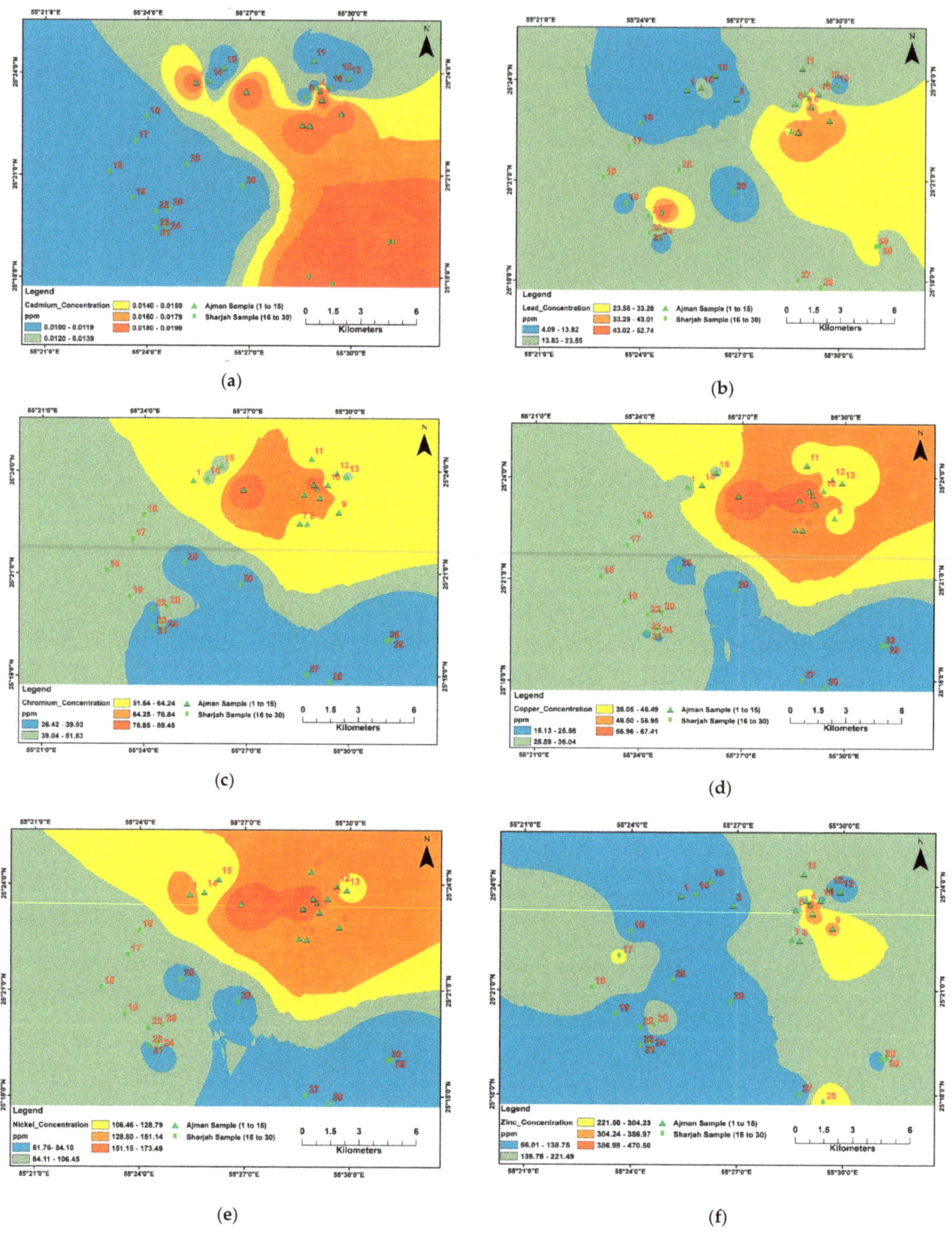

Figure 7. Maps showing the concentrations of (**a**) Cd, (**b**) Pb, (**c**) Cr, (**d**) Co, (**e**) Ni (**f**) Zn in the study area.

The composition of dust collected from industrial areas presents much higher concentrations of Zn and Ni than other metals. The highest concentration of Zn was found in samples 4, 6, and 9 (400.49, 377.30, and 316.49 mg/Kg, respectively), collected from the Ajman industrial area. The high zinc concentrations result from the steel processing activities, tire abrasion, and the corrosion of metallic parts of cars. The highest concentrations of Ni were contained by samples 7, 5, and 8 (173.49, 167.21, and 165.65 mg/Kg, respectively), collected from the Ajman industrial area. Nickel could originate from natural sources, but its presence in the air results from fuel combustion or metal plating activity.

The copper concentrations at sites 18, 28, 22, and 27 are the highest (67.41, 66.76, 65.71, and 61.75 mg/Kg). Site 18 is a bus station, and the presence of a high concentration of Cu can be attributed to traffic, tire abrasion, and the corrosion of metallic parts of cars. Site 22 is located in the Sharjah industrial area. Thus, Cu's presence can be attributed to industrial activities. The other two sites (27 and 28) are located at the University of Sharjah, where the heavy traffic can explain the high pollution.

The heavy metals concentrations in the collected dust samples from the study area were compared with those in selected cities in the world and the world reference values (Table 5). Based on the values of the pollutants' concentrations reported in different studies, the study zone occupies the first place for Cr pollution, the second one (after Hawaii) for Ni pollution, and the third for Zn pollution. These values indicate that the dust content is an issue in the area of Sharjah and Ajman.

Since each city has its specific combination of elemental compositions and the observed similarities may not reflect the actual natural and anthropogenic diversity among the different urban settings, it is necessary to establish a standard procedure to analyze the urban dust samples and draw conclusions based on the experiments [24,53].

Table 5. Heavy metals concentration in dust in different cities around the world, (mg/kg).

Location	Cr	Ni	Cu	Zn	Cd	Pb	Reference
Study area	89.44	173.48	67.91	470.49	0.018	52.73	This study
Beijing	69.33	25.97	72.13	219.20	0.64	202.82	[24]
Ottawa	43.30	15.20	65.84	112.50	0.37	39.05	[54]
Hawaii	273.0	177.0	167.0	434.0	-	106.0	[55]
Birmingham	-	41.1	466.9	534.0	1.62	48.0	[56]
Hong Kong	-	28.60	110.0	3840.0	-	120.0	[57]
Background values	70	50	30	90	0.35	35	[58]

The pollutants' concentrations recorded at different sites are not essentially influenced by wind transportation.

This conclusion results from comparing the wind rose and the metals concentrations in the samples collected at opposite sites, such as 25 and 28 or 27 and 30. We also remark that sites 29 and 30 are close to each other, but the concentrations of Zn differ. The same is valid for sites 25 and 26. This is due to the existence of small factories situated in the neighborhood of 25 and 29.

3.2. Health Risk Assessment

First, the non-carcinogenic effect on health was assessed by calculating the average daily doses (*ADD*) values, then the hazard quotient (*HQ*). The minimum, mean, and maximum levels of *ADD* and total *ADD* for adults via ingestion, dermal, and inhalation contact routes in the study area are listed in Table 6.

Table 6. Average daily dose (*ADD*) and total *ADD* for heavy metals through different pathways.

	Metal	Cd	Pb	Cr	Cu	Ni	Zn
ADD_{ing}	Mean	1.84×10^{-8}	2.75×10^{-5}	6.96×10^{-5}	5.07×10^{-5}	1.53×10^{-4}	2.20×10^{-4}
	Min.	6.85×10^{-9}	5.58×10^{-6}	3.62×10^{-5}	2.07×10^{-5}	8.46×10^{-5}	7.66×10^{-5}
	Max.	2.47×10^{-8}	7.22×10^{-5}	1.23×10^{-4}	9.23×10^{-5}	2.38×10^{-4}	6.45×10^{-4}
ADD_{derm}	Mean	4.47×10^{-11}	6.70×10^{-8}	1.69×10^{-7}	1.23×10^{-7}	3.72×10^{-7}	5.34×10^{-7}
	Min.	1.67×10^{-11}	1.36×10^{-8}	8.81×10^{-8}	5.04×10^{-8}	2.06×10^{-7}	1.87×10^{-7}
	Max.	6.00×10^{-11}	1.76×10^{-7}	2.98×10^{-7}	2.25×10^{-7}	5.78×10^{-7}	1.57×10^{-6}
ADD_{inh}	Mean	2.78×10^{-12}	4.17×10^{-9}	1.05×10^{-8}	7.68×10^{-9}	2.31×10^{-8}	3.33×10^{-8}
	Min	1.04×10^{-12}	8.46×10^{-10}	5.48×10^{-9}	3.14×10^{-9}	1.28×10^{-8}	1.16×10^{-8}
	Max.	3.74×10^{-12}	1.09×10^{-8}	1.86×10^{-8}	1.40×10^{-8}	3.60×10^{-8}	9.77×10^{-8}
Total ADD	Mean	1.84×10^{-8}	2.76×10^{-5}	6.97×10^{-5}	5.08×10^{-5}	1.53×10^{-4}	2.20×10^{-4}
	Min.	6.87×10^{-8}	5.60×10^{-6}	3.63×10^{-5}	2.08×10^{-5}	8.48×10^{-5}	7.68×10^{-5}
	Max.	2.47×10^{-8}	7.24×10^{-5}	1.23×10^{-4}	9.26×10^{-5}	2.38×10^{-4}	6.46×10^{-4}

The highest *ADD* values are those for Ni and Zn, corresponding to absorption by ingestion, while the lowest are those for Cd. The main pathway the pollutants enter the organism is ingestion. Indeed, ADD_{ing} is about 10^3 times higher than ADD_{derm} and 10^4 times higher than ADD_{inh}.

The ADD_{ing}, ADD_{derm}, and ADD_{inh} are lower than the *RfD* for the studied heavy metals, which preliminarily indicates no significant effect on the health.

The mean levels of total *ADD* (*ADD total*) (in mg/kg-day) are 1.84×10^{-8} for Cd, 2.76×10^{-5} for Pb, 6.97×10^{-5} for Cr, 5.08×10^{-5} for Cu, 1.53×10^{-4} for Ni, and 2.20×10^{-4} for Zn. The mean values of total *ADD* for adults are ordered decreasingly as follows: Zn > Ni> Cr >Cu > Pb > Cd.

The minimum, mean, and maximum levels of *HQ* and total *HQ* for adults through ingestion, dermal, and inhalation contact pathways are presented in Table 7.

Table 7. *HQ* for heavy metals through different pathways and *HI*.

	Metal	Cd	Pb	Cr	Cu	Ni	Zn
HQ_{ing}	Mean	1.84×10^{-5}	7.87×10^{-3}	1.39×10^{-2}	1.37×10^{-3}	7.64×10^{-3}	7.32×10^{-4}
	Min	6.85×10^{-6}	1.60×10^{-3}	7.24×10^{-3}	5.60×10^{-4}	4.23×10^{-3}	2.55×10^{-4}
	Max	2.47×10^{-5}	2.06×10^{-2}	2.45×10^{-2}	2.50×10^{-3}	1.19×10^{-2}	2.15×10^{-3}
HQ_{derm}	Mean	8.94×10^{-7}	1.28×10^{-4}	6.77×10^{-4}	1.13×10^{-4}	3.72×10^{-4}	8.91×10^{-6}
	Min	3.33×10^{-7}	2.59×10^{-5}	3.52×10^{-4}	4.63×10^{-5}	2.06×10^{-4}	3.11×10^{-6}
	Max	1.20×10^{-6}	3.35×10^{-4}	1.19×10^{-3}	2.06×10^{-4}	5.78×10^{-4}	2.61×10^{-5}
HQ_{inh}	Mean	2.78×10^{-9}	1.19×10^{-6}	3.69×10^{-4}	1.91×10^{-7}	1.12×10^{-6}	1.04×10^{-7}
	Min	1.04×10^{-9}	2.40×10^{-7}	1.92×10^{-4}	7.81×10^{-8}	6.22×10^{-7}	3.63×10^{-8}
	Max	3.74×10^{-9}	3.11×10^{-6}	6.49×10^{-4}	3.48×10^{-7}	1.75×10^{-6}	3.05×10^{-7}
Total HQ	Mean	1.93×10^{-5}	8.00×10^{-3}	1.50×10^{-2}	1.48×10^{-3}	8.01×10^{-3}	7.41×10^{-4}
	Min	7.18×10^{-6}	1.62×10^{-3}	7.78×10^{-3}	6.06×10^{-4}	4.44×10^{-3}	2.59×10^{-4}
	Max	2.59×10^{-5}	2.10×10^{-2}	2.63×10^{-2}	2.70×10^{-3}	1.25×10^{-2}	2.18×10^{-3}

$HQ \leq 1$ indicates no adverse health effects, while $HQ > 1$ indicates likely negative health effects [59]. All the studied heavy metals had total HQs below 1 (Table 7). Accordingly, the health risk estimation of Cd, Pb, Cr, Cu, Ni, and Zn suggests a low level of non-carcinogenic harmful health risk in all samples taken from the Ajman and Sharjah studied areas. The average hazard index *HI* is 3.32×10^{-2}. It shows a negligible non-carcinogenic risk to residents' health.

3.3. Site Clustering

Clustering has been performed for grouping the observation sites' function of the pollution impact on the population health, based on the health indexes.

The series containing the pollutants concentrations recorded at each site were normalized, and the silhouette and elbow methods (Figure 8) were used to determine the optimal number of clusters. Based on them, *k* was found to be 2 and 4.

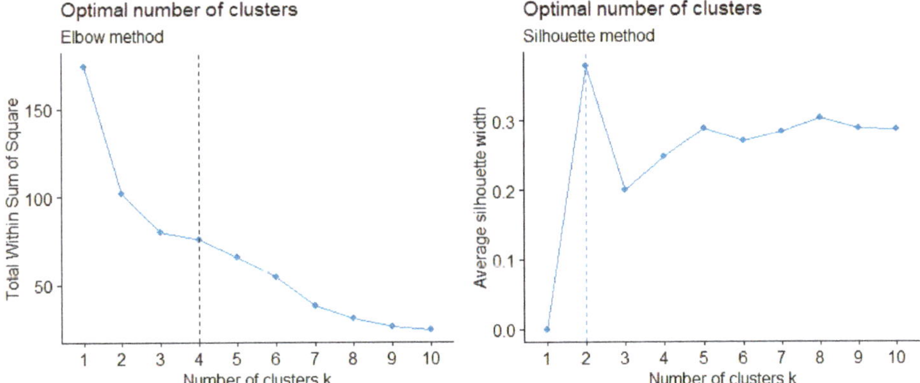

Figure 8. The elbow method for the normalized pollutants dataset.

Running the *k*-means algorithm with $k = 2$, all the sites, but the first one, are contained in the same cluster. Running the *k*-means algorithm with $k = 4$, the following sites have been included in the clusters: (1) 1; (2) 2–4, 6, 8, 9, 11–13; (3) 14–31; (4) 5, 7, 10 (Figure 9).

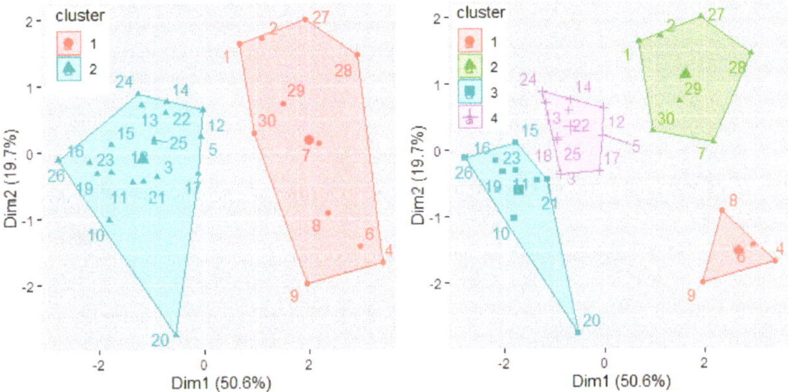

Figure 9. Clusters determined by the k-means algorithm with $k = 2$ (in the left) and $k = 4$ (in the right), based on the pollutants' concentrations.

Using $k = 2$, it resulted that the sites with the highest concentrations of Zn (4, 6, 9), Ni (7, 8), and Cu (27, 28) are in the same cluster. Still, sites 5, 18, and 22 with high concentrations of Ni and Zn are contained in the second cluster. Using $k = 4$, the sites with the highest concentrations of Zn (4, 6, 9) and Ni (8) are in the first cluster.

Samples 27 and 28 (high concentration of Cu) are kept in another cluster, while the samples with the lowest concentrations are in Cluster 3. Comparing the clustering based on the sum of squares of the distances between the groups (SSD) over the total sum of distances (TSD), the best clustering is the second (SSD/TSD = 41.5% when $k = 2$, and 60.8% for $k = 4$).

All the indices previously computed were utilized for clustering the sites. The procedure was performed using the *k*-means algorithm with $k = 2$ and $k = 4$ (determined by the elbow method). Figure 10 shows the sites' clustering (based on the health indexes).

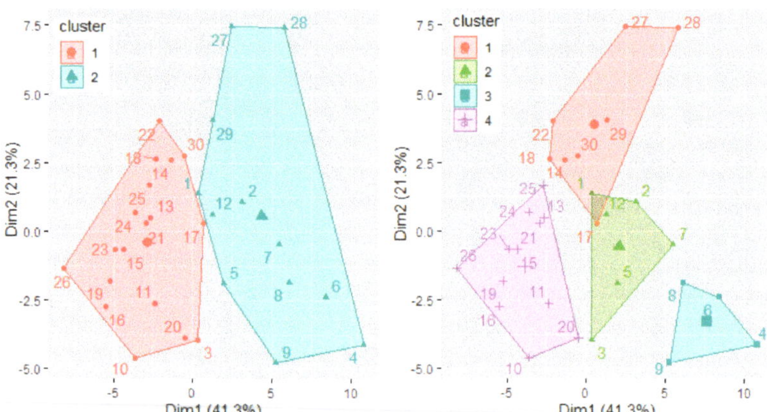

Figure 10. Clusters determined by the k-means algorithm with $k = 2$ (in the left), and $k = 4$ (in the right) based on the health risk indicators.

For $k = 2$, the sites with the highest concentrations with Ni (5, 7, 8), Zn (4, 6, 9), 27, and 28 are in the same cluster, confirming an increased risk impact on health due to high pollution with different elements in the air. For $k = 4$, the samples with the highest values of the health indices are mainly situated in clusters 1 (sites 18, 22, 27, 28), 2 (5, 7), and 3 (4, 5, 8, 9). For $k = 4$, the samples with the largest values of the health indices are mainly situated in clusters 1 (sites 18, 22, 27, 28), 2 (5, 7), and 3 (4, 5, 8, 9). The best clustering corresponds to $k = 4$ because SSD/TSD is 29.2% for $k = 2$, compared to 54.9%, for $k = 4$.

Cluster 3 from Figure 10 and Cluster 1 from Figure 9 have the same members, so the highest health risk is due to high concentrations of Zn and Ni. The sites 27–30 belong to the same cluster in Figures 9 and 10, showing a similar effect of the same pollutants on human health. Cluster 4 in Figure 10 contains the sites with the lowest impact on population health.

4. Conclusions

This study investigated the existence of heavy metals in the samples of atmospheric dust collected in the Sharjah and Ajman emirates, of the United Arab Emirates. It assessed the impact of pollutants on human health. This type of study is very significant for the residents of the UAE since the economic, industrial, and commercial development has increased the volume of exhausted gases and dust in the environment, which is severely impacting air pollution within the country.

The results show that the average concentration of heavy metals in the collected and analyzed dust samples can be ordered in decreasing order as follows: Zn > Ni > Cr > Cu > Pb > Cd. Compared with the recommended maximum allowable limits, Zn, Ni, and Cr concentrations exceeded the admissible concentrations at some locations—mainly situated in the industrial zones—indicating anthropogenic pollution. Still, at this stage of the research, the contribution of desert sand to the heavy metals pollution cannot be distinguished from that produced by anthropogenic sources.

Hazard quotient values for single and hazard index values for all studied metals are lower than the safe level for adults, indicating a non-significant non-carcinogenic. The mean values of HI through ingestion, dermal contacts, and inhalation adsorption showed a low non-carcinogenic risk to residents' health.

The clustering of the sites based on raw data and computed indices indicated four locations with the highest risks for human health (mainly due to the high concentrations of Zn and Ni).

Since many health issues of the population have been linked to air pollution with heavy metals, some measures have been proposed and are necessary to prevent such health risks [60]. They include developing detection protocols, guidelines and practices, and legislation to reduce emissions, particularly in areas with high levels of heavy metal pollution.

Since the Sharjah and Ajman cities are continuously developing, a monitoring program should be implemented. Automatic stations that record the concentrations of the most important pollutants should be placed in crowded areas and industrial zones. These should provide real-time information to the population, through electronic devices placed on visible displays. They also might be connected to a system that sends alerts to the population when the admissible pollution limit is exceeded.

Furthermore, engineering solutions are critical to both minimize pollution and prevent occupational exposure. An essential stage towards prevention is the early monitoring of human exposure to environmental pollution for a prompt action to reduce emissions and, by consequence, the adverse health effects. National collaborative efforts are needed to shape effective strategies, policies, and practices to control and prevent heavy metal toxicity.

Author Contributions: Conceptualization, Y.N. and F.H.; methodology, A.B.; software, M.S., C.M.X.; validation, A.B., F.E.K. and C.S.D.; formal analysis, A.B.; investigation, N.B.O. and A.E.B.; resources, F.H.; data curation, C.M.X., Y.N. and A.B.; writing—original draft preparation, A.A.A.-T.; writing—review and editing, A.B. and C.S.D.; visualization, J.I.; supervision, A.B.; project administration, Y.N.; funding acquisition, F.H. All authors have read and agreed to the published version of the manuscript.

Funding: The APC was funded by Zayed University, Abu Dhabi, and RIF project R20115 funded by Zayed University.

Institutional Review Board Statement: Not applicable.

Informed Consent Statement: Not applicable.

Data Availability Statement: Data will be available on request.

Conflicts of Interest: The authors declare no conflict of interest.

References

1. Ali, H.; Khan, E.; Ilahi, I. Environmental chemistry and ecotoxicology of hazardous heavy metals: Environmental persistence, toxicity, and bioaccumulation. *J. Chem.* **2019**, *2019*, 6730305. [CrossRef]
2. Briffa, J.; Sinagra, E.; Blundell, R. Heavy metal pollution in the environment and their toxicological effects on humans. *Heliyon* **2020**, *6*, e04691. [CrossRef] [PubMed]
3. 3. Herawati, N.; Suzuki, S.; Hayashi, K.; Rivai, I.F.; Koyoma, H. Cadmium, copper and zinc levels in rice and soil of Japan, Indonesia and China by soil type. *Bull. Environ. Contam. Toxicol.* **2000**, *64*, 33–39. [CrossRef] [PubMed]
4. He, Z.L.; Yang, X.E.; Stoffella, P.J. Trace elements in agroecosystems and impacts on the environment. *J. Trace Elem. Med. Biol.* **2005**, *19*, 125–140. [CrossRef]
5. Kok, J.F.; Parteli, E.J.; Michaels, T.I.; Karam, D.B. The physics of wind-blown sand and dust. *Rep. Prog. Phys.* **2012**, *75*, 106901. [CrossRef] [PubMed]
6. Jin, Y.; O'Connor, D.; Ok, Y.S.; Tsang, D.C.W.; Liu, A.; Hou, D. Assessment of sources of heavy metals in soil and dust at children's playgrounds in Beijing using GIS and multivariate statistical analysis. *Environ. Int.* **2019**, *124*, 320–328. [CrossRef] [PubMed]
7. Hou, D.; O'Connor, D.; Nathanail, P.; Tian, L.; Ma, Y. Integrated GIS and multivariate statistical analysis for regional scale assessment of heavy metal soil contamination: A critical review. *Environ. Pollut.* **2017**, *231*, 1188–1200. [CrossRef]
8. Tchounwou, P.B.; Yedjou, C.G.; Patlolla, A.K.; Sutton, D.J. Heavy metal toxicity and the environment. *Exp. Suppl.* **2012**, *101*, 133–164. [CrossRef] [PubMed]
9. Bradl, H. *Heavy Metals in the Environment: Origin, Interaction and Remediation*; Academic Press: London, UK, 2002; Volume 6.
10. Bărbulescu, A.; Dumitriu, C.S. Assessing the water quality by statistical methods. *Water* **2021**, *13*, 1026. [CrossRef]
11. Bărbulescu, A.; Barbeș, L.; Dumitriu, C.S. Statistical Assessment of the Water Quality Using Water Quality Indicators—Case study from India. In *Water Safety, Security and Sustainability. Advanced Sciences and Technologies for Security Applications*; Vaseashta, A., Maftei, C., Eds.; Springer: Amsterda, The Netherlands, 2021; pp. 599–613. [CrossRef]

12. Al-Taani, A.A.; Nazzal, Y.; Howari, F.; Iqbal, J.; Bou-Orm, N.; Xavier, C.M.; Bărbulescu, A.; Sharma, M.; Dumitriu, C.S. Contamination Assessment of Heavy Metals in Agricultural Soil, in the Liwa Area (UAE). *Toxics* **2021**, *9*, 53. [CrossRef]
13. Nazzal, Y.H.; Bărbulescu, A.; Howari, F.; Al-Taani, A.A.; Iqbal, J.; Xavier, C.M.; Sharma, M.; Dumitriu, C.Ș. Assessment of metals concentrations in soils of Abu Dhabi Emirate using pollution indices and multivariate statistics. *Toxics* **2021**, *9*, 95. [CrossRef]
14. Mihăilescu, M.; Negrea, A.; Ciopec, M.; Negrea, P.; Duțeanu, N.; Grozav, I.; Svera, P.; Vancea, C.; Bărbulescu, A.; Dumitriu, C.S. Full factorial design for gold recovery from industrial solutions. *Toxics* **2021**, *9*, 111. [CrossRef] [PubMed]
15. Aonofriesei, F.; Bărbulescu, A.; Dumitriu, C.-S. Statistical analysis of morphological parameters of microbial aggregates in the activated sludge from a wastewater treatment plant for improving its performances. *Rom. J. Phys.* **2021**, *66*, 809.
16. WHO/IAEA/FAO—World Health Organization, International Atomic Energy Agency & Food and Agriculture Organization of the United Nations. Trace Elements in Human Nutrition and Health. 1996. Available online: https://apps.who.int/iris/handle/10665/37931 (accessed on 21 August 2021).
17. Tchounwou, P.B.; Centeno, J.A.; Patlolla, A.K. Arsenic toxicity, mutagenesis, and carcinogenesis: A health risk assessment and management approach. *Mol. Cell. Biochem.* **2004**, *255*, 47–55. [CrossRef] [PubMed]
18. Al-Taani, A.A.; Nazzal, Y.; Howari, F.M. Assessment of heavy metals in roadside dust along the Abu Dhabi–Al Ain National Highway, UAE. *Environ. Earth Sci.* **2019**, *78*, 411. [CrossRef]
19. National Strategy and Action Plan for Environmental Health for the United Arab Emirates. 2010. Available online: https://sph.unc.edu/wp-content/uploads/sites/112/2013/07/report.pdf (accessed on 19 August 2021).
20. The United Arab Emirates Unified Aerosol Experiment. 2006. Available online: http://sonmi.weebly.com/uploads/2/4/7/4/24749526/the_united_arab_emirates_unified_aerosol_experiment_uae2_2006.pdf (accessed on 21 August 2021).
21. Nazzal, Y.; Bărbulescu, A.; Howari, F.; Yousef, A.; Al-Taani, A.A.; Al Aydaroos, F.; Naseem, M. New insights on sand dust storm from historical records, UAE. *Arab. J. Geosci.* **2019**, *12*, 396. [CrossRef]
22. Suryawanshi, P.V.; Rajaram, B.S.; Bhanarkar, A.D.; Chalapati Rao, C.V. Determining heavy metal contamination of road dust in Delhi, India. *Atmósfera* **2016**, *29*, 221–234. [CrossRef]
23. Bărbulescu, A.; Șerban, C.; Caramihai, S. Assessing the soil pollution using a genetic algorithm. *Rom. J. Phys.* **2021**, *66*, 806.
24. Du, Y.; Gao, B.; Zhou, H.; Ju, X.; Hao, H.; Yin, S. Health risk assessment of heavy metals in road dusts in urban parks of Beijing, China. *Procedia Environ. Sci.* **2013**, *18*, 299–309. [CrossRef]
25. Barbulescu, A.; Nazzal, Y. Statistical analysis of dust storms in the United Arab Emirate. *Atmos. Resear.* **2020**, *231*, 104669. [CrossRef]
26. Nazzal, Y.; Ghrefat, H.; Rose, M.A. Application of multivariate geostatistics in the investigation of heavy metal contamination of roadside dusts from selected highways of the Greater Toronto Area, Canada. *Environ. Earth Sci.* **2014**, *71*, 1409–1419. [CrossRef]
27. Pant, P.; Harrison, R.M. Estimation of the contribution of road traffic emissions to particulate matter concentrations from field measurements: A review. *Atmos. Environ.* **2013**, *77*, 78–97. [CrossRef]
28. Thorpe, A.; Harrison, R.M. Sources and properties of non-exhaust particulate matter from road traffic: A review. *Sci. Total Environ.* **2008**, *400*, 270–282. [CrossRef] [PubMed]
29. Apeagyei, E.; Bank, M.S.; Spengler, J.D. Distribution of heavy metals in road dust along an urban-rural gradient in Massachusetts. *Atmos. Environ.* **2011**, *45*, 2310–2323. [CrossRef]
30. Kelly, J.; Thornton, I.; Simpson, P.R. Urban geochemistry: A study of the influence of anthropogenic activity on heavy metal content of soils in traditionally industrial and nonindustrial areas of Bristol. *Appl. Geochem.* **1996**, *11*, 363–370. [CrossRef]
31. Gabarron, M.; Faz, A.; Acosta, J.A. Effect of different industrial activities on heavy metal concentrations and chemical distribution in topsoil and road dust. *Environ. Earth Sci.* **2017**, *76*, 129. [CrossRef]
32. Losacco, C.; Perillo, A. Particulate matter air pollution and respiratory impact on humans and animals. *Environ. Sci. Pollut. Res. Int.* **2018**, *25*, 33901–33910. [CrossRef]
33. Kabata-Pendias, A. *Trace Elements in Soil and Plants*, 4th ed.; Taylor & Francis: Boca Raton, FL, USA, 2011.
34. Caspah, K.; Mathuthu, M.; Madhuku, M. Health risk assessment of heavy metals in soils from Witwatersrand Gold Mining Basin, South Africa. *Int. J. Environ. Res. Public Health* **2016**, *13*, 663.
35. Taylor, A.A.; Tsuji, J.S.; Garry, M.R.; McArdle, M.E.; Goofellow, W.L., Jr.; Adams, W.J.; Menzie, C.A. Critical Review of Exposure and Effects: Implications for Setting Regulatory Health Criteria for Ingested Copper. *Environ. Manag.* **2020**, *65*, 131–159. [CrossRef]
36. Ogwuegbu, M.O.C.; Muhanga, W. Investigation of lead concentration in the blood of people in the copper belt province of Zambia. *J. Environ.* **2005**, *1*, 66–75.
37. Plum, L.M.; Rink, L.; Haase, H. The Essential Toxin: Impact of Zinc on Human Health. *Int. J. Environ. Res. Public Health* **2010**, *7*, 1342–1365. [CrossRef] [PubMed]
38. Chen, Q.Y.; Brocato, J.; Laulicht, F.; Costa, M. Mechanisms of nickel carcinogenesis. In *Essential and Non-Essential Metals. Molecular and Integrative Toxicology*; Mudipalli, A., Zelikoff, J.T., Eds.; Springer: New York, NY, USA, 2017; pp. 181–197.
39. Sinicropi, M.S.; Caruso, A.; Capasso, A.; Palladino, C.; Panno, A.; Saturnino, C. Heavy metals: Toxicity and carcinogenicity. *Pharmacologyonline* **2010**, *2*, 329–333.
40. Genchi, G.; Carocci, A.; Lauria, G.; Sinicropi, M.S.; Catalano, A. Nickel: Human Health and Environmental Toxicology. *Int. J. Environ. Res. Public Health* **2020**, *17*, 679. [CrossRef] [PubMed]
41. Genchi, G.; Sinicropi, M.S.; Lauria, G.; Carocci, A.; Catalano, A. The Effects of Cadmium Toxicity. *Int. J. Environ. Res. Public Health* **2020**, *17*, 3782. [CrossRef] [PubMed]

42. ATSDR. Agency for Toxic Substances and Disease Registry. Environmental Health and Medical Education. Chromium Toxicity. Available online: https://www.atsdr.cdc.gov/csem/chromium/physiologic_effects_of_chromium_exposure.html (accessed on 7 September 2021).
43. Ministry of Presidential Affairs, National Centre of Metrology-Climate History-Sharjah. Available online: www.ncm.ae. (accessed on 2 October 2020).
44. Dewan, A.M.; Yamaguchi, Y.; Rahman, M.Z. Dynamics of land use/cover changes and the analysis of landscape fragmentation in Dhaka Metropolitan, Bangladesh. *GeoJournal* **2012**, *77*, 315–330. [CrossRef]
45. Azapagic, A.; Chalabi, Z.; Fletcher, T.; Grundy, C.; Jones, M.; Leonardi, G.; Osammor, O.; Sharifi, V.; Swithenbank, J.; Tiwarya, A.; et al. An integrated approach to assessing the environmental and health impacts of pollution in the urban environment: Methodology and a case study. *Process. Saf. Environ.* **2013**, *91*, 508–520. [CrossRef]
46. Wei, X.; Gao, B.; Wang, P.; Zhou, H.; Lu, J. Pollution characteristics and health risk assessment of heavy metals in street dusts from different functional areas in Beijing, China. *Ecotoxicol. Environ. Saf.* **2015**, *112*, 186–192. [CrossRef]
47. U.S. EPA. *Exposure Factors Handbook (1997, Final Report)*; EPA/600/P-95/002F a–c; US Environmental Protection Agency: Washington, DC, USA, 1997. Available online: https://cfpub.epa.gov/si/si_public_record_report.cfm?Lab=NCEA&dir,EntryId=12464 (accessed on 2 October 2020).
48. Kamani, H.; Ashrafi, S.D.; Isazadeh, S.; Jaafari, J.; Hoseini, M.; Mostafapour, F.K.; Bazrafshan, E.; Nazmara, S.; Mahvi, A.H. Heavy metal contamination in street dusts with various land uses in Zahedan, Iran. *Bull. Environ. Contam. Toxicol.* **2015**, *94*, 382–386. [CrossRef]
49. U.S. EPA. Risk Assessment Guidance for Superfund, Volume 1: Human Health Evaluation Manual. EPA/540/1-89/002. 1989. Available online: https://www.epa.gov/sites/default/files/2015-09/documents/rags_a.pdf (accessed on 15 October 2020).
50. U.S. EPA. Risk Assessment Guidance for Superfund: Volume III—Part A, Process for Conducting Probabilistic Risk Assessment. 2001. Available online: https://www.epa.gov/risk/risk-assessment-guidance-superfund-rags-volume-iii-part (accessed on 15 October 2020).
51. Cluster Analysis: Basic Concepts and Algorithms. Available online: https://www-users.cs.umn.edu/~{}kumar001/dmbook/ch8.pdf (accessed on 10 November 2019).
52. Everitt, B.S.; Landau, S.; Leese, M.; Stahl, D. *Cluster Analysis*, 5th ed.; Wiley: Chichester, UK, 2011.
53. Duzgoren-Aydin, N.S.; Wong, C.S.C.; Aydin, A.; Song, Z.; You, M.; Li, X.D. Heavy metal concentrations and distribution in the urban environment of Guangzhou, SE China. *Environ. Geochem. Health* **2006**, *28*, 375–391. [CrossRef]
54. Rasmussen, P.E.; Subranmanian, K.S.; Jessiman, B.J. A multi-element profile of house dust in relation to exterior dust and soils in the city of Ottawa, Canada. *Sci. Total Environ.* **2001**, *267*, 125–140. [CrossRef]
55. Sutherland, R.A.; Tolosa, C.A. Multi-element analysis of road-deposited sediment in an urban drainage basin, Honolulu, Hawaii. *Environ. Pollut.* **2000**, *110*, 483–495. [CrossRef]
56. Charlesworth, S.; Everett, M.; McCarthy, R. A comparative study of heavy metal concentration and distribution in deposited street dusts in a large and a small urban area: Birmingham and Coventry, West Midlands, UK. *Environ. Int.* **2003**, *29*, 563–573. [CrossRef]
57. Yeung, Z.L.L.; Kwok, R.C.W.; Yu, K.N. Determination of multi-element profiles of street dust using energy dispersive X-ray fluorescence (EDXRF). *Appl. Radiat. Isot.* **2003**, *58*, 339–346. [CrossRef]
58. China National Environment Monitoring Centre. *Background Values of Soil Elements in China*; China Environmental Science Press: Beijing, China, 1990.
59. U.S. EPA. Superfund Public Health Evaluation Manual EPA/540/1-86. 1986. Available online: https://nepis.epa.gov/Exe/ZyNET.exe/2000DATB.TXT?ZyActionD=ZyDocument&Client=EPA&Index=1986+Thru+1990&Docs=&Query=&Time=&EndTime=&SearchMethod=1&TocRestrict=n&Toc=&TocEntry=&QField=&QFieldYear=&QFieldMonth=&QFieldDay=&IntQFieldOp=0&ExtQFieldOp=0&XmlQuery=&File=D%3A%5Czyfiles%5CIndex%20Data%5C86thru90%5CTxt%5C00000000%5C2000DATB.txt&User=ANONYMOUS&Password=anonymous&SortMethod=h%7C-&MaximumDocuments=1&FuzzyDegree=0&ImageQuality=r75g8/r75g8/x150y150g16/i425&Display=hpfr&DefSeekPage=x&SearchBack=ZyActionL&Back=ZyActionS&BackDesc=Results%20page&MaximumPages=1&ZyEntry=1&SeekPage=x&ZyPURL (accessed on 15 October 2020).
60. Jaishankar, M.; Tseten, T.; Anbalagan, N.; Mathew, B.B.; Beeregowda, K.N. Toxicity, mechanism and health effects of some heavy metals. *Interdiscip. Toxicol.* **2014**, *7*, 60–72. [CrossRef]

Article

The Influence of Transport on PAHs and Other Carbonaceous Species' (OC, EC) Concentration in Aerosols in the Coastal Zone of the Gulf of Gdansk (Gdynia)

Joanna Klaudia Buch, Anita Urszula Lewandowska *, Marta Staniszewska, Kinga Areta Wiśniewska and Karolina Venessa Bartkowski

Institute of Oceanography, University of Gdansk, Al. Marszałka J. Piłsudskiego 46, 81-378 Gdynia, Poland; jnna.bu@gmail.com (J.K.B.); marta.staniszewska@ug.edu.pl (M.S.); kinga.wisniewska@phdstud.ug.edu.pl (K.A.W.); karolina.bartkowski@gmail.com (K.V.B.)

* Correspondence: anita.lewandowska@ug.edu.pl; Tel.: +48-523-68-37

Abstract: The aim of this study was to determine the influence of transport on the concentration of carbon species in aerosols collected in the coastal zone of the Gulf of Gdansk in the period outside the heating season. Elemental carbon (EC), organic carbon (OC), and the $\Sigma PAHs_5$ concentrations were measured in aerosols of two size: <3 μm (respirable aerosols) and >3 μm in diameter (inhalable aerosols). Samples were collected between 13 July 2015 and 22 July 2015 (holiday period) and between 14 September 2015 and 30 September 2015 (school period). In both periods samples were taken only during the morning (7:00–9:00 a.m.) and afternoon (3:00–5:00 p.m.) road traffic hours. The highest mean values of the $\Sigma PAHs_5$ and EC were recorded in small particles during the school period in the morning road traffic peak hours. The mean concentration of OC was the highest in small aerosols during the holiday period. However, there were no statistically significant differences between the concentrations of organic carbon in the morning and afternoon peak hours. Strict sampling and measurement procedures, together with the analysis of air mass backward trajectories and pollutant markers, indicated that the role of land transport was the greatest when local to regional winds prevailed, bringing pollution from nearby schools and the beltway.

Keywords: aerosols; PAHs; OC and EC; transport sources; urbanized coastal station

Citation: Buch, J.K.; Lewandowska, A.U.; Staniszewska, M.; Wiśniewska, K.A.; Bartkowski, K.V. The Influence of Transport on PAHs and Other Carbonaceous Species' (OC, EC) Concentration in Aerosols in the Coastal Zone of the Gulf of Gdansk (Gdynia). *Atmosphere* **2021**, *12*, 1005. https://doi.org/10.3390/atmos12081005

Academic Editor: Alina Barbulescu

Received: 13 July 2021
Accepted: 1 August 2021
Published: 5 August 2021

Publisher's Note: MDPI stays neutral with regard to jurisdictional claims in published maps and institutional affiliations.

Copyright: © 2021 by the authors. Licensee MDPI, Basel, Switzerland. This article is an open access article distributed under the terms and conditions of the Creative Commons Attribution (CC BY) license (https://creativecommons.org/licenses/by/4.0/).

1. Introduction

Coastal cities tend to have cleaner air than inland cities. However, even in their atmosphere the concentration of pollutants may increase, especially in the immediate vicinity of their emission sources [1]. In urbanized coastal cities, transport plays an important role in shaping poor air quality, in addition to the municipal and housing sector. The first source appears most clearly during the heating season. In the warmer months of the year, communication can take over the role of the dominant emitter of pollutants. The term 'communication' is usually understood as road (heavy and passenger), rail, and air transport. Land-based road pollution also enters the atmosphere as a result of the abrasion of tires and brakes and re-suspension of road dust [2]. In this way, large particles with a diameter of 2.5–10 μm are emitted, while fine aerosols (<2.5 μm) are present in the atmosphere mainly as a result of fuel combustion. The quality of the atmosphere in coastal cities is also negatively affected by sea transport (e.g., ferries, container ships, bulk carriers, chemical tankers) and the proximity of ports (e.g., transshipment activities). In coastal or port regions, emissions from ships can significantly increase the concentration of NO_x, SO_2, PM_x, and their components [3,4]. The largest increase in the concentration of aerosols and their components is recorded along the traffic routes. This is manifested mainly by a high concentration of elemental carbon (EC), which is the basic indicator of air pollution from transport [5]. In the atmosphere of urbanized cities, as much as

80% of EC in aerosols come from exhaust emissions, 14% from heating houses, 4% from maritime transport, and only 2% from the energy industry and refinery activity [6]. In addition to elemental carbon, the composition of aerosols emitted in the transport sector is dominated by organic carbon, polycyclic aromatic hydrocarbons, and sulfur and nitrogen compounds [7–9]. Organic aerosols from road traffic can be released directly from the exhaust due to incomplete combustion of fuels and lubricating oil or can be formed in the atmosphere by the oxidation of traffic generated VOCs such as aromatics [10]. Lang et al., (2017) found a very high correlation coefficient (r^2) between the annual average OC concentration with vehicular OC emissions (r^2 = 0.95) and VOC emissions (r^2 = 0.9) to the atmospheric OC level [11]. In turn Zhang (2006) found that the average content of OC and EC in fine (2.5 μm) particles is 38% and 4% from gasoline cars and even higher from diesel cars (58% and 16%, respectively) [12]. Studies conducted in China by Cai et al. (2017) showed similar results for diesel vehicles in the case of OC (56.9%) in $PM_{2.5}$ [13]. However, EC content in $PM_{2.5}$ was 17.6% for heavy duty diesel, 17.7% for light duty diesel, and 8% on average for gasoline. Of course, burning fossil fuels such as gasoline and diesel releases carbon dioxide, a greenhouse gas, into the atmosphere [14]. Considering the health aspects, carbon aerosols currently require the greatest attention. Since the 1990s, it has been indicated that the presence of road pollutants in aerosols is associated with human exposure. Vehicle emissions contribute to the formation of ground level ozone, which together with other chemicals emitted by various means of transport, can trigger human health problems such as aggravated asthma, reduced lung capacity, and increased susceptibility to respiratory illnesses, including pneumonia and bronchitis [15]. The increase in air pollution from transport emission contributes also to the increase in the incidence of cardiovascular diseases and cancer. This, in turn, leads to a higher mortality, especially in urbanized areas [16–19]. Diesel particulate matter is of particular concern because long-term exposure is likely to cause lung cancer.

There is a direct relationship between the exposure to human health and life and the particle size and chemical composition. Larger particles, 2.5 to 10 μm in diameter, are retained in the upper respiratory tract, while the smallest (<2.5 μm) reach the lungs and alveoli, and even the bloodstream [20–22]. Long-term exposure of the brain to the traffic pollution slows down the maturation processes of this organ and causes changes in its functioning. This is manifested by decreased brain activity when viewing and listening. In turn, in the youngest children (up to 5 years of age) whose mothers experienced longer exposure to traffic pollution during pregnancy, structural changes in the brain were found. It has also been observed that in the left hemisphere of the brain there was a reduction in the volume of white matter, which is responsible for supporting memory [15,23–27]. Long-term exposure to polluted air also reduces the volume of brain tissue in the elderly [26,27]. Fetuses, new-born children, elder people, and people with chronic illnesses are especially susceptible to the effects of air pollutants from transport sources.

Due to the constant development of transport routes, the motorization of the population is increasing, along with the number of passenger cars. In the Gdynia region in 2005, it amounted to nearly 101,000. Ten years later it was already 57% higher (178,146 units) [28]. This phenomenon results in increased traffic, which in turn leads to the increased emissions of transport pollutants into the air. So far, it has been proven that the increase of these components is directly correlated with the proximity of traffic routes [6]. People who live, work, or attend school near major roads appear to have an increased incidence and severity of health problems associated with air pollution exposure related to roadway traffic. Children, the elderly, people with pre-existing cardiopulmonary disease, and people of low socioeconomic status are among those at higher risk of health impacts resulting from the air pollution near roadways [14]. Taking the above into account, the aim of this study was to determine the influence of transport on the concentration of carbon compounds (PAHs, OC, EC) in aerosols collected in the urbanized coastal zone of the southern Baltic Sea (Gdynia station) outside the heating season, in the morning and afternoon hours of the road traffic peak. In addition to the above, the aim of the research was to determine which period (school or vacation) and which meteorological conditions increase the role of transport in shaping high concentrations of the analyzed carbon compounds, especially in small aerosols (<3 μm in diameter), which are the most dangerous to human health.

2. Materials and Methods

2.1. Location of the Measuring Station

Aerosol samples were collected in Gdynia, at the Faculty of Oceanography and Geography of the University of Gdansk (54°30′ N, 18°32′ E). The building is located in the urbanized part of the city, about 600 m from the shoreline of the Baltic Sea (Gdansk Bay). The research station is surrounded by many traffic routes (Figure 1). The largest of them is the Tri-City ring road located to the south-west, 6000 m away from the IO UG. Moreover, heavy traffic characterizes Władysław IV Street (500 m) and Silesia Street (635 m). At a distance of 600 m from the station there is also a fast city rail. The closest is Pilsudski street (15 m), where the measurements were carried out. The Port of Gdynia is located north-west of the station, 3000 m away.

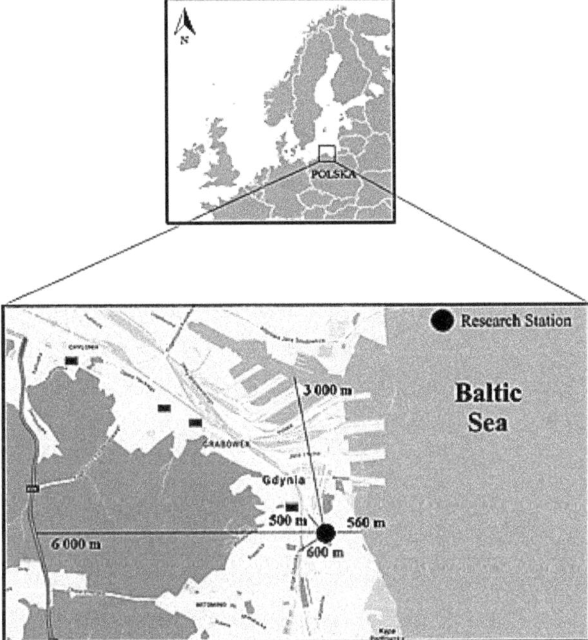

Figure 1. Location of the measuring station along with the surrounding traffic routes.

In the vicinity of the measuring station, there is increased traffic in the morning and afternoon hours, which is mainly related to the presence of numerous schools to which children are transported (Figure 2). The closest of them is located 128 m and the farthest 690 m away from the measuring station.

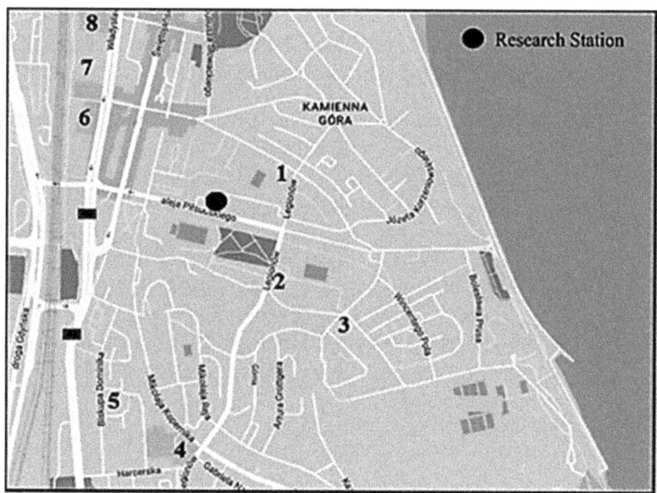

Figure 2. Location of the measuring station along with the surrounding schools.

2.2. *Aerosol Sampling*

Aerosol samples were collected in the period between 13 July 2015 and 22 July 2015 (holiday period) and between 14 September 2015 and 30 September 2015 (school period). In both cases, samples were taken outside of the heating period. It was aimed at eliminating the source related to the communal-living sector, which plays a great role in shaping the air quality in the research area. Measurements were carried out in two-hour cycles during the morning and afternoon rush hours (7:00–9:00 a.m.; 3:00–5:00 p.m.).

Aerosols were collected using a Tisch Environmental, Inc. high-flow impactor (TEI) (model: TE-235). It operates at a nominal flow of 1.132 $m^3 \cdot min^{-1}$ (40 scfm; 68 $m^3 \cdot h^{-1}$) at a pressure of 760 mm Hg and a temperature of 25 °C. Aerosols were collected on TE-QMA Micro Quartz filters, 14.3 cm × 13.7 cm in size (aerosols from 0.49 μm to 10 μm). The smallest particles, below 0.49 μm, were collected on a Whatman 41 filter, which had a size of 20.3 cm × 25.4 cm. Before use, all filters were preheated (580 °C, 6 h) and then conditioned in a desiccator for 24 h (Rh: 45% ± 5% and 20 °C ± 5 °C). All filters were weighed twice with an accuracy of 10^{-5} g on a vertical plate of a RADWAG microbalance AS 110.R2, adjusted to the size of the filters. After sampling, the filters were re-conditioned for 48 h in the desiccator and weighed twice again. All activities related to installing, removing, and weighing the filters were carried out in a laminar air flow chamber. The limit of quantification (LOQ) was set at 0.12 μg (20 replicates). The uncertainty of the method was <3.0% (at a certainty level of 99%).

2.3. *Analysis of Organic and Elemental Carbon and Polycyclic Aromatic Hydrocarbons*

The analysis of organic (OC) and elemental (EC) carbon in aerosols was performed by the thermo-optical method with the use of a thermo-optical analyzer (Sunset Laboratory Dual-Optical Carbonaceous Analyzer; protocol EUSAAR 2). For the analysis, a filter fragment with an area of 1.5 cm^2 was used. In addition to automatic calibration, an external standard (99.9% sugar solution) was analyzed every 10–15 samples [29,30]. The detection limit of the method was set to 0.1 $\mu g \cdot m^{-3}$ for both OC and EC (n = 12). The analytical error of the method was 4.5% at a confidence interval of 99% [22,29–32].

Concentrations of five PAHs (benzo(a)pyrene, benzo(a)anthracene, fluoranthene, pyrene and chrysene) were determined by means of high-performance liquid chromatography using a Dionex UltiMate 3000 analyzer with a fluorescence detector (benzo(a)pyrene λex. = 296 nm, λem. = 408 nm; fluoranthene and pyrene λex. = 270 nm, λem. = 440 nm; benzo(a)antracene and chrysene λex. = 275 nm, λem. = 380 nm). The isolation of PAHs was conducted by means of solvent extraction (acetonitrile: dichloromethane 3:1 v/v) in an ultrasonic bath [33]. The concentration values for the standard curve ranged from 0.125 to 10 $ng \cdot cm^{-3}$. The limit of quantification was 0.01 $ng \cdot cm^{-3}$. The recovery determined against the reference material (SRM-2585) was 83%, 78%, 91%, 91%, and 99% for BaP, FLA, PYR, B(a)A, and CHR, respectively [32–34].

2.4. Anion Analysis

Prior to chromatographic analysis, a fragment of 10.8 cm^2 was cut from quartz filters with dimensions of 14.3 cm × 13.7 cm, while a fragment of 3.8 cm^2 was cut from a filter with dimensions of 20.3 cm × 25.4 cm. Next, the cut filters were placed in polyethylene tubes and 12 cm^3 of milli-Q water was added. The next step was to sonicate the samples (20 min) in an ultrasonic bath (Sonic 6D, Sonic 10, Polsonic Palczyński, Warsaw, Poland) in order to bring the ions into solution. The extract obtained in this way was filtered through membrane filters with a pore diameter of 0.25 μm. The ions NO_3^- and SO_4^{2-} were determined by ion chromatography 881 Compact IC pro (Metrohm AG, Herisau, Switzerland) in accordance with Polish Standard PrPN-EN No 10304-1. For sulphates and nitrates, the limit of detection was 0.1 $\mu g \cdot m^{-3}$ and 0.2 $\mu g \cdot m^{-3}$, respectively, and the error of the method was 4.7% and 5.5%. In all cases, a confidence level of 99% was assumed [35].

2.5. Variation of Meteorological Parameters

Gdynia, where aerosol samples were collected, lies in the temperate climate zone, which is constantly modified by the influence of the vicinity of the Baltic Sea. Such a location determines the less severe winters and, at the same time, mild summers. The average annual temperature for the summer period is 14 °C, and for the winter period it is 2.3 °C. Average precipitation totals are 590 mm (1971–2000) with maximum values in July (13%). The dominant wind direction in Gdynia is westerly (1981–2010) [36].

During the research period, the highest average temperature value of 19.4 °C was recorded during the afternoon traffic rush during the summer holidays in July (with a maximum of 26.8 °C; 21 July; 3:00–5:00 p.m.) (Table 1).

The lowest temperature was also noted in the afternoon rush hour in September (school period), and amounted to 16.8 °C (with a minimum of 11.5 °C, 30.09; 3:00–5:00 p.m.) (Table 1). Relative air humidity ranged from 31% (17/7; 7:00–9:00 a.m.) to 83% (16/9; 3:00–5:00 p.m.). Higher RH values were recorded during the school period than during the holiday period. The mean wind speed values were slightly higher during the holiday season (3.2 $m \cdot s^{-1}$) than during the school season (2.0 $m \cdot s^{-1}$). The maximum wind speed was recorded on 17 July (vacation period) during the morning rush hour and it was 9.8 $m \cdot s^{-1}$. The lowest wind speed was recorded on 16 September between 7:00 a.m. and 9:00 a.m. and it was equal to 0.1 $m \cdot s^{-1}$ (Table 1). The mean atmospheric pressure was higher during the school period. The highest pressure was recorded on 29 September in the morning peak hours (1033 hPa), and the lowest on 16 September in the afternoon (999 hPa).

During the summer holidays, in the morning rush hour the westerly wind direction was dominant (80%), and in the afternoon south-westerly (46%) and southerly (34%) winds dominated. During the school period, in the morning traffic hours, the south-west direction had the highest share (84%), while in the afternoon traffic hours winds from the south-east direction were predominant (46%) (Table 1).

Table 1. Statistical characteristics of meteorological data during the research conducted in Gdynia in 2015.

	Holiday Period (13–21 July 2015)		**School Period (14–30 July 2015)**	
	7:00–9:00 a.m. Average (Min–Max)	**3:00–5:00 p.m. Average (Min–Max)**	**7:00–9:00 a.m. Average (Min–Max)**	**3:00–5:00 p.m. Average (Min–Max)**
T [°C]	18.9 ± 1.2 (15.3–21.3)	19. 4 ± 2.2 (16.8–26.8)	15.9 ± 2.4 (12.4–20.7)	16.8 ± 2.3 (11.5–20.0)
Rh [%]	54 ± 8 (31–72)	55 ± 7 (39–77)	64 ± 9 (44–78)	60 ± 8 (41–83)
Vw [m/s]	3.5 ± 0.8 (0.6–9.8)	2.8 ± 1.2 (0.2–8.5)	2.1 ± 0.9 (0.1–6.5)	1.9 ± 0.9 (0.3–6.4)
P [hPa]	1007 ± 4 (1001–1012)	1008 ± 3 (1003–1012)	1015 ± 13 (1002–1033)	1015 ± 14 (999–1033)
∑ precipitation [mm]	8		35	
Wind Direction	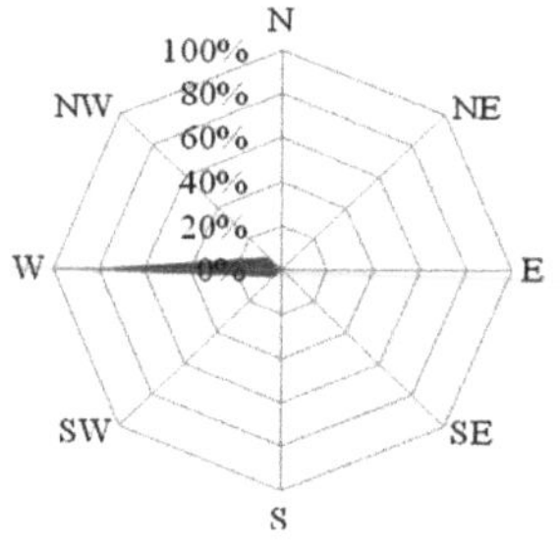 7:00–9:00 a.m. 3:00–5:00 p.m.	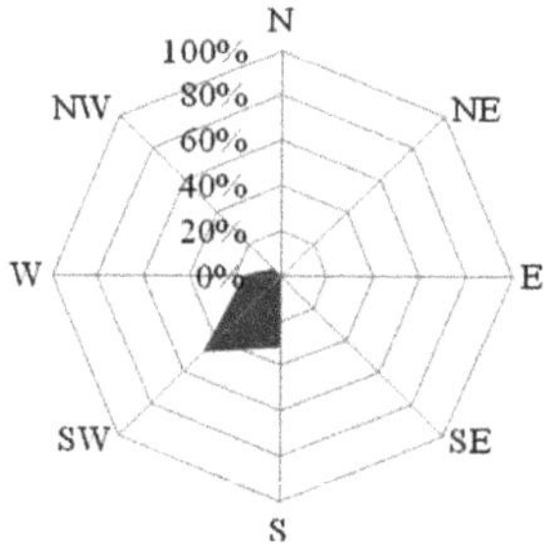	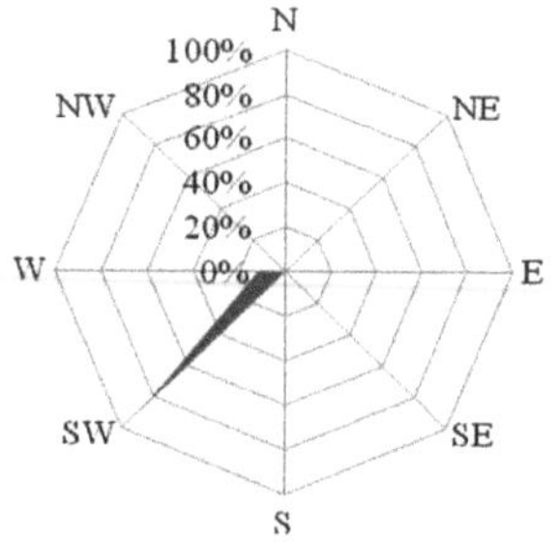 7:00–9:00 a.m. 3:00–5:00 p.m.	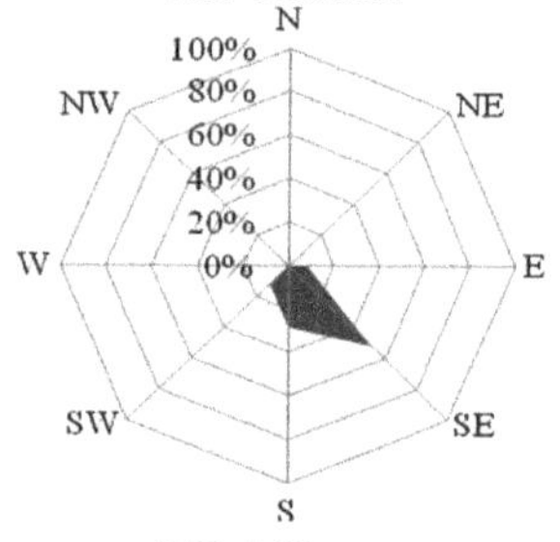

2.6. Atmospheric Air Pollution Indicators from Transport Sources Used in the Work

There are several indicators that allow us to estimate whether the chemical composition of aerosols in a given research area is determined by the emission from transport sources. While EC has a primary origin, OC can be both primarily emitted but also formed in the atmosphere through condensation to the aerosol phase of low vapor pressure compounds emitted primary as pollutants or formed in the atmosphere. Thereby, a large fraction of OC in the atmosphere has a secondary origin. Because of this, the OC/EC ratio in aerosol fractions differs widely, both in space and seasonally, and it could be a useful diagnostic ratio to investigate sources and processes happening in the atmosphere, which could lead to the formation of secondary organic compounds [37–40]. The value of the OC/EC ratio depends on the emission sources associated with different combustion processes. Higher concentrations of OC and EC occur during the heating season [41]. They are also increasing in areas of heavy traffic. In general, both OC and EC are characterized by higher concentrations near traffic routes than in rural or industrial areas [42–45]. When the OC/EC value is between 2.6 and 6.0, the organic carbon comes from the combustion of fossil fuels [43]. It is assumed that for biomass combustion, the coefficient exceeds 6 [46–48]. Pio et al., (2011) measured the OC and EC at both roadside and urban background sites in Portugal and the UK and obtained the lowest OC/EC ratio ranging from 0.3 to 0.4 for the road-generated aerosols. The results of Pio et al., (2011) are in agreement with the findings of Yu et al., (2011) [49]. On the other hand, they are lower than measured by Hildemann et al., (1991) [50] for particles emitted from gasoline (OC/EC = 2.2) and diesel vehicles (0.8). The latter results may be the consequence of using other methods of estimating OC and EC concentrations in the 1990s.

Polycyclic aromatic hydrocarbons (PAHs) have also been used as indicators of atmospheric pollution from transport sources in various areas of the world. For example, Masclet et al., (1986) [51] and Miguel et al., (1998) [52] found that the gasoline engine emissions were enriched in benzo(ghi)perylene and coronene and diesel exhausts emitted mainly chrysene, fluoranthene, and pyrene. In turn, Duan et al., (2016) [53] noted that fluoranthene, naphthalene, phenanthrene, pyrene, fluorene, chrysene, and benzo(a)pyrene are dominant PAHs emitted from coal-fired power plants. For a heavy oil and natural gas fueled-boiler, naphthalene, phenanthrene, fluoranthene, pyrene, fluorene, and benzo(b)chrysene were found to be the major PAHs. Sometimes relationships are found that allow us to determine the origin of PAHs in aerosols. For example, a B(a)A/chrysene ratio above 1 suggests that the source of the aerosols is fuel combustion. A similar source is indicated by a B(a)A/(B(a)A + CHR) ratio above 0.2 and a fluoranthene/pyrene ratio above 1. The ratio of fluoranthene/(pyrene + fluoranthene) within the range of 0.4–0.5 indicates combustion liquid fuels, and when its value is higher than 0.5, it implies burning coal and biomass. When the value of the above index falls below 0.4, the carbon source is oil combustion [7].

Another well-recognized marker is the aerosol nitrate to sulfate ratio. It is used to distinguish the air pollution coming from mobile sources from those emitted by stationary sources (point emitters, e.g., power plants, refineries). When nitrate ions dominate over sulphate ions in aerosols, meaning that the NO_3^-/SO_4^{2-} ratio is above 1, this indicates that transport is the main source of pollutants [54,55].

In addition to the chemical indicators listed above, the analysis of meteorological data facilitates the identification of aerosol sources. For this purpose, wind roses are plotted to determine potential local and regional sources of pollution (Table 1). In order to determine the movement of air masses from distant sources, the HYSPLIT model developed by NOAA can be used [56]. A detailed description of their trajectories has been presented in previous publications [57,58].

2.7. Statistical Treatment of the Data

To verify the significance of the impact of the analyzed factors (e.g., distance from the street, level of traffic), two tests were applied. The non-parametric U Mann–Whitney Test was applied to examine differences between two sets of independent data and the Kruskal–Wallis test was used for more than two groups of independent variables. Analogous tests were applied to determine the influence of selected factors on the deposition of organic carbon. For all dependencies presented in the publication, the levels of tests' significance have been considered to be important only when the p value was less than 0.05. All the statistical analysis was performed using STATISTICA® Software (Dell Inc., software.dell.com, Tulusa, OK, USA, Version 13).

3. Results

The research conducted in Gdynia in 2015 was aimed at determining the extent to which transport related to driving children to school contributes to air pollution with carbon compounds. For this reason, measurements were carried out only in the non-heating period, which was divided into two cycles. The first one covered summer holidays (July 2015), when there is no traffic related to transporting children to school. September (2015) was selected as the school period. In both measurement cycles, samples were taken during road traffic peak hours (7:00–9:00 a.m. and 3:00–5:00 p.m.). During the measurements in September, the traffic volume in Gdynia ranged from 37,000 to 45,000 vehicles a day and was on average one third higher than during the summer holidays [59]. This could have resulted in slightly higher concentrations of PAHs (21.4 ng·m^{-3}) and EC (0.5 µg·m^{-3}) in PM10 aerosols during the school period than during the holiday season (20.3 ng·m^{-3} and 0.3 µg·m^{-3}, for PAHs and EC, respectively). Among carbon compounds, only the concentration of organic carbon in PM10 was higher in July than in September 2015 (6.1 µg·m^{-3} and 4.3 µg·m^{-3}, respectively), which could be a consequence of the increased vegetation of plants on land and in the sea at that time. However, the Mann–Whitney U test did not confirm a statistically significant difference in the concentrations of all analyzed carbon compounds (PAHs, OC and EC) between the school and holiday periods (test, $p > 0.05$). In order to better interpret the sources of origin of the analyzed carbon compounds in the discussed periods, the results were divided into two size classes: up to 3 µm in diameter (respirable aerosols) and from 3 µm to 10 µm in diameter (inhalable aerosols) (Table 2). Additionally, the study takes into account the ionic components of aerosols (nitrates and sulphates) as a supplement to the information on air pollution from stationary and mobile sources (NO_3^-/SO_4^{2-} factor) [54,55].

Table 2. Statistical characteristics of PAHs, OC, and EC concentrations and selected ionic aerosol components during the morning and afternoon traffic peak during school and holiday periods.

Parameter	Aerosol Size [µm]	Holiday Period (13–21 July 2015)		School Period (14–30 September 2015)	
		7:00–9:00 a.m. Average (Min–Max)	3:00–5:00 p.m. Average (Min–Max)	7:00–9:00 a.m. Average (Min–Max)	3:00–5:00 p.m. Average (Min–Max)
ΣPAH$_5$ [ng·m^{-3}]	<3	15.44 ± 9.15 (3.12–26.66)	13.68 ± 8.21 (3.54–25.61)	11.04 ± 4.99 (5.77–18.42)	7.96 ± 4.67 (4.28–17.81)
	3–10	6.24 ± 6.88 (0.74–19.90)	5.46 ± 3.98 (0.52–10.55)	20.15 ± 22.9 (0.64–57.71)	4.95 ± 3.58 (0.38–8.90)
Benzo(a)anthracene B(a)A [ng·m^{-3}]	<3	0.04 ± 0.02 (0.02–0.06)	0.04 ± 0.06 (LD-0.14)	0.03 ± 0.02 (0.01–0.1)	0.02 ± 0.01 (0.01–0.02)
	3–10	0.02 ± 0.03 (LD-0.06)	0.03 ± 0.02 (LD-0.06)	0.14 ± 0.07 (0.09–0.20)	0.02 ± 0.01 (0.01–0.03)
Benzo(a)pyrene B(a)P [ng·m^{-3}]	<3	0.04 ± 0.04 (0.01–0.11)	0.03 ± 0.02 (LD-0.07)	0.02 ± 0.01 (0.01–0.03)	0.02 ± 0.01 (0.02–0.03)
	3–10	0.01 ± 0.01 (LD-0.03)	0.01 ± 0.01 (LD-0.02)	0.08 ± 0.00 (0.08–0.08)	0.07 ± 0.01 (0.06–0.10)
Chrysen CHR [ng·m^{-3}]	<3	0.12 ± 0.06 (0.07–0.23)	0.10 ± 0.05 (0.05–0.20)	0.09 ± 0.03 (0.06–0.14)	0.04 ± 0.02 (0.01–0.07)
	3–10	0.04 ± 0.04 (0.01–0.11)	0.08 ± 0.06 (0.03–0.20)	0.11 ± 0.11 (0.02–0.39)	0.04 ± 0.02 (0.02–0.07)
Fluoranthene FLU [ng·m^{-3}]	<3	15.17 ± 9.06 (2.96–26.27)	13.41 ± 8.05 (3.45–25.03)	10.87 ± 4.94 (5.56–18.12)	4.87 ± 3.53 (0.34–8.75)
	3–10	6.13 ± 6.75 (0.71–19.52)	5.29 ± 4.00 (0.19–10.37)	19.89 ± 22.13 (0.61–57.03)	7.83 ± 4.62 (4.19–17.59)
Pyrene PYR [ng·m^{-3}]	<3	0.08 ± 0.03 (0.05–0.13)	0.11 ± 0.08 (0.03–0.24)	0.05 ± 0.02 (0.04–0.09)	0.02 ± 0.01 (0.01–0.04)
	3–10	0.05 ± 0.06 (0.01–0.17)	0.06 ± 0.05 (0.01–0.16)	0.07 ± 0.06 (0.01–0.17)	0.02 ± 0.01 (0.01–0.04)
Nitrates NO_3^- [µg·m^{-3}]	<3	0.7 ± 0.5 (LD-1.5)	0.5 ± 0.5 (0.1–1.4)	2.0 ± 2.1 (0.5–5.8)	2.5 ± 0.7 (1.8–3.3)
	3–10	0.2 ± 0.1 (LD-0.2)	0.1 ± 0.2 (LD-0.2)	1.0 ± 0.4 (LD-1.4)	1.1 ± 1.2 (0.1–3.1)

Table 2. *Cont.*

Parameter	Aerosol Size [μm]	Holiday Period (13–21 July 2015)		School Period (14–30 September 2015)	
		7:00–9:00 a.m. Average (Min–Max)	3:00–5:00 p.m. Average (Min–Max)	7:00–9:00 a.m. Average (Min–Max)	3:00–5:00 p.m. Average (Min–Max)
Sulphates SO_4^{2-} [μg·m^{-3}]	<3	0.6 ± 0.3 (0.3–1.1)	0.6 ± 0.4 (0.2–1.2)	2.5 ± 2.0 (0.7–5.4)	2.2 ± 1.0 (1.0–4.0)
	3–10	0.1 ± 0.0 (LD-0.1)	0.1 ± 0.1 (LD-0.2)	1.4 ± 1.1 (LD-3.2)	2.1 ± 3.3 (LD-9.3)
OC [μg·m^{-3}]	<3	4.6 ± 0.7 (3.6–5.6)	4.3 ± 1.3 (2.8–6.0)	3.3 ± 0.8 (1.8–3.9)	2.7 ± 0.6 (1.8–3.4)
	3–10	1.7 ± 0.5 (1.2–2.6)	1.6 ± 0.3 (1.1–2.1)	1.2 ± 0.2 (0.9–1.6)	1.3 ± 0.5 (0.8–2.0)
EC [μg·m^{-3}]	<3	0.5 ± 0.2 (LD-0.7)	0.3 ± 0.1 (LD-0.5)	0.6 ± 0.2 (LD-0.9)	0.4 ± 0.4 (LD-1.0)
	3–10	0.2 ± 0.0 (LD-0.2)	0.2 ± 0.2 (LD-0.7)	0.4 ± 0.3 (LD-0.7)	0.3 ± 0.1 (LD-0.4)

The concentration of total PAHs was always higher in particles smaller than 3 μm in diameter. Only in the morning road traffic peak, during the school period, was there a reverse tendency that the mean the concentration of $\sum PAH_5$ was higher in particles with a diameter of 3 to 10 μm. It was also the only case where the concentration of $\sum PAH_5$ was higher during the school period than during the holiday period. At the same time, regardless of the particle size and duration of measurements (school and holiday period), the concentration of $\sum PAH_5$ was always higher in the morning than in the afternoon (Table 2). Among the analyzed PAHs, the highest concentration values in both fractions, both during school and holiday periods, as well as during the morning and afternoon road traffic peak, were exhibited by fluoranthene (Table 2). The concentrations of other PAHs were at a similar level. The lowest values were found for B(a)A (from <LD of the analytical method to 0.2 ng·m^{-3}). Apart from benzo(a)pyrene, which belongs to the five-ring hydrocarbons, the remaining analyzed PAHs are classified as tetracyclic (pyrene, chrysene, fluoranthene, benzo(a)anthracene). Two-and three-ring PAHs have a low molecular weight (LMW), four-ring PAHs have an average molecular weight (MMW), while five- and six-ring PAHs have a high molecular weight (HMW). The physical and chemical properties of PAHs change with the molecular weight and chemical structure. Low molecular weight compounds, which were not analyzed in this study, have a higher vapor pressure and are present in the environment in gaseous form. In addition, they are less hydrophobic than medium and high molecular weight hydrocarbons and therefore dissolve more easily in water. PAHs of medium and high molecular weights are more difficult to degrade, and thus more persistent in the natural environment. PAHs with four or more aromatic rings are hydrophobic and typically non-polar compounds. This determines their behavior in the natural environment. In general, PAHs with a higher molecular weight exhibit sorption properties on smaller aerosols [60], which could explain why the concentration of PAHs was higher in particles smaller than 3 μm in diameter. The high concentrations of PAHs in large aerosols obtained in the morning hours during the school period could have resulted from the prevailing weather conditions. The process of PAHs sorption on aerosols is more intensive under higher air humidity. In the discussed period of time, the average air humidity was 64 ± 9% and was higher than in other research periods (Table 1). At that time, the atmospheric pressure was also characterized by the highest range of values (from 1002 to 1033 hPa). The increase in atmospheric pressure reduces the speed of air circulation and prevents the transfer of PAHs from aerosol to gaseous form [61]. For this reason, during the morning hours of school period, when the

air humidity and atmospheric pressure were higher, higher concentrations of the analyzed PAHs in aerosols >3 μm of diameter could be recorded. In the afternoon hours of the school period, the pressure periodically dropped below 1000 hPa, and the humidity was several % lower than in the morning hours (Table 1).

The mean concentration of organic (OC) and elemental (EC) carbon was always higher in aerosols below 3 μm in diameter. In a similar manner to the concentration of $\sum PAH_5$, the concentration of these compounds was also higher in the morning hours than in the afternoon traffic peak hours. However, while the concentration of EC was higher or at a similar level during the school period, higher values of OC were observed during the summer holidays. Organic carbon during the holiday season accounted for as much as 94% in the total carbon fraction in particles smaller than 3 μm and 92% in particles larger than 3 μm in diameter. Its share decreased during the school period, when the EC concentration increased. At that time, OC constituted 88% of the TC mass in particles <3 μm in diameter, and 84% of TC in particles >3 μm in diameter.

The average concentration of nitrates and sulphates, as well as the EC concentration, was higher during the school period than during the holiday period (Table 2). Regardless of the season and the time of day, it was always greater in aerosols <3 μm in diameter.

4. Discussion

4.1. The Origin of Carbon Compounds during the Holiday Season

Measurements carried out in Gdynia during the holiday season did not show statistically significant differences in concentrations depending on the time of sampling in the case of $\sum PAH_5$, both forms of carbon (OC and EC), or basic ionic components (NO_3^-, SO_4^{2-}) (Figure 3).

The median concentrations of all compounds were very similar in the morning and afternoon hours. This may be due to the fact that during the holidays in the area of the Tri-City agglomeration, the traffic volume is largely determined by tourism. Therefore, it is not at its highest during peak traffic hours. Rather, it falls in the late morning hours, when tourists head for the beach and the early afternoon hours, when tourists come down for lunch. Additionally, some residents are on vacation during the summer months. Of course, driving children to school is also eliminated. However, the obtained value of the PAHs origin index described by the relationship PIR/B(a)P was high and amounted to 6.9. This indicates that the dominant source of these compounds in aerosols over Gdynia during the summer holidays was combustion in diesel engines [62,63]. The same source was found both for small particles in the morning hours (6.1) and in the afternoon (5.7) and for large particles during both road traffic peak hours (6.2 and 11.4, respectively in the morning and afternoon). The high concentration of fluoranthene in relation to pyrene (mean value Flu/Pyr = 159) also indicated the communicative source of PAHs during the holidays. This source played a more significant role in the morning (Flu/Pyr = 337) than in the afternoon (Flu/Pyr = 138). The more than two times higher concentration of fluoranthene in small aerosols as compared to large particles also proves the impact of combustion in diesel engines. This trend was recorded both in the morning and in the afternoon (Table 2). The same source of aerosols during the holiday season was indicated by the value of the NO_3^-/SO_4^{2-} ratio. Again, its greater importance was established in the morning (7:00–9:00 a.m.) when the mean value of the ratio was 1.3 (1.2 and 1.4, respectively, in particles <3 μm and >3 μm in diameter). In the afternoon (3:00–5:00 p.m.) the NO_3^-/SO_4^{2-} ratio was slightly lower and averaged 1.1 (1.0 and 1.2, respectively, in particles <3 μm and >3 μm in diameter) [54,55].

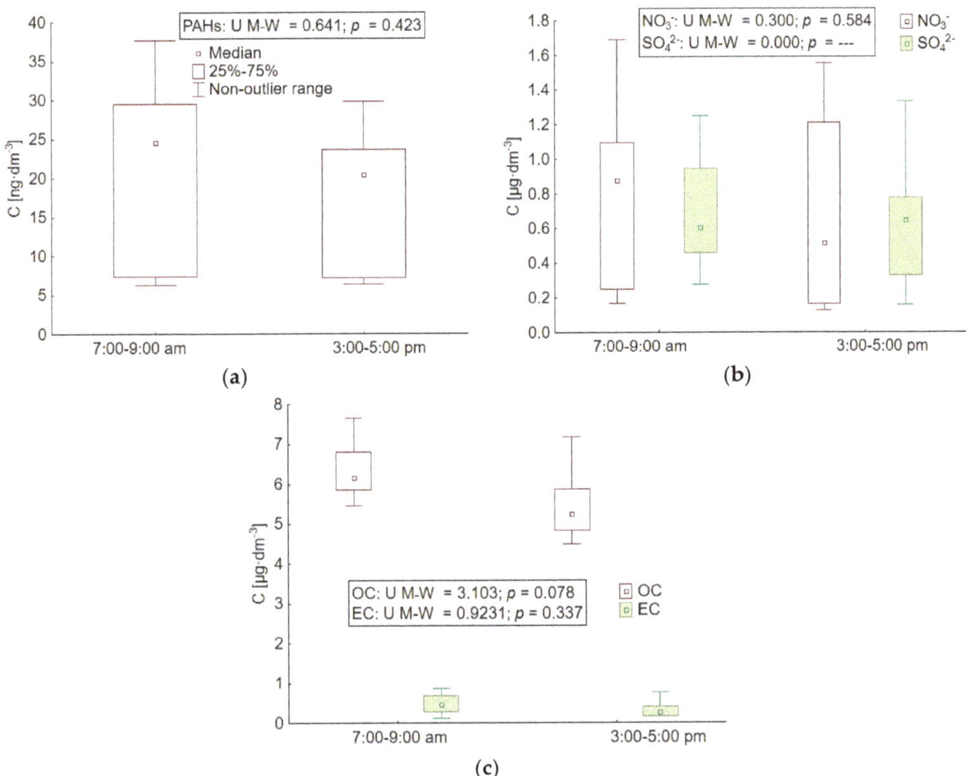

Figure 3. Statistical characteristics of the concentration of (**a**) PAH_5, (**b**) nitrate and sulphate ions, (**c**) and OC and EC in the morning and afternoon rush hours in the atmosphere over Gdynia during the holiday season of 2015.

Another indicator, the OC/EC ratio, was at the level of 14.0 during the holiday period (with an average of 11.4 in the entire measurement period), which proves the significant role of vegetation in forming the high concentrations of organic carbon at that time [62]. This compound could be present in aerosols as a consequence of naturally occurring processes, i.e., emission of plant spores, pollen, vegetation debris, microorganisms, and organic matter from the soil surface and the nearby sea [32,57,64–68]. The value of the coefficient was always higher in the smaller aerosols, both in the morning and afternoon hours (12.3 and 23.4, respectively) than in the aerosols with a diameter of 3 to 10 µm (8.2 and 7.1 in the morning and afternoon, respectively). It was also found that in aerosols <3 µm in diameter, the source of OC and EC origin during the summer holidays was always common, as indicated by the Pearson correlation coefficient between the concentrations of OC and EC ($r = 0.8$ and $r = 0.95$, respectively in the morning hours and afternoon). In particles >3 µm in diameter during the holiday season, no common source of OC and EC origin was established during any of the road traffic peak hours (Pearson correlation $r < 0.5$). This could be due to the fact that as much as 67% of the EC concentrations measured in these particles in the morning hours and 43% of the concentrations in the afternoon hours were below the limit of quantification of the method (Table 2). This suggests a different source of organic carbon, apart from the transport sector, is large aerosols, despite lower OC/EC values compared to small particles. Organic carbon, apart from plant vegetation, could be present at that time in large aerosols as a consequence of biomass combustion during food processing. The research was conducted in the summer, when both residents and tourists often grill [69–71]. It could also be a component of secondary aerosols resulting from the physical or chemical adsorption of gases on particles, which led to an increase in its concentration [72,73].

The influence of transport was noticeable during the summer holidays in small particles, especially in the morning hours, when the wind dominated from the Tri-City ring road. Its force was then up to 10 $m \cdot s^{-1}$ (Table 1, Figure 1). During this time, the OC/EC ratio in particles <3 μm in diameter was almost two times lower than in the afternoon (23.4 and 12.3, respectively), as a consequence of the increase in EC concentration [6]. The ring road connects all the cities of the Tri-City (Gdansk, Sopot, Gdynia) and at the same time a route leading to the Hel Peninsula, which is one of the places most visited by tourists on the Polish Baltic coast during the holidays. Its significance for the increase in EC concentration in aerosols has already been reported in this area of research [34,58,61,74].

In the afternoon hours, the average wind speed was 2.8 $m \cdot s^{-1}$ and was lower than in the morning hours (Table 1). The highest value of OC/EC recorded at that time in particles smaller than 3 μm (23.4) was the result of two times lower EC concentrations compared to the morning hours. However, since the source of OC and EC origin was common at that time (Pearson's correlation $r = 0.95$), the influence of transport on the concentration of both compounds cannot be ruled out. At that time, the road leading through Gdynia, located 600 m south-west of the measuring station, can be indicated as a potential carbon compound source from the transport sector (Figure 1). On the other hand, the high values of OC concentrations in small particles present in the atmosphere over Gdynia during the afternoon hours are probably a consequence the presence of secondary organic carbon in them or/and of biomass combustion during food processing [69–71].

4.2. The Origin of Carbon Compounds during the School Period

During the school period (September 2015), the difference in the concentrations obtained in the morning and afternoon traffic rush hours was more pronounced than in the holiday season (Figure 4). This relationship was confirmed by the U Mann–Whitney test for PAHs ($p = 0.05$) and for elemental carbon ($p = 0.04$). Statistical significance was not confirmed for nitrates, sulphates, or organic carbon ($p > 0.05$).

Higher median concentrations of $\sum PAH_5$ and EC, as well as OC, were recorded in the morning from 7:00 a.m. to 9:00 a.m. (Table 2), when children are transported to school and adults are going to work. Then, the traffic intensity in the study area increases, which could have generated an increase in air pollution from transport sources [6,75]. In the morning, when class starts at 8:00 a.m. or 9:00 a.m., dozens of cars dropping off children and running their engines are observed in front of schools. In the afternoon, high levels of traffic are spread over time. This is due to the different times that the classes end for particular groups of students and the additional activities they perform (extracurricular activities). For this reason, a large proportion of children return home on foot, without the need for a car. These factors determined the differences in the concentrations of traffic pollution in aerosols measured during the school period in the atmosphere over Gdynia in the morning and afternoon hours. In the morning, the value of the PIR/B(a)P ratio pointed to the transport source of PAHs related to combustion in diesel engines, which was almost twice as high as in the afternoon (4.0 and 2.2, respectively) [60,63]. The value of this coefficient was several times higher in small aerosol particles (4.4 and 3.4, respectively in the morning and afternoon traffic rush hours) than in large particles (1.4 and 0.8, respectively in the morning and afternoon traffic rush hours). It was found that the value of the coefficient was not affected by the concentration of B(a)P, which during the school period did not show statistically significant differences depending on the time of day (U Mann–Whitney test, $p = 0.2$). Both in the morning and afternoon hours, the concentration of this compound was also at a similar level in large and small particles (Table 2). Pyrene was the PAH that differentiated the PIR/B(a)P ratio during the school period. Its median differed statistically significantly depending on the time of day, both in small and large particles (Table 2), which was confirmed by the U Mann–Whitney test ($p = 0.03$). In the case of small aerosols, the concentration of pyrene in the morning traffic rush hours was twice as high, and in the case of large aerosols it was even three times higher than in the afternoon hours (Table 2). Such high concentrations of pyrene indicated that in the morning there was an additional source of PAHs in aerosols, apart from transport sector, probably related to the combustion of fuels for heating purposes [60,63,76]. Taking into account the beginning of autumn and the cooling that prevailed in Poland at that time, it is possible that users of detached houses in the mornings heated them more intensively using solid fuels for this purpose. This would also explain the high concentrations of fluoranthene, which is a congener of PAHs, which, in addition to transport emissions, also result from the combustion of coal and wood [60,77]. In the morning hours, the median concentration of this compound was 25.62 ng·dm^{-3}, and it was almost two times higher than in the afternoon (13.71 ng·dm^{-3}). In addition, high concentrations of fluoranthene in the morning were also recorded in large particles, which may confirm their non-transport source of origin (Table 2). A similar dependence was shown in large aerosols of benzo(a)anthracene, whose median concentration in the morning traffic rush hours was 0.14 ng·dm^{-3} and was seven times higher than that obtained in the afternoon (0.02 ng·dm^{-3}).

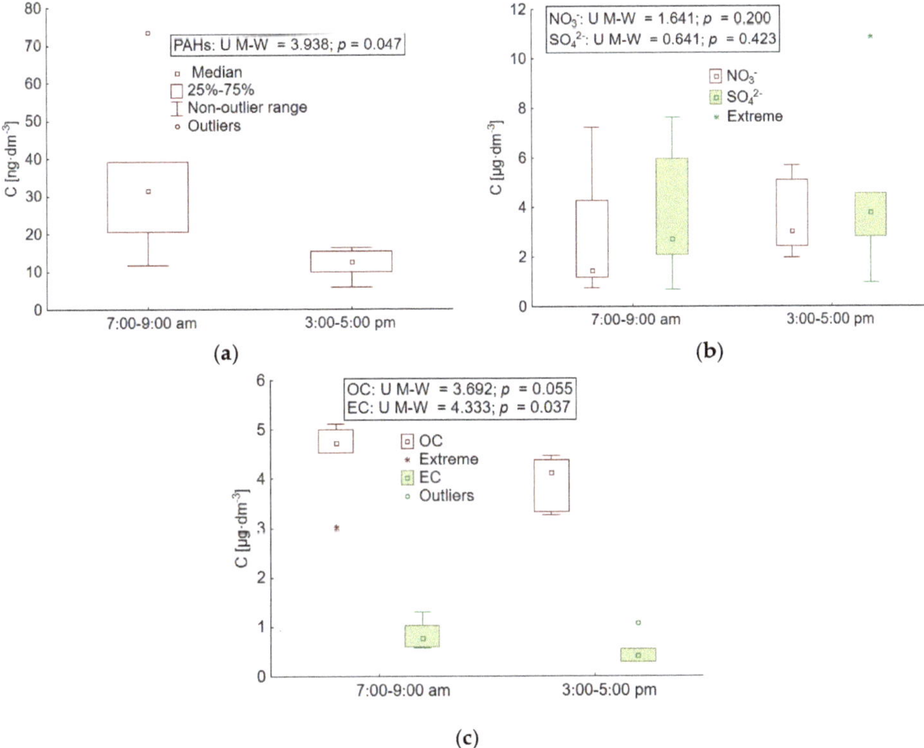

Figure 4. Statistical characteristics of the concentration of: (**a**) PAHs, (**b**) nitrate and sulphate ions, (**c**) and OC and EC in the morning and afternoon rush hours in the atmosphere over Gdynia during the school period in 2015.

During the school period, the greater impact of road transport in the morning was also confirmed by the NO_3^-/SO_4^{2-} ratio, which on average amounted to 1.2 during this time. In the afternoon, the value of the coefficient decreased to 0.9, suggesting that some of the pollutants in the atmosphere above Gdynia could come from emissions from stationary sources [54,55]. The OC/EC value during the school period was set at 6.8 and was lower in the morning than in the afternoon (5.8 and 9.1, respectively). This indicates a greater importance of transport in the emission of elemental carbon to the atmosphere during the hours of transporting children to school [64,78,79]. Similar results were obtained by Querol et al., (2013) [64] conducting research in the years 1999–2011 at 78 research stations located throughout Spain. The researchers considered areas with varying degrees of urbanization, agricultural areas, and background stations away from large cities. The lowest value of the OC/EC ratio was obtained by Querol et al., (2013) [64], similarly to this study, in the morning. It always corresponded to a marked increase in the volume of traffic.

4.3. Selected Episodes with the Highest Influence of Land and Maritime Transport

Both during the holiday and school periods, there were several interesting cases in which the concentration of the analyzed compounds was determined by meteorological conditions and the time of day. The first episode took place on 16 July 2015. Then, air masses were transported over the station from the north-west (from the North Sea), but the wind direction changed significantly with the time of day (Figure 5). The value of the NO_3^-/SO_4^{2-} ratio in PM10 was then the highest in the entire holiday season, and amounted to 1.9 on average. The concentration of $\sum PAH_5$ was also very high (29.17 ng·dm^{-3}). The average wind speed equal to 2.4 m·s^{-1} indicated a local to regional source of aerosols [58,80].

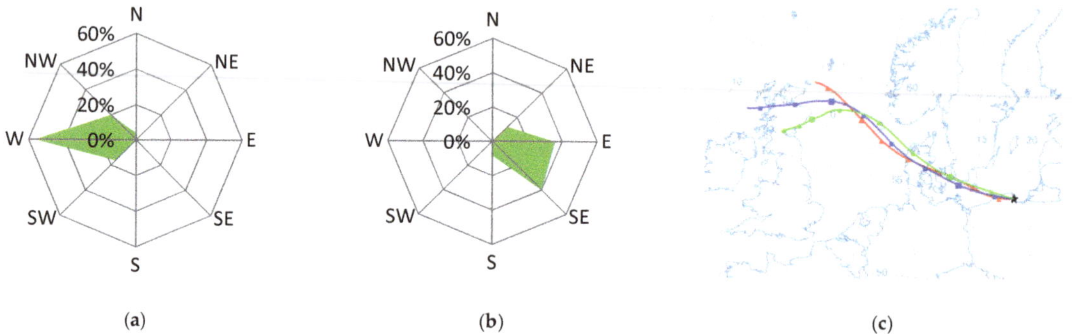

Figure 5. Dominant wind direction in the morning (**a**) and the afternoon (**b**) and the dominant air masses (**c**) on 16 July 2015 in Gdynia.

As the discussed situation occurred during the holiday season, and the air temperature that day reached 20 °C, the increase in the concentration of pollutants in the air over Gdynia could have been caused by increased tourist traffic towards the beaches of Tri-City. In the morning hours (7:00–9:00 a.m.), when the wind direction from the ring road dominated (Figure 5a) and the wind speed was up to 7 m·s^{-1}, the value of the NO_3^-/SO_4^{2-} coefficient in PM10 aerosols measured in Gdynia increased to 2.2. This indicated the transport source of the pollution origin at that time [5,55]. This was confirmed by the low value of the OC/EC ratio compared to the average for the entire holiday period (11.5 and 19.6, respectively) [64]. Additionally, on the morning of 16 July the maximum $\sum PAH_5$ concentration was recorded (37.7 ng·dm^{-3}), which also proves the significant influence of transport [75]. In the afternoon, the winds were blowing from east to south-east from the streets around the station. The wind speed was already low (1.4 m·s^{-1}), which led to less dispersion of pollutants. For this reason, the concentration of $\sum PAH_5$ remained at a high level (20.7 ng·dm^{-3}). Additionally, the value of the NO_3^-/SO_4^{2-} coefficient was similar to that in the morning (1.9).

The next episode took place on 20 July 2015, when the wind was coming from the north-west (Figure 6). On that day, the concentration of nitrates was very low, with an average of 0.7 µg·dm^{-3}, while the concentration of sulfates was high and amounted to 1.1 µg·dm^{-3} (mean 0.67 µg·dm^{-3}). Consequently, the NO_3^-/SO_4^{2-} ratio had a very small value of 0.1. Taking into account the proximity of the port and the incoming wind from its region, the probable sources of pollution in the atmosphere above the research station were day shipping and port activity [81]. The value of the OC/EC ratio was high (18.4) and did not indicate a large share of elemental carbon in aerosols at that time. However, the high concentration of ∑PAH$_5$ (15.8 ng·dm^{-3}) and the obtained value of the PIR/B(a)P ratio at the level of 16.1 may suggest the presence of carbon compounds emitted to the atmosphere from combustion in diesel engines. As on that day the wind speed reached 8.5 m·s^{-1}, the pollutants were well dispersed. On the next measuring day, the concentrations had halved.

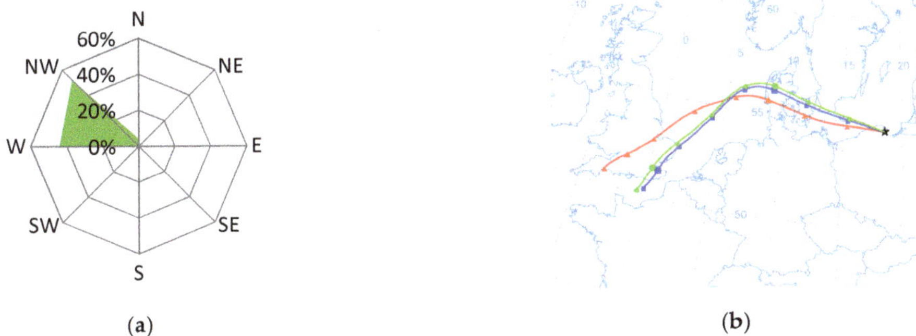

Figure 6. Dominant wind direction (**a**) and dominant air masses (**b**) on 20 July 2015 in Gdynia.

Another situation was recorded on 16 September 2015, when the NO_3^-/SO_4^{2-} ratio reached its maximum value during the entire measurement period, equal to 2.7. The likely cause of the increase in nitrate concentration was the film festival taking place in Gdynia and the related increased traffic volume. On that day, the wind direction from the east (in the morning) and from the south-east (afternoon hours) was recorded (Figure 7). The wind speed was low (on average 1.4 m·s^{-1}), which led to the accumulation of pollutants close to the emission source and their poor dispersion [5,80]. The obtained value of the PIR/B(a)P ratio, amounting to an average of 2.2, indicated the role of petrol-powered cars in the formation of high concentrations of ∑PAH$_5$ (20.75 ng·dm^{-3}).

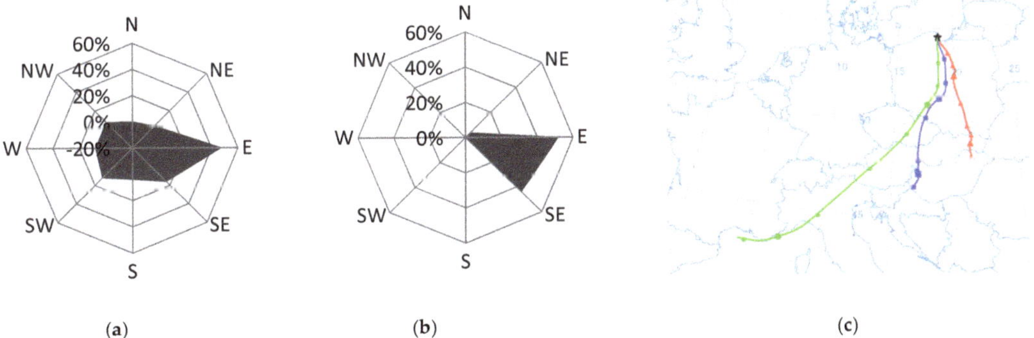

Figure 7. Dominant wind direction in the morning (**a**) and the afternoon (**b**) and dominant air masses (**c**) on 16 September 2015 in Gdynia.

The role of transport on that day was clear, both in the morning and in the afternoon rush hours. From 7:00 a.m. to 9:00 a.m. the NO_3^-/SO_4^{2-} ratio was equal to 2.3, and from 3:00 to 5:00 p.m. it adopted the highest value over the entire measurement period, amounting to 2.9. It was found that the source of air pollution, determined using the PIR/B(a)P ratio, was related to the emissions from combustion in gasoline engines, both in the morning and in the afternoon (1.7 and 2.5, respectively). In the morning, when the average wind speed was very low and averaged 0.7 m·s^{-1}, the concentration of ∑PAH$_5$

reached 31.5 ng·dm^{-3}. In the afternoon, when the wind speed doubled (1.7 m·s^{-1} on average), the concentration of $\sum$PAH$_5$ decreased to 10.0 ng·dm^{-3}. In the morning, in the vicinity of the research station, there could be a greater accumulation of local pollutants. In turn, in the afternoon, pollutants could be transported from the important roads located on the southeast and east of the measuring station [80,82]. Many of them are access roads to nearby schools (Figure 2).

5. Conclusions

PM10 measurements conducted in the coastal zone of the Baltic Sea in 2015 indicated higher concentrations of nitrate, sulphate, and elemental carbon in the school period, while the concentration of organic carbon in aerosols was higher during the holidays. In the case of PAHs concentration, the difference between the school and vacation periods was not clear. While the concentrations of fluoranthene, chrysene, and pyrene were higher during the holidays, the concentration of B(a)P and B(a)A were higher in the school period.

The analysis of aerosol pollution markers suggested that during the holidays, the quality of the surrounding air was mainly determined by the combustion of diesel oil in transport related to tourism, passenger ships, and port activity. These sources apparently appeared in the morning hours (7:00–9:00 a.m.). During the school period, the main source of pollutants was gasoline combustion. At the beginning of autumn, due to the drop in air temperature, the role of the heating sector also cannot be ignored.

The highest mean values of the ΣPAHs$_5$ and EC were recorded in small particles (<3 µm in diameter) during the school period, which were found in the morning road traffic peak hours. The mean concentration of OC was also the highest in small aerosols during the holiday period. However, there were no statistically significant differences between the concentrations of organic carbon concentration in the morning and afternoon peak hours. Strict sampling and measurement procedure, together with analysis of air mass backward trajectories and pollutant markers, indicated that the role of land transport was the greatest when local to regional winds prevailed, bringing pollution from nearby schools and the beltway.

Author Contributions: Conceptualization, A.U.L.; data curation, J.K.B., K.A.W., and K.V.B.; formal analysis, J.K.B., K.A.W., and K.V.B.; funding acquisition, A.U.L. and M.S.; investigation, J.K.B. and K.V.B.; methodology, M.S. and A.U.L.; project administration, A.U.L.; resources, A.U.L., M.S., and K.V.B.; supervision, A.U.L.; validation, A.U.L. and M.S.; visualization, J.K.B. and K.A.W.; writing—original draft, J.K.B., A.U.L., and M.S.; writing—review and editing, A.U.L. All authors have read and agreed to the published version of the manuscript.

Funding: This research received no external funding.

Institutional Review Board Statement: Not applicable.

Informed Consent Statement: Not applicable.

Data Availability Statement: Not applicable.

Conflicts of Interest: The authors declare no conflict of interest.

References

1. Viana, M.; Kuhlbusch, T.A.J.; Querol, X.; Alastuey, A.; Harrison, R.M.; Hopke, P.K.; Winiwarter, W.; Vallius, M.; Szidat, S.; Prévôt, A.S.H.; et al. Source Apportionment of Particulate Matter in Europe: A Review of Methods and Results. *J. Aerosol Sci.* **2008**, *39*, 827–849. [CrossRef]
2. Jandacka, D.; Durcanska, D.; Bujdos, M. The Contribution of Road Traffic to Particulate Matter and Metals in Air Pollution in the Vicinity of an Urban Road. *Transp. Res. Part D Transp. Environ.* **2017**, *50*, 397–408. [CrossRef]
3. Donateo, A.; Gregoris, E.; Gambaro, A.; Merico, E.; Giua, R.; Nocioni, A.; Contini, D. Contribution of Harbour Activities and Ship Traffic to PM2.5, Particle Number Concentrations and PAHs in a Port City of the Mediterranean Sea (Italy). *Environ. Sci. Pollut. Res.* **2014**, *21*, 9415–9429. [CrossRef]
4. Merico, E.; Gambaro, A.; Argiriou, A.; Alebic-Juretic, A.; Barbaro, E.; Cesari, D.; Chasapidis, L.; Dimopoulos, S.; Dinoi, A.; Donateo, A.; et al. Atmospheric Impact of Ship Traffic in Four Adriatic-Ionian Port-Cities: Comparison and Harmonization of Different Approaches. *Transp. Res. Part D Transp. Environ.* **2017**, *50*, 431–445. [CrossRef]

5. Abu-Allaban, M.; Gillies, J.A.; Gertler, A.W.; Clayton, R.; Proffitt, D. Tailpipe, Resuspended Road Dust, and Brake-Wear Emission Factors from on-Road Vehicles. *Atmos. Environ.* **2003**, *37*, 5283–5293. [CrossRef]
6. Keuken, M.P.; Zandveld, P.; Jonkers, S.; Moerman, M.; Jedynska, A.D.; Verbeek, R.; Visschedijk, A.; Elshout, S.; Panteliadis, P.; Velders, G.J.M. Modelling Elemental Carbon at Regional, Urban and Traffic Locations in The Netherlands. *Atmos. Environ.* **2013**, *73*, 73–80. [CrossRef]
7. Arnott, W.P.; Zielinska, B.; Rogers, C.F.; Sagebiel, J.; Park, K.; Chow, J.; Moosmüller, H.; Watson, J.G.; Kelly, K.; Wagner, D.; et al. Evaluation of 1047-Nm Photoacoustic Instruments and Photoelectric Aerosol Sensors in Source-Sampling of Black Carbon Aerosol and Particle-Bound PAHs from Gasoline and Diesel Powered Vehicles. *Environ. Sci. Technol.* **2005**, *39*, 5398–5406. [CrossRef] [PubMed]
8. Geller, M.D.; Sardar, S.B.; Phuleria, H.; Fine, P.M.; Sioutas, C. Measurements of Particle Number and Mass Concentrations and Size Distributions in a Tunnel Environment. *Environ. Sci. Technol.* **2005**, *39*, 8653–8663. [CrossRef] [PubMed]
9. Piątkowski, P.; Bohdal, T. Testing of Ecological Properties of Spark Ignition Engine Fed with LPG Mixture. *Rocz. Ochr. Srodowiska* **2011**, *13*, 607–618.
10. Kanakidou, M.; Seinfeld, J.H.; Pandis, S.N.; Barnes, I.; Dentener, F.J.; Facchini, M.C.; Van Dingenen, R.; Ervens, B.; Nenes, A.; Nielsen, C.J.; et al. Organic Aerosol and Global Climate Modelling: A Review. *Atmos. Chem. Phys.* **2005**, *5*, 1053–1123. [CrossRef]
11. Lang, J.; Zhang, Y.; Zhou, Y.; Cheng, S.; Chen, D.; Guo, X.; Chen, S.; Li, X.; Xing, X.; Wang, H. Trends of PM2.5 and Chemical Composition in Beijing, 2000–2015. *Aerosol Air Qual. Res.* **2017**, *17*, 412–425. [CrossRef]
12. Zhang, Y.X. *Study on Speciation of Particulate Organic Matter from Combustion Sources*; Beking University: Beijing, China, 2006.
13. Cai, T.; Zhang, Y.; Fang, D.; Shang, J.; Zhang, Y.; Zhang, Y. Chinese Vehicle Emissions Characteristic Testing with Small Sample Size: Results and Comparison. *Atmos. Pollut. Res.* **2017**, *8*, 154–163. [CrossRef]
14. Available online: www.epa.gov (accessed on 30 June 2021).
15. Boothe, V.L.; Shendell, D.G. Potential Health Effects Associated with Residential Proximity to Freeways and Primary Roads: Review of Scientific Literature, 1999–2006. *J. Environ. Health* **2008**, *70*, 33–41. [PubMed]
16. Dockery, D.W.; Pope, C.A. Acute Respiratory Effects of Particulate Air Pollution. *Annu. Rev. Public Health* **1994**, *15*, 107–132. [CrossRef]
17. Dockery, D.W.; Pope, C.A.; Xu, X.; Spengler, J.D.; Ware, J.H.; Fay, M.E.; Ferris, B.G.; Speizer, F.E. An Association between Air Pollution and Mortality in Six U.S. Cities. *N. Engl. J. Med.* **1993**, *329*, 1753–1759. [CrossRef]
18. de Kok, T.M.C.M.; Driece, H.A.L.; Hogervorst, J.G.F.; Briedé, J.J. Toxicological Assessment of Ambient and Traffic-Related Particulate Matter: A Review of Recent Studies. *Mutat. Res. Rev. Mutat. Res.* **2006**, *613*, 103–122. [CrossRef] [PubMed]
19. Rogula-Kozłowska, W.; Rogula-Kopiec, P.; Klejnowski, K.; Błaszczyk, J. Influence of traffic emission on the concentration of two forms of carbon and their mass distribution in relation to the particle size in the atmospheric aerosol of an urban area. *Annu. Set Environ. Prot.* **2013**, *15*, 1623–1644. (In Polish)
20. Hassanvand, M.S.; Naddafi, K.; Faridi, S.; Nabizadeh, R.; Sowlat, M.H.; Momeniha, F.; Gholampour, A.; Arhami, M.; Kashani, H.; Zare, A.; et al. Characterization of PAHs and Metals in Indoor/Outdoor PM10/PM2.5/PM1 in a Retirement Home and a School Dormitory. *Sci. Total Environ.* **2015**, *527–528*, 100–110. [CrossRef] [PubMed]
21. Witkowska, A.; Lewandowska, A.U. Water Soluble Organic Carbon in Aerosols (PM1, PM2.5, PM10) and Various Precipitation Forms (Rain, Snow, Mixed) over the Southern Baltic Sea Station. *Sci. Total Environ.* **2016**, *573*, 337–346. [CrossRef] [PubMed]
22. Witkowska, A.; Lewandowska, A.U.; Saniewska, D.; Falkowska, L.M. Effect of Agriculture and Vegetation on Carbonaceous Aerosol Concentrations ($PM_{2.5}$ and PM_{10}) in Puszcza Borecka National Nature Reserve (Poland). *Air Qual. Atmos. Health* **2016**, *9*, 761–773. [CrossRef] [PubMed]
23. Peterson, B.S.; Rauh, V.A.; Bansal, R.; Hao, X.; Toth, Z.; Nati, G.; Walsh, K.; Miller, R.L.; Arias, F.; Semanek, D.; et al. Effects of Prenatal Exposure to Air Pollutants (Polycyclic Aromatic Hydrocarbons) on the Development of Brain White Matter, Cognition, and Behavior in Later Childhood. *JAMA Psychiatry* **2015**, *72*, 531–540. [CrossRef]
24. Chen, S.; Gao, C.; Tang, W.; Zhu, H.; Han, Y.; Jiang, Q.; Li, T.; Cao, X.; Wang, Z. Self-Powered Cleaning of Air Pollution by Wind Driven Triboelectric Nanogenerator. *Nano Energy* **2015**, *14*, 217–225. [CrossRef]
25. Wilker, E.H.; Preis, S.R.; Beiser, A.S.; Wolf, P.A.; Au, R.; Kloog, I.; Li, W.; Schwartz, J.; Koutrakis, P.; DeCarli, C.; et al. Long-Term Exposure to Fine Particulate Matter, Residential Proximity to Major Roads and Measures of Brain Structure. *Stroke* **2015**, *46*, 1161–1166. [CrossRef] [PubMed]
26. World Health Organization. *Effects of Air Pollution on Children's Health and Development: A Review of the Evidence*; WHO: Geneva, Switzerland, 2005.
27. Perera, F.P.; Wang, S.; Rauh, V.; Zhou, H.; Stigter, L.; Camann, D.; Jedrychowski, W.; Mroz, E.; Majewska, R. Prenatal Exposure to Air Pollution, Maternal Psychological Distress, and Child Behavior. *Pediatrics* **2013**, *132*, e1284–e1294. [CrossRef] [PubMed]
28. Available online: www.imgw.pl (accessed on 30 June 2021).
29. Schmid, H.; Laskus, L.; Jürgen Abraham, H.; Baltensperger, U.; Lavanchy, V.; Bizjak, M.; Burba, P.; Cachier, H.; Crow, D.; Chow, J.; et al. Results of the "Carbon Conference" International Aerosol Carbon Round Robin Test Stage I. *Atmos. Environ.* **2001**, *35*, 2111–2121. [CrossRef]
30. Cavalli, F.; Viana, M.; Yttri, K.E.; Genberg, J.; Putaud, J.-P. Toward a Standardised Thermal-Optical Protocol for Measuring Atmospheric Organic and Elemental Carbon: The EUSAAR Protocol. *Atmos. Meas. Tech.* **2010**, *3*, 79–89. [CrossRef]

31. Wiśniewska, K.; Lewandowska, A.U.; Witkowska, A. Factors Determining Dry Deposition of Total Mercury and Organic Carbon in House Dust of Residents of the Tri-City and the Surrounding Area (Baltic Sea Coast). *Air Qual. Atmos. Health* **2017**, *10*, 821–832. [CrossRef]
32. Wiśniewska, K.; Lewandowska, A.U.; Staniszewska, M. Air Quality at Two Stations (Gdynia and Rumia) Located in the Region of Gulf of Gdansk during Periods of Intensive Smog in Poland. *Air Qual. Atmos. Health* **2019**, *12*, 879–890. [CrossRef]
33. Staniszewska, M.; Graca, B.; Bełdowska, M.; Saniewska, D. Factors Controlling Benzo(a)Pyrene Concentration in Aerosols in the Urbanized Coastal Zone. A Case Study: Gdynia, Poland (Southern Baltic Sea). *Environ. Sci. Pollut. Res.* **2013**, *20*, 4154–4163. [CrossRef]
34. Lewandowska, A.U.; Staniszewska, M.; Witkowska, A.; Machuta, M.; Falkowska, L. Benzo(a)Pyrene Parallel Measurements in PM1 and PM2.5 in the Coastal Zone of the Gulf of Gdansk (Baltic Sea) in the Heating and Non-Heating Seasons. *Environ. Sci. Pollut. Res.* **2018**, *25*, 19458–19469. [CrossRef]
35. Falkowska, L.; Lewandowska, A. Sulphates in Particles of Different Sizes in the Marine Boundary Layer over the Southern Baltic Sea. *Oceanologia* **2004**, *46*.
36. Available online: https://bip.um.gdynia.pl/ (accessed on 30 June 2021).
37. Na, K.; Sawant, A.A.; Song, C.; Cocker, D.R. Primary and Secondary Carbonaceous Species in the Atmosphere of Western Riverside County, California. *Atmos. Environ.* **2004**, *38*, 1345–1355. [CrossRef]
38. Lonati, G.; Ozgen, S.; Giugliano, M. Primary and Secondary Carbonaceous Species in PM2.5 Samples in Milan (Italy). *Atmos. Environ.* **2007**, *41*, 4599–4610. [CrossRef]
39. Pio, C.; Cerqueira, M.; Harrison, R.M.; Nunes, T.; Mirante, F.; Alves, C.; Oliveira, C.; Sanchez de la Campa, A.; Artíñano, B.; Matos, M. OC/EC Ratio Observations in Europe: Re-Thinking the Approach for Apportionment between Primary and Secondary Organic Carbon. *Atmos. Environ.* **2011**, *45*, 6121–6132. [CrossRef]
40. Cesari, D.; Merico, E.; Dinoi, A.; Marinoni, A.; Bonasoni, P.; Contini, D. Seasonal Variability of Carbonaceous Aerosols in an Urban Background Area in Southern Italy. *Atmos. Res.* **2018**, *200*, 97–108. [CrossRef]
41. Gu, J.; Bai, Z.; Liu, A.; Wu, L.; Xie, Y.; Li, W.; Dong, H.; Zhang, X. Characterization of Atmospheric Organic Carbon and Element Carbon of PM2.5 and PM10 at Tianjin, China. *Aerosol Air Qual. Res.* **2010**, *10*, 167–176. [CrossRef]
42. Watson, J.G.; Chow, J.C.; Lowenthal, D.H.; Pritchett, L.C.; Frazier, C.A.; Neuroth, G.R.; Robbins, R. Differences in the Carbon Composition of Source Profiles for Diesel- and Gasoline-Powered Vehicles. *Atmos. Environ.* **1994**, *28*, 2493–2505. [CrossRef]
43. Cao, J.J.; Lee, S.C.; Ho, K.F.; Zou, S.C.; Fung, K.; Li, Y.; Watson, J.G.; Chow, J.C. Spatial and Seasonal Variations of Atmospheric Organic Carbon and Elemental Carbon in Pearl River Delta Region, China. *Atmos. Environ.* **2004**, *38*, 4447–4456. [CrossRef]
44. Chow, J.C.; Watson, J.G.; Kuhns, H.; Etyemezian, V.; Lowenthal, D.H.; Crow, D.; Kohl, S.D.; Engelbrecht, J.P.; Green, M.C. Source Profiles for Industrial, Mobile, and Area Sources in the Big Bend Regional Aerosol Visibility and Observational Study. *Chemosphere* **2004**, *54*, 185–208. [CrossRef]
45. Bautista, A.T.; Pabroa, P.C.B.; Santos, F.L.; Racho, J.M.D.; Quirit, L.L. Carbonaceous Particulate Matter Characterization in an Urban and a Rural Site in the Philippines. *Atmos. Pollut. Res.* **2014**, *5*, 245–252. [CrossRef]
46. Shen, Z.; Cao, J.; Arimoto, R.; Han, Y.; Zhu, C.; Tian, J.; Liu, S. Chemical Characteristics of Fine Particles (PM1) from Xi'an, China. *Aerosol Sci. Technol.* **2010**, *44*, 461–472. [CrossRef]
47. Watson, J.G.; Chow, J.C.; Chen, L.-W.A.; Lowenthal, D.H.; Fujita, E.M.; Kuhns, H.D.; Sodeman, D.A.; Campbell, D.E.; Moosmüller, H.; Zhu, D.; et al. Particulate Emission Factors for Mobile Fossil Fuel and Biomass Combustion Sources. *Sci. Total Environ.* **2011**, *409*, 2384–2396. [CrossRef] [PubMed]
48. Tiitta, P.; Vakkari, V.; Croteau, P.; Beukes, J.P.; van Zyl, P.G.; Josipovic, M.; Venter, A.D.; Jaars, K.; Pienaar, J.J.; Ng, N.L.; et al. Chemical Composition, Main Sources and Temporal Variability of PM_1 Aerosols in Southern African Grassland. *Atmos. Chem. Phys.* **2014**, *14*, 1909–1927. [CrossRef]
49. Yu, J.Z.; Huang, X.H.H.; Ho, S.S.H.; Bian, Q. Nonpolar Organic Compounds in Fine Particles: Quantification by Thermal Desorption–GC/MS and Evidence for Their Significant Oxidation in Ambient Aerosols in Hong Kong. *Anal. Bioanal. Chem.* **2011**, *401*, 3125–3139. [CrossRef] [PubMed]
50. Hildemann, L.M.; Markowski, G.R.; Cass, G.R. Chemical Composition of Emissions from Urban Sources of Fine Organic Aerosol. *Environ. Sci. Technol.* **1991**, *25*, 744–759. [CrossRef]
51. Masclet, P.; Mouvier, G.; Nikolaou, K. Relative Decay Index and Sources of Polycyclic Aromatic Hydrocarbons. *Atmos. Environ. (1967)* **1986**, *20*, 439–446. [CrossRef]
52. Miguel, A.H.; Kirchstetter, T.W.; Harley, R.A.; Hering, S.V. On-Road Emissions of Particulate Polycyclic Aromatic Hydrocarbons and Black Carbon from Gasoline and Diesel Vehicles. *Environ. Sci. Technol.* **1998**, *32*, 450–455. [CrossRef]
53. Duan, X.; Shen, G.; Yang, H.; Tian, J.; Wei, F.; Gong, J.; Zhang, J. Dietary Intake Polycyclic Aromatic Hydrocarbons (PAHs) and Associated Cancer Risk in a Cohort of Chinese Urban Adults: Inter- and Intra-Individual Variability. *Chemosphere* **2016**, *144*, 2469–2475. [CrossRef]
54. Wang, Z.; Zhang, X.; Chen, Z.; Zhang, Y. Mercury Concentrations in Size-Fractionated Airborne Particles at Urban and Suburban Sites in Beijing, China. *Atmos. Environ.* **2006**, *40*, 2194–2201. [CrossRef]
55. Lai, S.; Zou, S.; Cao, J.; Lee, S.; Ho, K. Characterizing Ionic Species in PM2.5 and PM10 in Four Pearl River Delta Cities, South China. *J. Environ. Sci.* **2007**, *19*, 939–947. [CrossRef]
56. Available online: www.arl.noaa.gov/ready/hysplit4.html (accessed on 30 June 2021).

57. Lewandowska, A.; Falkowska, L.; Murawiec, D.; Pryputniewicz, D.; Burska, D.; Bełdowska, M. Elemental and Organic Carbon in Aerosols over Urbanized Coastal Region (Southern Baltic Sea, Gdynia). *Sci. Total Environ.* **2010**, *408*, 4761–4769. [CrossRef] [PubMed]
58. Lewandowska, A.U.; Falkowska, L.M. Sea salt in aerosols over the southern Baltic. Part 1. The generation and transportation of marine particles. *Oceanologia* **2013**, *55*, 279–298. [CrossRef]
59. Available online: https://bdl.stat.gov.pl (accessed on 30 June 2021).
60. Ravindra, K.; Sokhi, R.; Van Grieken, R. Atmospheric Polycyclic Aromatic Hydrocarbons: Source Attribution, Emission Factors and Regulation. *Atmos. Environ.* **2008**, *42*, 2895–2921. [CrossRef]
61. Skalska, K.; Lewandowska, A.U.; Staniszewska, M.; Reindl, A.; Witkowska, A.; Falkowska, L. Sources, Deposition Flux and Carcinogenic Potential of PM2.5-Bound Polycyclic Aromatic Hydrocarbons in the Coastal Zone of the Baltic Sea (Gdynia, Poland). *Air Qual. Atmos. Health* **2019**, *12*, 1291–1301. [CrossRef]
62. Ravindra, K.; Bencs, L.; Wauters, E.; de Hoog, J.; Deutsch, F.; Roekens, E.; Bleux, N.; Berghmans, P.; Van Grieken, R. Seasonal and Site-Specific Variation in Vapour and Aerosol Phase PAHs over Flanders (Belgium) and Their Relation with Anthropogenic Activities. *Atmos. Environ.* **2006**, *40*, 771–785. [CrossRef]
63. Tobiszewski, M.; Namieśnik, J. PAH Diagnostic Ratios for the Identification of Pollution Emission Sources. *Environ. Pollut.* **2012**, *162*, 110–119. [CrossRef]
64. Querol, X.; Alastuey, A.; Viana, M.; Moreno, T.; Reche, C.; Minguillón, M.C.; Ripoll, A.; Pandolfi, M.; Amato, F.; Karanasiou, A.; et al. Variability of Carbonaceous Aerosols in Remote, Rural, Urban and Industrial Environments in Spain: Implications for Air Quality Policy. *Atmos. Chem. Phys.* **2013**, *13*, 6185–6206. [CrossRef]
65. Duan, F.; Liu, X.; Yu, T.; Cachier, H. Identification and Estimate of Biomass Burning Contribution to the Urban Aerosol Organic Carbon Concentrations in Beijing. *Atmos. Environ.* **2004**, *38*, 1275–1282. [CrossRef]
66. Kim, W.; Lee, H.; Kim, J.; Jeong, U.; Kweon, J. Estimation of Seasonal Diurnal Variations in Primary and Secondary Organic Carbon Concentrations in the Urban Atmosphere: EC Tracer and Multiple Regression Approaches. *Atmos. Environ.* **2012**, *56*, 101–108. [CrossRef]
67. Lewandowska, A.U.; Śliwińska-Wilczewska, S.; Woźniczka, D. Identification of Cyanobacteria and Microalgae in Aerosols of Various Sizes in the Air over the Southern Baltic Sea. *Mar. Pollut. Bull.* **2017**, *125*, 30–38. [CrossRef]
68. Wiśniewska, K.A.; Śliwińska-Wilczewska, S.; Lewandowska, A.U. The First Characterization of Airborne Cyanobacteria and Microalgae in the Adriatic Sea Region. *PLoS ONE* **2020**, *15*, e0238808. [CrossRef] [PubMed]
69. Chow, J.C.; Watson, J.G.; Chen, L.-W.A.; Rice, J.; Frank, N.H. Quantification of $PM_{2.5}$ Organic Carbon Sampling Artifacts in US Networks. *Atmos. Chem. Phys.* **2010**, *10*, 5223–5239. [CrossRef]
70. Minguillón, M.C.; Perron, N.; Querol, X.; Szidat, S.; Fahrni, S.M.; Alastuey, A.; Jimenez, J.L.; Mohr, C.; Ortega, A.M.; Day, D.A.; et al. Fossil versus Contemporary Sources of Fine Elemental and Organic Carbonaceous Particulate Matter during the DAURE Campaign in Northeast Spain. *Atmos. Chem. Phys.* **2011**, *11*, 12067–12084. [CrossRef]
71. Mohr, C.; DeCarlo, P.F.; Heringa, M.F.; Chirico, R.; Slowik, J.G.; Richter, R.; Reche, C.; Alastuey, A.; Querol, X.; Seco, R.; et al. Identification and Quantification of Organic Aerosol from Cooking and Other Sources in Barcelona Using Aerosol Mass Spectrometer Data. *Atmos. Chem. Phys.* **2012**, *12*, 1649–1665. [CrossRef]
72. Marenco, F.; Bonasoni, P.; Calzolari, F.; Ceriani, M.; Chiari, M.; Cristofanelli, P.; D'Alessandro, A.; Fermo, P.; Lucarelli, F.; Mazzei, F.; et al. Characterization of Atmospheric Aerosols at Monte Cimone, Italy, during Summer 2004: Source Apportionment and Transport Mechanisms. *J. Geophys. Res. Atmos.* **2006**, *111*. [CrossRef]
73. Cerqueira, M.; Pio, C.; Legrand, M.; Puxbaum, H.; Kasper-Giebl, A.; Afonso, J.; Preunkert, S.; Gelencsér, A.; Fialho, P. Particulate Carbon in Precipitation at European Background Sites. *J. Aerosol Sci.* **2010**, *41*, 51–61. [CrossRef]
74. Lewandowska, A.U.; Bełdowska, M.; Witkowska, A.; Falkowska, L.; Wiśniewska, K. Mercury Bonds with Carbon (OC and EC) in Small Aerosols (PM_1) in the Urbanized Coastal Zone of the Gulf of Gdansk (Southern Baltic). *Ecotoxicol. Environ. Saf.* **2018**, *157*, 350–357. [CrossRef]
75. Alves, C.A.; Vicente, A.M.P.; Gomes, J.; Nunes, T.; Duarte, M.; Bandowe, B.A.M. Polycyclic Aromatic Hydrocarbons (PAHs) and Their Derivatives (Oxygenated-PAHs, Nitrated-PAHs and Azaarenes) in Size-Fractionated Particles Emitted in an Urban Road Tunnel. *Atmos. Res.* **2016**, *180*, 128–137. [CrossRef]
76. Tolis, E.I.; Saraga, D.E.; Filiou, K.F.; Tziavos, N.I.; Tsiaousis, C.P.; Dinas, A.; Bartzis, J.G. One-Year Intensive Characterization on PM2.5 Nearby Port Area of Thessaloniki, Greece. *Environ. Sci. Pollut. Res.* **2015**, *22*, 6812–6826. [CrossRef]
77. Bernalte, E.; Marín Sánchez, C.; Pinilla Gil, E.; Cereceda Balic, F.; Vidal Cortez, V. An Exploratory Study of Particulate PAHs in Low-Polluted Urban and Rural Areas of Southwest Spain: Concentrations, Source Assignment, Seasonal Variation and Correlations with Other Air Pollutants. *Water Air Soil Pollut.* **2012**, *223*, 5143–5154. [CrossRef]
78. Zielinska, B.; Sagebiel, J.; McDonald, J.D.; Whitney, K.; Lawson, D.R. Emission Rates and Comparative Chemical Composition from Selected In-Use Diesel and Gasoline-Fueled Vehicles. *J. Air Waste Manag. Assoc.* **2004**, *54*, 1138–1150. [CrossRef] [PubMed]
79. El Haddad, I.; Marchand, N.; Dron, J.; Temime-Roussel, B.; Quivet, E.; Wortham, H.; Jaffrezo, J.L.; Baduel, C.; Voisin, D.; Besombes, J.L.; et al. Comprehensive Primary Particulate Organic Characterization of Vehicular Exhaust Emissions in France. *Atmos. Environ.* **2009**, *43*, 6190–6198. [CrossRef]
80. Michalski, M.-C.; Michel, F.; Sainmont, D.; Briard, V. Apparent ζ-Potential as a Tool to Assess Mechanical Damages to the Milk Fat Globule Membrane. *Colloids Surf. B Biointerfaces* **2002**, *23*, 23–30. [CrossRef]

81. Evangelos, K. The Impact of Vegetation on Ohe Characteristics of the Flow in an Inclined Open Channel Using the Piv Method. *J. Water Resour. Ocean Sci.* **2012**, *1*, 1. [CrossRef]
82. Falkowska, L.; Lewandowska, A. *Aerosols and Gases in the Earth's Atmosphere-Global Changes*; University of Gdańsk: Gdańsk, Poland, 2009; ISBN 978-83-7326-624-7.

MDPI

Article

The Impact of Air Pollution on Pulmonary Diseases: A Case Study from Brasov County, Romania

Carmen Maftei [1,*], Radu Muntean [1] and Ionut Poinareanu [2,3,4,*]

1 Faculty of Civil Engineering, Transilvania University of Brasov, 500152 Brasov, Romania; radu.m@unitbv.ro
2 Faculty of Medicine, Ovidius University of Constanta, 900527 Constanta, Romania
3 Sacele Public Hospital, 505600 Brasov, Romania
4 Faculty of Materials Science and Engineering, Transilvania University of Brasov, 500036 Brasov, Romania
* Correspondence: cemaftei@gmail.com (C.M.); ionut.poinareanu@univ-ovidius.ro (I.P.)

Abstract: Air pollution is considered one of the most significant risk factors for human health. To ensure air quality and prevent and reduce the harmful impact on human health, it is necessary to identify and measure the main air pollutants (sulfur and nitrogen oxides, PM_{10} and $PM_{2.5}$ particles, lead, benzene, carbon monoxide, etc.), their maximum values, as well as the impact they have on mortality/morbidity rates caused by respiratory diseases. This paper aims to assess the influence of air pollution on respiratory diseases based on an analysis of principal pollutants and mortality/morbidity data sets. In this respect, four types of data are used: pollution sources inventory, air quality data sets, mortality/morbidity data at the local and national level, and clinical data of patients diagnosed with different forms of lung malignancies. The results showed an increased number of deaths caused by respiratory diseases for the studied period, correlated with the decreased air quality due to industrial and commercial activities, households, transportation, and energy production.

Keywords: air quality index; polluting agents; respiratory diseases; monitoring stations

Citation: Maftei, C.; Muntean, R.; Poinareanu, I. The Impact of Air Pollution on Pulmonary Diseases: A Case Study from Brasov County, Romania. *Atmosphere* **2022**, *13*, 902. https://doi.org/10.3390/atmos13060902

Academic Editor: Daniele Contini

Received: 12 May 2022
Accepted: 30 May 2022
Published: 2 June 2022

Publisher's Note: MDPI stays neutral with regard to jurisdictional claims in published maps and institutional affiliations.

Copyright: © 2022 by the authors. Licensee MDPI, Basel, Switzerland. This article is an open access article distributed under the terms and conditions of the Creative Commons Attribution (CC BY) license (https://creativecommons.org/licenses/by/4.0/).

1. Introduction

Air pollution represents the principal risk to human health. In 2021, European Commission adopted the EU Action Plan called "Towards a zero pollution for air, water, and soil". The main objective is to improve the air quality and reduce the premature mortality caused by air pollution by 55% [1].

Several studies have demonstrated the adverse effects of air pollution on the environment and human health, especially on respiratory diseases. In 2018, Dumitru et al. [2] published a retrospective study on the influence of air pollution over the respiratory infections in Romania, covering a period of ten years. Similar studies were developed in different countries and for different periods of time: Rodríguez-Villamizar et al. [3] studied the influence of air pollution on respiratory and circulatory morbidity in Colombia, Nhung et al. [4] in Hanoi, the capital city of Vietnam, while Al-Taani et al. [5] and Nazzal et al. [6,7] focused their research in Sharjah and Ajman Emirates (UAE). Dastoorpoor et al. [8] studied the short-term effects of air pollution in Iran while Stafoggia et al. [9] conducted similar research for the Southern Europe. Carlsten et al. [10] recommended several strategies to minimize personal exposure to ambient air pollution, while Barbulescu et al. [11,12] used statistical methods for modeling and assessing the influence of different pollutants. According to Eurostat [13], 339,000 deaths were caused by respiratory diseases in EU-27, an equivalent of 75 death per 100,000 habitants (SDR–Standardized Death Rate). EEA (European Environment Agency) reported in 2020 that air pollution caused 400,000 premature deaths in Europe [14]. In a recent study, Schraufnagel et al. [15,16] demonstrated that air pollution can affect the respiratory tract and every organ in the body.

To secure the air quality, and prevent and reduce harmful impact on human health, the Directive 2008/50/EC [17] sets some measures: (i) quality standards under which the quality evaluation is based on threshold values of different pollutants such as sulfur dioxide, nitrogen dioxide and oxides of nitrogen, particulate matter (PM_{10} and $PM_{2.5}$), lead, benzene and carbon monoxide; (ii) establishment of air quality management and evaluation area are mandatory for all member states and (iii) improving health.

Even if Romania has registered progress in reducing the emissions of pollutants during the 1990–2016 period, the air quality represents a significant concern and the authority shall endeavor to achieve the new limits proposed in the new Directive 2016/2284/UE [18]. Statistics from Eurostat [13] shows that Romania had an SDR greater than the EU-27 average in 2017 for all diseases associated with respiratory function (Table 1). At the same time, from the analysis concerning the average length of hospital stays for in-patients treated for respiratory disease, results show that in Romania, the average hospital stay in 2018 is 6.7 days, less than the EU-27 average (7.0). Asthma patients spent a highest number of days in hospital (7.9 days), more than the EU average. EEA estimated for 2018 that 29,200 premature deaths in Romania are due to particulate matter and NO_2 concentrations, representing 6.5% of EU-27 countries [19]. According to with European Public Health Alliance report [20], Romania has the highest cost per capita caused by air pollution (1810 euros/capita).

Table 1. Standardized death rates—respiratory disease (2017).

	Influenza	Pneumonia	Chronic Lower Respiratory Diseases	Asthma and Status Asthmaticus	Other Lower Respiratory Diseases	Other Diseases of the Respiratory System
EU −27 [1]	1.0	24.3	34.2	1.2	33.0	22.0
Romania	0.1	42.5	39.2	1.4	37.8	11.7

[1] source Eurostat [13].

Currently, the most quality air index used in Europe is CAQI (Common Air Quality Index), calculated on a three-time level (hourly, daily and annual) developed under the project CITEAIR [21]. The pollutants used in CAQI evaluation are CO (Carbone monoxide), NO_2 (Nitrogen Dioxide), O_3 (Ozone), SO_2 (Sulfur dioxide), $PM_{2.5}$ (Fine Particle Matter), and PM_{10} (Particle Matter). This index does not have the flexibility to aggregate all the pollutants [14]; the air quality is given by the worst value of the contaminants included in the determination. Stieb [22] introduced a new index, AQHI (Air Quality Health Index), based on the "sum of excess mortality risk associated with individual pollutants." Olstrup [23] calculated this index for Stockholm during 2015–2017 and concluded that it could be an efficient tool to estimate air quality based on the combined effect of multiple pollutants. Still, the meta coefficient is available only for a local AQHI evaluation. In a recent review, the authors [20] investigated 19 methods for AQHI evaluation and concluded that most of them could not include in the estimation a new pollutant whether it's designed for a specific number of contaminants or the aggregation function has not the possibility to aggregate a new pollutant.

In this context, this paper aims to assess the influence of air pollution on respiratory diseases based on an analysis of principal pollutants and mortality/morbidity data sets. The main primary histopathological forms of lung malignancies are analyzed, including their association with environmental factors and the primary pollutants [1,2].

2. Materials and Methods

2.1. Study Data

Brasov County is situated in the center part of Romania (Figure 1) at 45°38′ north latitude and 25°35′ east longitude. The elevation increases from north to south (Figure 1). The region is located at the junction of three large natural units: the Eastern Carpathians and the Southern Carpathians, some places exceeding 2000 m, and Transilvania Plateau. The average altitude is 625 m. The climate is temperate with 8.8 °C multiannual average temperature, and annual precipitation is around 654 mm.

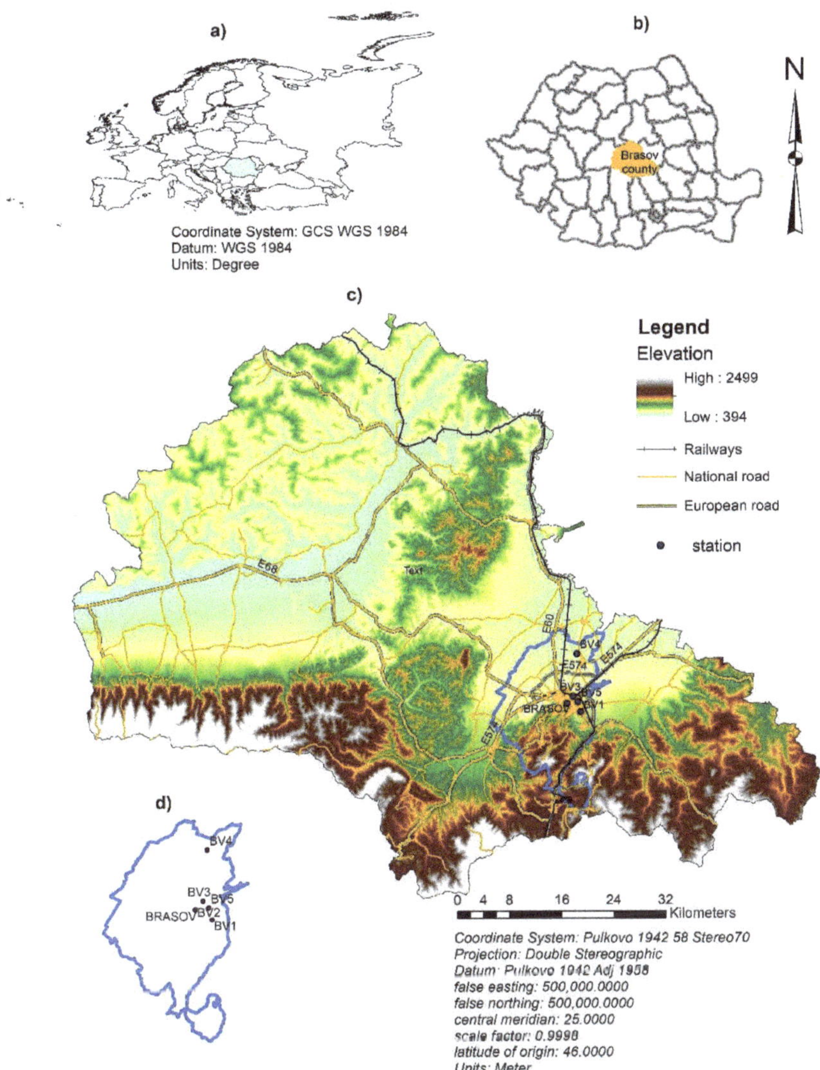

Figure 1. Location of Brasov County and air quality measurement stations: (**a**) map of Europe and the border of Romania; (**b**) map of Romania and the border of Brasov County; (**c**) relief, roads and railways traffic map of Brasov County and the location of monitoring stations; (**d**) Brasov metropolitan area.

The land use distribution is 52% agricultural and 48% non-agricultural, from which 38% are represented by forests.

From an administrative point of view, Brasov County is a part of the Center Development Region and has 58 localities with a population of 627,597 (according to 2011 census). The public road network of Brasov County has a length of 1659 km. It should be mentioned that Brasov is crossed by the European Corridor 4 and the European road E 60 (Figure 1). Brasov County has a railway network with a total length of 353 km, of which 184 km is electrified. Currently, an airport is under construction. In Brasov County, the machine-building industry, the metal processing industry, the pharmaceutical, food, and wood processing industry, and the field of construction, transport, and services have developed. Brașov has a long tradition in tourism, being the most popular ski and winter sports destination in Romania and the resorts in the Prahova Valley.

2.2. Data and Methodology Used

Four types of data are used in this study: pollution sources inventory, air quality data sets, data about mortality/morbidity at the Brasov County level, and clinical data of patients diagnosed histopathologically in Sacele Brasov Municipal Hospital.

Air quality data sets are obtained from the National/Local network of quality Air Monitoring (RLMCA Brasov). The air quality is monitored in five automatic stations in urban, suburban, and industrial areas (Figure 1 and Table 2). The sixth station is a reference station for air quality assessment situated in the mountain area (EMI-Fundata station). The period investigated is 2016–2020, and daily time series data are used. Notice that the Air Quality Monitoring National System was establish in 2011 by Law no 104, which transpose in national legislation Directive 2008/50/CE and 2004/107/CE provisions.

Table 2. Monitoring stations in Brasov.

Station Indicative	Location	Type	Elevation [m]
BV1	Calea Bucureşti Blvd.	traffic	593
BV2	Castanilor street	urban	593
BV3	Garii Blvd.	traffic	593
BV4	Sânpetru village	suburban	560
BV5	Vlahuta street	industrial	593
EMI	Fundata village	regional	1360

SDR and morbidity annually data are obtained from National Institute of Public Health. The clinical data were obtained from Sacele Brasov Municipal Hospital, Pathology Department.

The methodology used in this paper refers to:

1. Understanding pollution sources; to achieve this analysis, we investigated the pollution inventory recorded by the Environment Pollution Agency of Brasov (APM) during 2016–2020. The objective of this activity is to identify the principal economic activity responsible for pollution in Brasov County and the prevalent pollutants that contribute to air quality reduction.
2. Analysis of mortality/morbidity data; the objective of this analysis is to identify the mortality/morbidity caused by respiratory disease and its evolution.
3. Analysis of the current status of principal air quality data and air quality index assessment. The analysis consists of (i) descriptive statistics to establish the frequency of occurrence for each pollutant and station; (ii) air index calculation based on daily, monthly, and annual data sets. For each daily value and pollutant is assigned a scale from "1", excellent, to "6" severe with respect to the calculation grid [24]. The aggregated air quality index is calculated using the worst value of all pollutants used for each station.
4. Analysis of clinical data of patients diagnosed with primary lung malignancies; the patient's ages, sex, and domicile were quantified to identify a correlation between pollutants and histological forms of cancer. The diagnosed tissue specimens came from the thoracic Surgery Clinic of the Brasov Military Hospital's surgery rooms and the Pneumoftiziology Clinic of the Respiratory Diseases Hospital Brasov. Thus, tumorectomies, lung specimens (lobes, segments), endobronchial biopsies, pleural fluids, aspirates, and bronchial lavages were diagnosed. The histopathological specimens were subjected to pathological processing techniques (fixation in 10% buffered formalin, dehydration, and paraffin impregnation by automatic processing, sectioning, staining of sections by hematoxylin-eosin stain. The examined liquids were centrifuged, the sediment being examined both directly and in Papanicolaou and Giemsa

staining, and by inclusion in paraffin, in the form of a cytoblock prepared with neutral proteins. For histological confirmation of the microscopic diagnosis, immunohistochemistry was performed, using the panel of mom and polyclonal antibodies specific to primary lung malignancies (TTF1 clone SP141, Napsin A clone MRQ-60, anti-p40 clone BC28, Anti-Pan Keratin clone AE1/AE3/PCK26). Immunohistochemistry was performed automatically using Benchmark Ventana Gx equipment. The microscopic study was performed using a Zeiss Primo Star microscope and capturing images from the paper was performed using an AxioCam 105 color microscopy camera. The Pathology Department owns the medical equipment. To establish the post-surgical treatment, some specimens, depending on the tumor stage, were investigated by molecular biology techniques to develop the prognostic factors.

3. Results

3.1. Analysis of Pollution Sources

The analysis of pollution sources was based on the emission of pollutants inventory recorded by APM (Environment Pollution Agency) of Brasov during the 2016-2020 period and it is presented in Table 3. Figure 2 shows the distribution of $PM_{2.5}$ and PM_{10}, NOx, and SOx on activities type. As example, 61% of $PM_{2.5}$ and 48% of PM_{10} emitting, respectively, are produced by households (1.A.4.b.i), followed by asphalting works (2.A.6), transport activity (1.A.3.b.i, 1.A.3.b.ii, 1.A.3.b.ii, 1.A.3.b.iv and 1.A.3.c) and cement production (2.A.1). The codes in brackets comply with NFR (Nomenclature for Reporting) code [25]. It can be concluded that the main sectors contributing to the emission of air pollutants in Brasov are: commercial, institutional, and households, transport (road and rail), industrial processes, and energy production and distribution (Table 3). Metal production (iron and steel production) is under 1%.

Table 3. Industrial sectors contributing to emission of air pollutants in Brasov.

NFR Source Categories	NOx	$PM_{2.5}$	PM_{10}	SOx
commercial and households	9%	61.2%	48.2%	23.8%
transport (road and rail)	54.8%	11.8%	10.6%	-
energy production	9%	0.3%	0.3%	0.2%
manufacturing industries and construction	26.7%	14.2%	11.1%	75.8%
agriculture	0.5%	0.2%	0.2%	0.2%
mineral products	-	12.3%	29.6%	-

3.2. Analysis of SDR and Morbidity Data Sets

This analysis is carried out based on data sets of the National Institute of Statistics [26] and the National Institute of Public Health (INSP) during the 1990–2020 period. According to INSP, the main causes of death in Romania, in descending order, are the disease of the circulatory systems, malignant tumors, respiratory system, and digestive system diseases. Of these, the deaths caused by the respiratory disease are investigated, the leading cause of death from the respiratory illness being primary lung malignancy.

The population of Brasov County is 627,597and the average age of the population is 37.4 years (36.7 for men and 39 for women), continuing to increase (Figure 3).

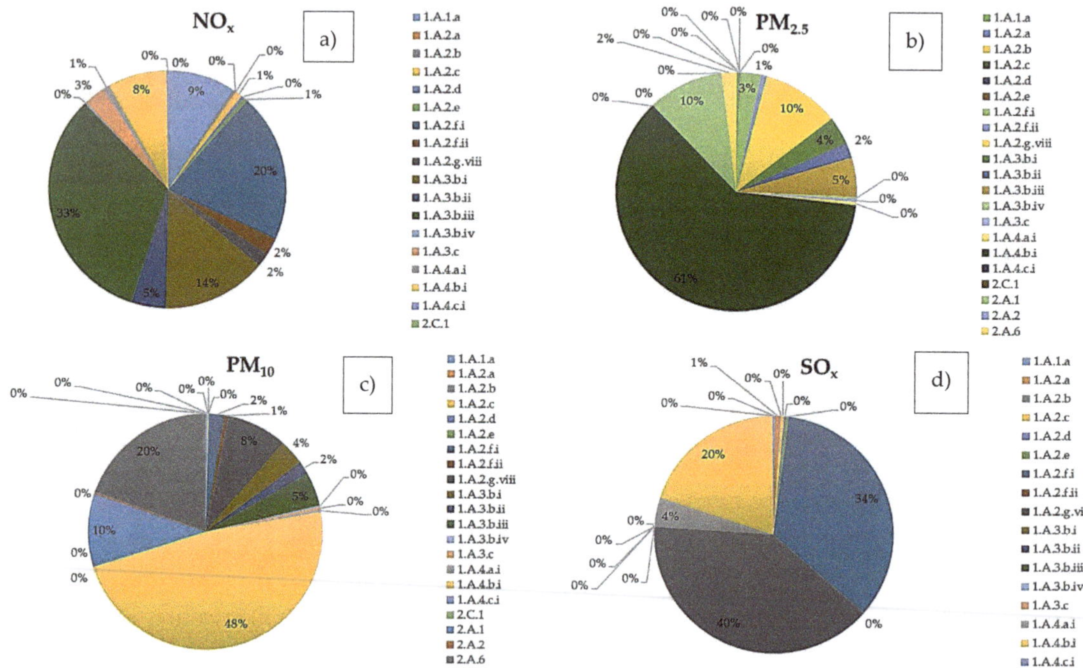

Figure 2. Distribution of principal pollutant emitting. (**a**) NOx; (**b**) $PM_{2.5}$; (**c**) PM_{10}; (**d**) SOx; in respect with NFR code. Legend: 1.A.1.a-Public electricity and heat production; 1.A.2.a Combusting in manufacturing industry -iron and steel; 1.A.2.b-Combusting in manufacturing industry-nonferrous metal; 1.A.2.c Combusting in manufacturing industry-chemical; 1.A.2.d-Combusting in manufacturing industry-; 1.A.2.e Combusting in manufacturing industry -food, drink and tobacco; 1.A.2.f.i-Combusting in manufacturing industry-nonferrous mineral 1.A.2.f.ii-Combusting in manufacturing industry –mobile equipment and machinery; 1.A.2.g.viii-Stationary combustion in manufacturing industry and construction; 1.A.3.b.i-Road transport, passenger cars; 1.A.3.b.ii-Road transport, light duty vehicles; 1.A.3.b.iii-Road transport Heavy duty vehicles; 1.A.3.b.iv-Road transport, mopeds and motorcycles; 1.A.3.c-Railways; 1.A.4.a.i-Comercial hold; 1.A.4.b.i-Household residential; 1.A.4.c.i-Agriculture/Forestry/Fishing; 2.C.1-Iron and steel production; 2.A.1-Cement production; 2.A.2-Lime production; 2.A.6-Road paving with asphalt.

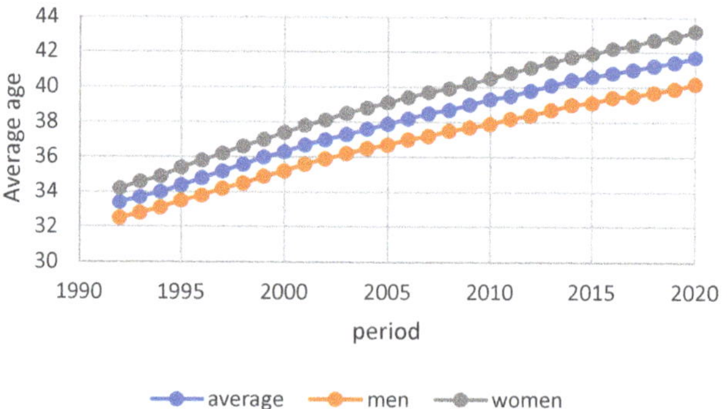

Figure 3. Evolution of average age for the investigated period.

Several deaths in Brasov County are presented in Figure 4. As can be noted, the number of deaths varies from a minimum of 5541 to a maximum of 6526, the average being 5973 people (Figure 4a) and representing 9% of the population, on average. However, more important is that starting with 2013, the number of deaths exceeded the multiannual average (5973 people). Similar behavior can be noticed in the evolution of the number of deaths caused by respiratory disease (Figure 4b). During the 1990–2008 period, the values vary near the average; after 2008, the values decreased, and starting with 2014, an increase can be observed, reaching a value of 481 in 2019. Moreover, the number of deaths caused by respiratory disease represents 4.8% of the total number of deaths in Brasov County. The 2020 data were discarded from this analysis because they are temporarily affected by COVID-19 (Figure 4a,b in red).

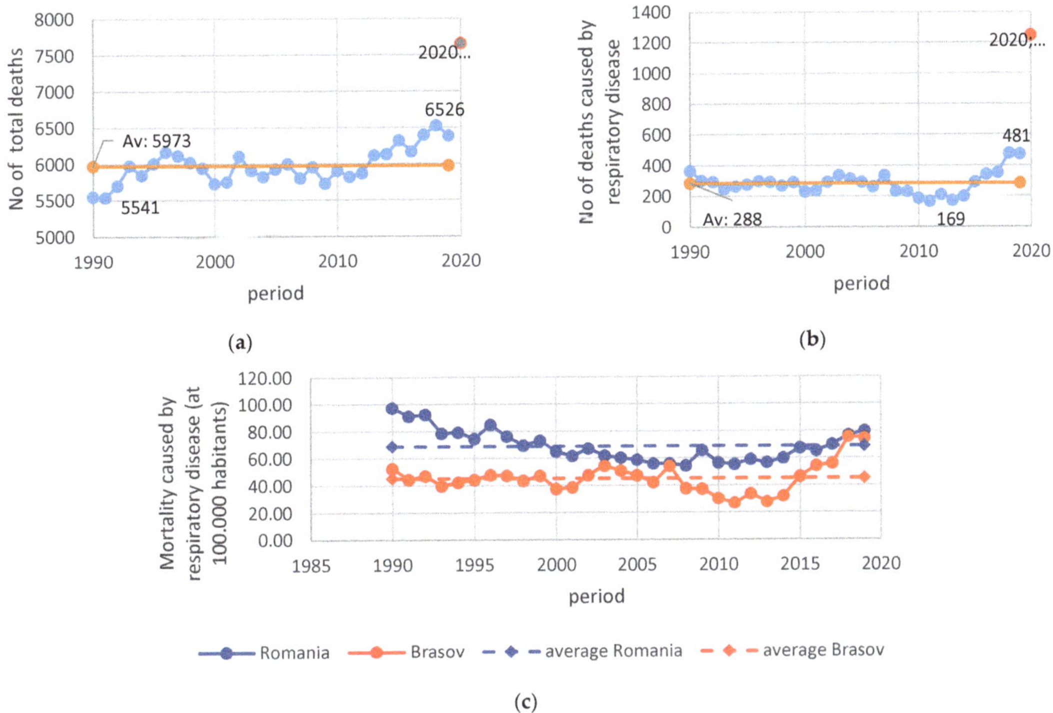

Figure 4. The evolution of the number of deaths and SDR in Brasov County (**a**) Number of total deaths; (**b**) Number of deaths caused by respiratory diseases; (**c**) SDR caused by respiratory diseases.

The standardized Death Rate (SDR) caused by respiratory diseases during the investigated period is presented in Figure 4c. During the 1990–2008 period, a decrease in SDR can be observed at the national level (blue line—Figure 4c). More than that, in the period between 2000 and 2016, the SDR is under the multiannual average (69.14 deaths at 100,000 habitants). Starting with 2018–2019, this index is increasing. On the contrary, the evolution of the same parameter for Brasov Country (red line—Figure 4c) is different, and the period investigated could be divided into three sections (i) 1990–2006—there are no significant variations from the multiannual SDR average (45.5); (ii) 2007–2014 when the values of SDR are 0.6 to 1.2 times under the multiannual average and (iii) 2015–2019 when the values are increasing, reaching the national value in 2018 (75.73).

The most common diseases in neoplasm mortality is malignant neoplasm of bronchus and lungs [27]. In Brasov County, SDR caused by tumors is increasing; starting with 2005, the SDR value is over the multiannual average. Unfortunately, there is no information concerning the number of deaths caused by malignant neoplasm of the bronchus and lungs.

There are few data concerning the morbidity at Brasov County-level caused by respiratory disease. INSP communicates several healthy profile statuses starting with 2014 generally based on official statistics. Starting with 2017, three kinds of morbidity index are calculated: incidence rate (number of new cases at 100,000 habitants), prevalence rate (number of cases of a disease existing in a population), and hospitalized morbidity. The data are presented below (Table 4).

Table 4. Morbidity for Brasov County per 100,000 habitants according to INSP data.

	Year	Incidence	Prevalence	Hospitalized
Malign tumors from which neoplasm of the bronchus and lungs	2017	330.3/30.3 *	2474/86.6 *	798.6/37.0 *
	2018	253.6/25.9 *	2453.8/319.2 *	763.9/28.8 *
	2019	110.7/11.6 *	1928.6/259.2 *	745.6/42.0 *
Chronic obstructive pulmonary disease (COPD)	2017	189.6	1940.8	331.7
	2018	125.8	1940.6	310.9
	2019	157.2	2254.7	221.2
Asthma	2017	139.9		25.2
	2018	150.1		23.9
	2019	184.7		24.1

* the "/" symbols mean "from which" (e.g., 330.3/30.3 = 330.3 total malign tumors from which 30.3 are neoplasm of bronchus and lungs.

Due to this lack of data, we could not draw any conclusions. The morbidity rate for malign tumors, including neoplasm of the bronchus and lungs, is decreasing (except for the value for hospitalized morbidity in 2019). The prevalence for COPD morbidity is increasing, while the hospitalized morbidity is decreasing. Asthma incidence is rising.

3.3. The Current Status of Principal Air Quality Data

As previously stated, the air quality index is monitored at six stations from which one is situated at above 1000 m altitude, and it is a regional station.

Table 5 presents some statistical information on the daily values monitored. Generally, the values registered vary in significant limits. The standard daily limits for NOx and $PM_{2.5}$ (50 mg/m^3) exceed 67% of cases during the investigated period, while the SO_2 values are under the standard daily limit (125 mg/m^3). The values of PM_{10} exceed the level of 50 mg/m^3 in 14% of cases (on average). It is known that the number of daily averages above the standard limit for PM_{10} must not exceed 35 days in a year [23]. This analysis has highlighted that, on average, this limit is exceeded for BV2 and BV3 stations (43 and 72 days, respectively). For BV4, which is situated in a suburban area, the value is not exceeded.

The values measured at the EMI station do not exceed the standard limit, but the number of observations is under 500 values (less than 100 observations per year). For this reason, we decided to continue the analysis without this station. $PM_{2.5}$ values are measured only on the BV2 station.

Based on the daily average, for the investigated period (2016–2020), the NOx index of frequency of occurrence is presented in the following table (Table 6).

For the BV4 station situated in a residential area, the criterion is excellent, with a 94% frequency of occurrence. For BV3 located in a heavy traffic area, the index is 11% frequency of occurrence for excellent criterion (Table 6). Similarly, the frequency of occurrence is calculated for each pollutant and station. Similar behavior was observed.

The aggregated air quality index (AQI) is calculated (Figure 5). The overall index is between "fine" (2) and "moderate" (3) for each station investigated and each year, with several exceptions (BV5 in 2016) due to lack of data for PM_{10} and BV4 situated in suburban area. For BV4, the index is between "excellence" and "fine."

Table 5. Statistical analysis of daily principal pollutants investigated.

Pollutant	Station	Observation	Obs. without Missing Data	Minimum	Maximum	Mean	Std. Deviation	No of Days Over Limit/Year
NOx	BV1	1827	1710	8.12	565.52	73.58	61.09	237
	BV2	1827	1547	9.62	628.90	69.96	63.63	187
	BV3	1827	1692	7.97	681.83	94.57	63.57	300
	BV4	1827	1682	2.14	142.60	17.24	13.89	19
	BV5	1827	1235	12.21	641.04	78.78	64.72	121
	EMI	1827	466	2.39	26.60	7.44	2.24	0
SO_2	BV1	1827	1076	0.47	17.02	5.73	2.07	0.00
	BV2	1827	1588	0.46	24.39	6.07	1.83	0.00
	BV3	1827	1227	1.40	21.56	6.87	2.66	0.00
	BV4	1827	1682	2.14	142.60	17.24	13.89	0.00
	BV5	1827	1483	0.00	19.85	5.61	2.04	0.00
	EMI	1827	485	0.09	18.11	5.47	2.67	0.00
PM_{10}	BV1	1827	1636	2.74	179.23	27.96	18.82	27
	BV2	1827	1305	2.36	255.93	28.84	22.27	44
	BV3	1827	1697	2.91	216.48	31.93	21.13	73
	BV4	1827	1223	1.09	200.95	23.82	21.00	14
	BV5	1827	659	0.43	272.09	25.24	27.24	15
	EMI	1827	397	0.73	66.86	9.23	8.37	0
$PM_{2.5}$	BV2	1827	1560	1.09	198.31	19.22	17.17	10

Table 6. NOx index frequency of occurrence.

Rating	Index Value for NOx	BV1	BV2	BV3	BV4	BV5
excellent	0–40	31%	39%	11%	94%	24%
fine	40–90	41%	33%	41%	5%	45%
moderate	90–120	12%	13%	26%	0%	14%
poor	120–230	12%	11%	17%	0%	12%
very poor	230–340	1%	2%	4%	0%	3%
severe	340–1000	3%	1%	1%	0%	1%

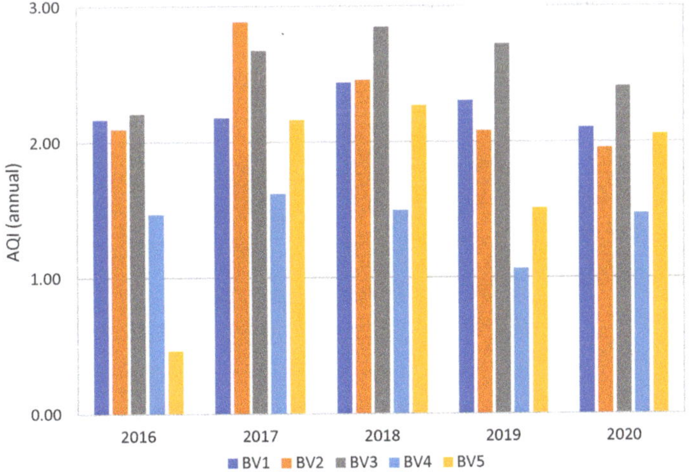

Figure 5. Aggregated Air Quality Index for each station.

There are also "severe" (6) and "very poor" (5) levels registered generally in the winter period or/and late autumn.

The variation of multiannual monthly average for each pollutant shows a certain behavior (Figure 6). In the winter and autumn months (October—February) the pollutant's values are greater than in spring and summer period, much more pronounced for the NOx variation than for PM. Only SO_2 di not have any variation.

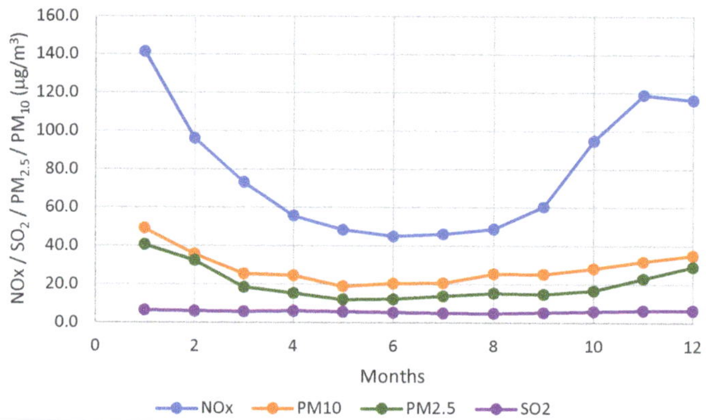

Figure 6. Variation of the multiannual monthly average per pollutant.

Figure 7 presents the aggregated index calculated for each station based on multiannual monthly values of each pollutant. Indeed, between November and February, the air quality index value is high. The value 4 represents poor quality. For BV4 (suburban station), the value of the index is 2, meaning a "fine" quality, without May, June, and July when the air quality index is 1.

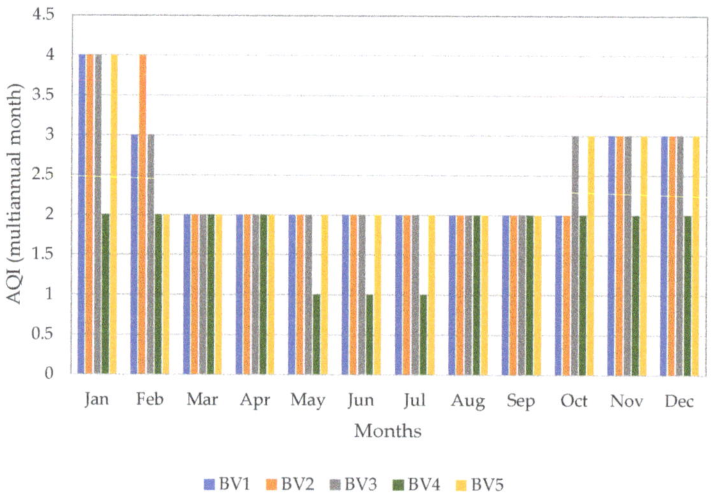

Figure 7. Variation of the multiannual monthly average per station.

3.4. *Analysis of Clinical Data of Patients Diagnosed with Primary Histological Forms of Lung Malignancies*

All cases of microscopically diagnosed primary lung tumors from 2018–2021 were studied retrospectively. The study represented the establishment of a database that contains minimal patient identification forms (age, sex, and domicile) and the microscopic form of malignant tumors diagnosed in the hospital Pathology Department.

Table 7 shows the distribution of cases according to age, sex, domicile, and histological forms of primary lung malignancy. 104 primary lung malignancies were examined, and malignant histological forms were pulmonary adenocarcinoma and squamous cell carcinoma.

Table 7. The distribution of cases.

Year of Study	Cases	Sex Ratio	Cases	Histological Forms	Cases
2018	25	Men	70	Pulmonary Adenocarcinoma (ADK)	25
2019	47	Women	34	Squamous cell carcinoma (SCC)	79
2020	9				
2021	23				
Ages	**Cases**			**Domicile and histological forms**	**Cases**
54–60	18			Urban/ADK	20
61–70	54	**Domicile**	**Cases**	Urban/SCC	55
71–80	29	Urban	75	Rural/ADK	5
81–82	3	Rural	29	Rural/SCC	24

Figure 8 shows the microscopic aspects of the usual hematoxylin-eosin staining of pulmonary adenocarcinoma (ADK) and squamous cell carcinoma (SCC).

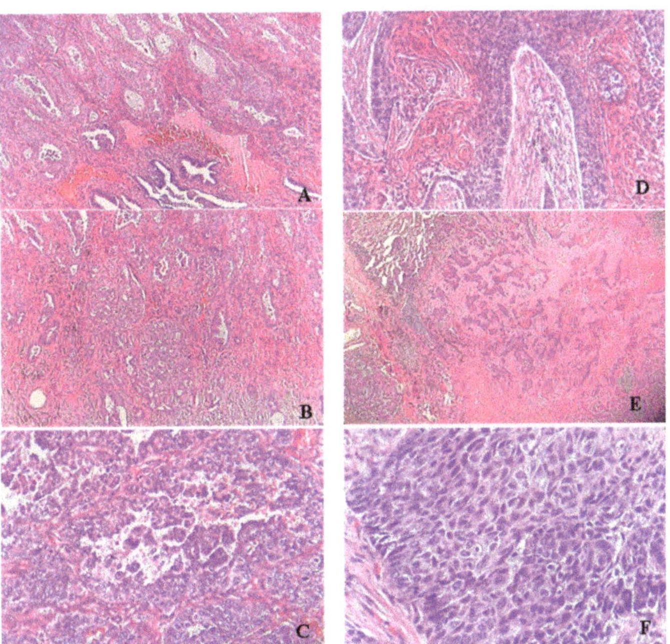

Figure 8. (**A**) Pulmonary adenocarcinoma moderate defined, bronchial origin (HE × 100); (**B**) Cribriform pulmonary adenocarcinoma (HE × 100); (**C**) Poor differentiated pulmonary adenocarcinoma (HE × 400); (**D**) Keratinized squamous cell carcinoma (HE × 200); (**E**) Not differentiated squamous cell carcinoma (HE × 100); (**F**) Atypical mitoses in squamous cell carcinoma (HE × 400).

4. Discussion

The analysis of clinical and histological data analysis showed that the most common form of lung cancer in patients included in the study is squamous cell carcinoma (76%) compared with pulmonary adenocarcinoma (24%). It should be noted that no primary neuroendocrine forms or small cell carcinomas were identified in the study. In addition, the incidence of the disease was much higher in urban areas than in rural areas. Probably the pollutants and environment have given this point of view major importance in the context of the influence of the environment on the development of cancers in general, and lung cancers in particular. Among the cases diagnosed in rural areas, the highest frequency was squamous cell carcinoma as well as in the cases from urban areas.

Regarding the distribution by sex ratio, the most common cases were in men (67%), while in women, the frequency was half (33%), according to medical literature. By analyzing the incidence by age groups, out of 104 diagnosed cases, the most common were in the age category 61–70 years (52%), the other categories being less affected.

The incidence of primary malignant lung tumors during the four years of the study shows that in 2019, 45% of cases were diagnosed. The patient's addressability to the doctor was also influenced by the SARS-CoV2 pandemic context, which explains the much lower number of cases in 2020 and the explosion recorded in 2021, approximately close to that of 2018 (over 20% of all tumors studied).

A report by John Hopkins University [28] shows that cigarettes cause 91% of squamous cell carcinoma, but the exposure to other toxic pollutants or radon are important risk factors. A study led in Korea shows that PM_{10} and NO_2 increase the number of lung cancer incidence [29]. It is worth mentioning that Brasov operated a thermal power plant till 2015, which is the most important pollutant activity in the area. In this context, even if we have not a specific tool to distinguish between the effects of NOx pollution and other pollutants on lung diseases, considering the results obtained especially for NOx pollutant, we conclude that the increases of lung cancer number in the latest period could be affected by the air pollution.

Our study's principal limitation consists in the number of pollution stations and their spatial distribution (Figure 1), which does not offer the possibility of realizing a spatial distribution of results or applying a Multicriteria Evaluation (MCE) method integrated with GIS. In this context, maybe modeling air quality based on wind, rainfall, or other climacteric parameters will be possible to continue this work. The second impediment is the lack of clinical data before 2018. This is because Sacele municipal hospital was closed in 2011 by a government decision. After its opening in 2017, the new Pathology Department started developing a research database related to malignant tumors.

The study represents an association of laboratory medical findings made on the group of patients who addressed the pulmonology services with malignant tumor suspicion that was confirmed histopathologically, with the level of air pollution in the metropolitan area of Brasov. The association found that the level of PM_{10} air pollutant detected in the respirated air in the metropolitan area is associated with the presence of squamous lung malignancies in patients, compared to other histological forms of bronchopulmonary cancer. The idea of developing the study started from the initially superficial analysis on the type of cases examined medically in the pathological anatomy service, due to the increased frequency of malignant lung tumor pathology in Brasov County compared to other counties in Romania. Thus, the analysis of the environmental factors in the territory was deepened and it was revealed that the polluting particles from the breathed air influence the malignant transformation of the respiratory epithelium by squamous metaplasia at bronchial level. The analysis and purpose of our study is to trigger an alarm signal that in the region, the risk of developing pulmonary squamous cell carcinoma is associated with the presence and levels of PM_{10} and NO_2 pollutants.

Author Contributions: Conceptualization, C.M. and I.P.; methodology, C.M. and I.P.; validation, R.M., C.M. and I.P.; formal analysis, R.M.; investigation, C.M. and I.P.; resources, C.M. and I.P.;

data curation, R.M. and I.P.; writing—original draft preparation, C.M. and I.P.; writing—review and editing, C.M., R.M and I.P. visualization, C.M. and I.P.; supervision, C.M. and RM; All authors have read and agreed to the published version of the manuscript.

Funding: This research received no external funding.

Institutional Review Board Statement: Not applicable.

Informed Consent Statement: Not applicable.

Conflicts of Interest: The authors declare no conflict of interest.

References

1. European Commission Directorate-General for Environment; Kantor, E.; Klebba, M.; Richer, C.; Kubota, U.; Zeisl, Y.; Dittrich, M.; Blanes Guardia, N.; Fons Estevez, J.; Salomons, E.; et al. *Assessment of Potential Health Benefits of Noise Abatement Measures in the EU: Phenomena Project*; Publications Office of the European Union: Luxembourg, 2021.
2. Dumitru, I.M.; Lilios, G.; Arbune, M. Respiratory Infections and Air Pollution, Retrospective Study Over the Past 10 Years. *J. Environ. Prot. Ecol.* **2018**, *19*, 1445–1451.
3. Rodríguez-Villamizar, L.A.; Rojas-Roa, N.Y.; Blanco-Becerra, L.C.; Herrera-Galindo, V.M.; Fernández-Niño, J.A. Short-term effects of air pollution on respiratory and circulatory morbidity in colombia 2011–2014: A multi-city, time-series analysis. *Int. J. Environ. Res. Public Health* **2018**, *15*, 1610. [CrossRef] [PubMed]
4. Nhung, N.T.T.; Schindler, C.; Dien, T.M.; Probst-Hensch, N.; Perez, L.; Künzli, N. Acute effects of ambient air pollution on lower respiratory infections in Hanoi children: An eight-year time series study. *Environ. Int.* **2018**, *110*, 139–148. [CrossRef] [PubMed]
5. Al-Taani, A.A.; Nazzal, Y.; Howari, F.M.; Iqbal, J.; Bou Orm, N.; Xavier, C.M.; Bărbulescu, A.; Sharma, M.; Dumitriu, C.-S. Contamination Assessment of Heavy Metals in Agricultural Soil, in the Liwa Area (UAE). *Toxics* **2021**, *9*, 53. [CrossRef] [PubMed]
6. Nazzal, Y.; Orm, N.B.; Barbulescu, A.; Howari, F.; Sharma, M.; Badawi, A.E.; Al-Taani, A.; Iqbal, J.; El Ktaibi, F.; Xavier, C.M.; et al. Study of Atmospheric Pollution and Health Risk Assessment: A Case Study for the Sharjah and Ajman Emirates (UAE). *Atmosphere* **2021**, *12*, 1442. [CrossRef]
7. Nazzal, Y.; Bărbulescu, A.; Howari, F.; Al-Taani, A.A.; Iqbal, J.; Xavier, C.M.; Sharma, M.; Dumitriu, C.Ș. Assessment of Metals Concentrations in Soils of Abu Dhabi Emirate Using Pollution Indices and Multivariate Statistics. *Toxics* **2021**, *9*, 95. [CrossRef] [PubMed]
8. Dastoorpoor, M.; Khanjani, N.; Bahrampour, A.; Goudarzi, G.; Aghababaeian, H.; Idani, E. Short-term effects of air pollution on respiratory mortality in Ahvaz, Iran. *Med. J. Islam. Repub. Iran* **2018**, *32*, 173–181. [CrossRef] [PubMed]
9. Stafoggia, M.; Samoli, E.; Alessandrini, E.; Cadum, E.; Ostro, B.; Berti, G.; Faustini, A.; Jacquemin, B.; Linares, C.; Pascal, M.; et al. Short-term Associations between Fine and Coarse Particulate Matter and Hospitalizations in Southern Europe: Results from the MED-PARTICLES Project. *Environ. Health Perspect.* **2013**, *121*, 1026–1033. [CrossRef] [PubMed]
10. Carlsten, C.; Salvi, S.; Wong, G.W.K.; Chung, K.F. Personal strategies to minimise effects of air pollution on respiratory health: Advice for providers, patients and the public. *Eur. Respir. J.* **2020**, *55*, 1902056. [CrossRef] [PubMed]
11. Bărbulescu, A.; Dumitriu, C.Ș. Assessing Water Quality by Statistical Methods. *Water* **2021**, *13*, 1026. [CrossRef]
12. Bărbulescu, A.; Dumitriu, C.S.; Ilie, I.; Barbeş, S.-B. Influence of Anomalies on the Models for Nitrogen Oxides and Ozone Series. *Atmosphere* **2022**, *13*, 558. [CrossRef]
13. Respiratory Diseases Statistics. Available online: http://ec.europa.eu/eurostat/statistics-explained/index.php/Respiratory_diseases_statistics (accessed on 4 September 2021).
14. European Environment Agency. *Healthy Environment, Healthy Lives: How the Environment Influences Health and Well-Being in Europe*; European Environment Agency: Copenhagen, Denmark, 2020.
15. Schraufnagel, D.E.; Balmes, J.R.; Cowl, C.T.; De Matteis, S.; Jung, S.-H.; Mortimer, K.; Perez-Padilla, R.; Rice, M.B.; Riojas-Rodriguez, H.; Sood, A.; et al. Air Pollution and Noncommunicable Diseases: A Review by the Forum of International Respiratory Societies' Environmental Committee, Part 1: The Damaging Effects of Air Pollution. *Chest* **2019**, *155*, 409–416. [CrossRef] [PubMed]
16. Schraufnagel, D.E.; Balmes, J.R.; Cowl, C.T.; De Matteis, S.; Jung, S.-H.; Mortimer, K.; Perez-Padilla, R.; Rice, M.B.; Riojas-Rodriguez, H.; Sood, A.; et al. Air Pollution and Noncommunicable Diseases: A Review by the Forum of International Respiratory Societies' Environmental Committee, Part 2: Air Pollution and Organ Systems. *Chest* **2019**, *155*, 417–426. [CrossRef] [PubMed]
17. Directive 2008/50/EC of the European Parliament and of the Council of 21 May 2008 on Ambient Air Quality and Cleaner Air for Europe. Available online: http://news.cleartheair.org.hk/wp-content/uploads/2013/02/LexUriServ.pdf (accessed on 10 May 2022).
18. European Parliament and Council. DIRECTIVE (EU) 2016/2284 on the Reduction of National Emissions of Certain Atmospheric Pollutants, Amending Directive 2003/35/EC and Repealing Directive 2001/81/EC. Available online: https://eur-lex.europa.eu/legal-content/RO/TXT/HTML/?uri=CELEX:32016L2284&from=RO (accessed on 2 September 2021).
19. Romania—Air Pollution Country Fact Sheet—European Environment Agency. Available online: https://www.eea.europa.eu/themes/air/country-fact-sheets/2020-country-fact-sheets/romania (accessed on 2 September 2021).

20. De Bruyn, S.; de Vries, J. *Health Costs of Air Pollution in European Cities and the Linkage with Transport*; CE Delft: Delft, The Netherlands, 2020.
21. CITEAIR—Comparing Urban Air Quality across Borders. Available online: http://airqualitynow.eu/download/CITEAIR-Comparing_Urban_Air_Quality_across_Borders.pdf (accessed on 3 September 2021).
22. Stieb, D.M.; Burnett, R.T.; Smith-Doiron, M.; Brion, O.; Shin, H.H.; Economou, V. A New Multipollutant, No-Threshold Air Quality Health Index Based on Short-Term Associations Observed in Daily Time-Series Analyses. *J. Air Waste Manag. Assoc.* **2008**, *58*, 435–450. [CrossRef] [PubMed]
23. Olstrup, H. An Air Quality Health Index (AQHI) with Different Health Outcomes Based on the Air Pollution Concentrations in Stockholm during the Period of 2015–2017. *Atmosphere* **2020**, *11*, 192. [CrossRef]
24. Order 1818 02/10/2020. Available online: http://legislatie.just.ro/Public/DetaliiDocument/231536 (accessed on 7 September 2021).
25. European Environment Agency. EMEP/EEA Air Pollutant Emission Inventory Guidebook. 2019. Available online: https://www.eea.europa.eu/publications/emep-eea-guidebook-2019 (accessed on 4 February 2022).
26. TEMPO Online. Available online: http://statistici.insse.ro:8077/tempo-online/#/pages/tables/insse-table (accessed on 4 September 2021).
27. INSP-CNSISP. Mortalitatea Generală. 2019. Available online: https://cnsisp.insp.gov.ro/wp-content/uploads/2021/01/MORTALITATEA-GENERALA-2019.pdf (accessed on 5 September 2021).
28. Ettinger, D.S.; Akerley, W.; Borghaei, H.; Chang, A.C.; Cheney, R.T.; Chirieac, L.R.; D'Amico, T.A.; Demmy, T.L.; Govindan, R.; Grannis, F.W.; et al. Non-Small Cell Lung Cancer, Version 2.2013. *J. Natl. Compr. Cancer Netw.* **2013**, *11*, 645–653. [CrossRef] [PubMed]
29. Lamichhane, D.K.; Kim, H.-C.; Choi, C.-M.; Shin, M.-H.; Shim, Y.M.; Leem, J.-H.; Ryu, J.-S.; Nam, H.-S.; Park, S.-M. Lung Cancer Risk and Residential Exposure to Air Pollution: A Korean Population-Based Case-Control Study. *Yonsei Med. J.* **2017**, *58*, 1111. [CrossRef] [PubMed]

Article

Levels, Sources, and Health Damage of Dust in Grain Transportation and Storage: A Case Study of Chinese Grain Storage Companies

Pengcheng Cui [1], Tao Zhang [1], Xin Chen [1] and Xiaoyi Yang [2,*]

1 Academy of National Food and Strategic Reserves Administration, Beijing 100037, China; cpc@ags.ac.cn (P.C.); zt@ags.ac.cn (T.Z.); chx@ags.ac.cn (X.C.)
2 School of Emergency Management and Safety Engineering, China University of Mining and Technology-Beijing, Beijing 100083, China
* Correspondence: yangxyi6@126.com; Tel.: +86-136-8305-3139

Abstract: A large amount of mixed dust exists in grain, which can easily stimulate the respiratory system and cause diseases. This study explored contamination levels and health effects of this grain dust. A total of 616 dust samples from different stages and types of grain were collected in China—in Hefei (Anhui), Shenzhen (Guangdong), Chengdu (Sichuan), Changchun (Jilin), and Shunyi (Beijing)—and analyzed using the filter membrane method and a laser particle size analyzer. A probabilistic risk assessment model was developed to explore the health effects of grain dust on workers in the grain storage industry based on the United States Environmental Protection Agency risk assessment model and the Monte Carlo simulation method. Sensitivity analysis methods were used to analyze the various exposure parameters and influencing factors that affect the health risk assessment results. This assessment model was applied to translate health risks into disability-adjusted life years (DALY). The results revealed that the concentration of dust ranged from 25 to 70 mg/m^3, which followed normal distribution and the proportion of dust with a particle size of less than 10 μm exceeded 10%. Workers in the transporting stage were exposed to the largest health risk, which followed a lognormal distribution. The average health risks for workers in the entering and exiting zones were slightly below 2.5×10^{-5}. The sensitivity analysis indicated that average time, exposure duration, inhalation rate, and dust concentration made great contributions to dust health risk. Workers in the grain storage and transportation stage had the health damage, and the average DALY exceeded 0.4 years.

Keywords: grain dust; health risk assessment; Monte Carlo simulation; disability adjusted life year

Citation: Cui, P.; Zhang, T.; Chen, X.; Yang, X. Levels, Sources, and Health Damage of Dust in Grain Transportation and Storage: A Case Study of Chinese Grain Storage Companies. *Atmosphere* **2021**, *12*, 1025. https://doi.org/10.3390/atmos12081025

Academic Editor: Alina Barbulescu

Received: 16 July 2021
Accepted: 9 August 2021
Published: 11 August 2021

Publisher's Note: MDPI stays neutral with regard to jurisdictional claims in published maps and institutional affiliations.

Copyright: © 2021 by the authors. Licensee MDPI, Basel, Switzerland. This article is an open access article distributed under the terms and conditions of the Creative Commons Attribution (CC BY) license (https://creativecommons.org/licenses/by/4.0/).

1. Introduction

Grain is an important foundation of national security. The scale of grain storage and the function of facilities determine the national grain circulation capacity. There are many occupational hazards that can affect the health of workers in the process of grain storage. Grain contains a large amount of mixed dust, including grain husks, bacteria, pests, microorganisms, and mixed fine sand [1]. Grain dust is a companion in the whole life cycle of modern grain, from purchase to storage, transportation, and processing. It is produced due to friction, collision, extrusion, crushing, etc. [2]. The particle size distribution of dust is approximately normal, ranging in size from 0 to 9.6×10^5 nm, with a true density within the range of 1.1–1.8 g/m^3 [3]. During the storage process, grain is constantly tumbling, and the dust particles are continuously separated under the influence of air flow. The instantaneous contact mass concentration can reach up to 1000 mg/m^3, and the time-weighted average allowable concentration is about 40 mg/m^3. Workers are often exposed to inhalable dust of >10 mg/m^3, and higher exposures can be found at the grain in-warehousing and out-warehousing stages [4].

Dust can cause serious environmental pollution, grain quality decline, mechanical equipment wear, and, more importantly, it also threatens the health of grain warehouse operators. Studies have shown that long-term exposure to high-concentration grain storage dust can easily irritate the respiratory system [5]. When accumulated after inhalation, it can cause allergic reactions [6] and cereal fever syndrome [7]. In severe cases, it may cause respiratory diseases such as pneumoconiosis, hardening of the respiratory organs, and recurrent nocturnal asthma [8]. Dust can also enter the blood with human cells, which can cause further cardiovascular and cerebrovascular diseases [9]. Due to its potential hazards, monitoring grain dust, studying its distribution rules, and quantifying the health damage are important steps in effectively recognizing and measuring the health hazards of dust.

Researchers have recently begun to pay more attention to the physical and chemical properties [10], sources [11], explosive characteristics [12–14], prevention measures [15], and mycotoxin content [16,17] of grain dust. When conducting occupational hazard analysis, the toxicological characteristics [18], the mechanism of action on the respiratory tract [19], and the clinical manifestations of diseases [20] related to grain dust have mostly been analyzed from a medical perspective, and there is a lack of quantitative evaluation of its health damage.

Dust health damage assessment has been carried out with reference to coal mines, construction, atmospheric environment, and other fields using deterministic analysis methods. For example, Behrooz et al. assessed the carcinogenic and non-carcinogenic health risks for three exposure pathways in airborne dust samples (TSP and $PM_{2.5}$) in Zabol, Iran during the summer dust period [21]; Guney et al. characterized contaminated soils (n = 6) and mine tailings samples (n = 3) for As, Cu, Fe, Mn, Ni, Pb, and Zn content and assessed elemental lung bioaccessibility in fine fraction [22]; Donghua et al. proposed an occupational health hazard risk assessment matrix method to rank the hazards of various risk factors in mining and mineral processing [23]; Lim et al. investigated the contamination levels and dispersion patterns of heavy metals and assessed the risk of health effects on the residence in the vicinity of the abandoned Songcheon Au-Ag mine, Korea [24]; Kan et al. calculated the health damage of air particulate pollution based on the epidemiological exposure–response function [25]; Zhang M. et al. assessed particulate pollution risk and quantified the public health damage caused by air emissions in Beijing in the period 2000–2004 [26]; Liu E. et al. labeled the indicative metals relating to non-exhaust traffic emissions and assessed anthropogenic sources of metals in TR dusts, combining their spatial pollution patterns, principal component analysis, and Pb isotopic compositions [27]; Zhang Y. et al. analyzed eight heavy metals (Cr, Ni, Cu, Cd, Pb, Zn, Mn, and Co) in the PM2.5, collected during four different seasons in Taiyuan, a typical coal-burning city in northern China [28]; Chen X. et al. established a health risk assessment system based on on-site measurements and assessed the health risks for a tunnel machine employee [29]; Zheng et al. investigated the heavy metal contamination in the street dust due to metal smelting in the industrial district of Huludao city and elucidated the spatial distribution of Hg, Pb, Cd, Zn, and Cu in the street dust [30]; Tahir et al. described the estimation of particulate matter (cotton dust) with different sizes in small-scale weaving industry (power looms) situated in district Hafizabad, Punjab, Pakistan, and the assessment of health problems of workers associated with these pollutants [31]; Zazouli et al. investigated the mineralogy, micro-morphology, chemical characteristics, and oxidation toxicity of respirable dusts generated in underground coal mines and assessed the health risk by EPA's health risk model [32]; and Li X et al. introduced the disability-adjusted life year (DALY) model for damage to human health caused by construction dust, to evaluate the environmental impact during the construction process [33]. Scholars have studied the health risks of dust from the perspective of components of heavy metals and the construction of a risk assessment model, which have laid the foundation for the risk assessment of grain dust. However, there are few studies evaluating the health damage caused by grain dust, despite the importance of grain storage and transport in the process of ensuring grain security. This study is particularly timely, as the circulation of grain in the market is accelerating, so a large amount of grain dust will be generated in specific areas. Inhalation is considered to be the main route of human exposure to grain dust [34]. The present research is of great significance for understanding the health status of grain warehouse personnel, as there is a need to evaluate the hazard to human health posed by the grain dust present in granaries.

Current research on quantitative occupational health risk assessment in other industries has yielded instructive conclusions and methods [35–37]. In fact, risk assessment methods include deterministic analysis methods and uncertainty analysis methods. The accuracy of the conclusions drawn from the deterministic method cannot be guaranteed, as the method calculates the health risk through the most probable and maximum value of human exposure parameters and pollutant content [38], resulting in large or small results [39]. However, the problem of uncertainty is inherent in health risk assessment and runs through the whole process [40,41]. The operation area, time, and season of the grain warehouse make the dust concentration uncertain. The uncertainty of human exposure parameters, temperature and humidity changes, and meteorological conditions in the relatively open working environment of the grain warehouse also causes uncertainty in the evaluation results. Hence, this study used the uncertainty analysis method (i.e., the Monte Carlo method) to evaluate the health damage caused by grain dust in warehouses.

To address the uncertainty in grain dust health hazard assessment, a total of 616 samples were collected in China in Hefei (Anhui), Shenzhen (Guangdong), Chengdu (Sichuan), Changchun (Jilin), and Shunyi (Beijing). Additionally, a risk assessment model for the health damage caused by grain dust through inhalation was proposed, considering the dust concentration and the uncertainty of the exposure parameters, and grouping them according to different operation processes and grain varieties (i.e., maize, rice, and wheat) based on the current health risk evaluation system of the United States Environmental Protection Agency (USEPA). The Monte Carlo simulation and Crystal Ball 11.1 software were used to quantify the health damage and to analyze the impact of various parameters on health risks. The results provide guidance for occupational health risk management in the grain storage industry. To our knowledge, this is one of the first studies to analyze and evaluate the hazards of grain dust to human health through inhalation by the uncertainty analysis method.

2. Materials and Methods

2.1. Dust Sampling and Analytical Method

Five grain storage warehouses in China—in Hefei (Anhui), Shenzhen (Guangdong), Chengdu (Sichuan), Changchun (Jilin), and Shunyi (Beijing)—were selected as samples due to their representative and relatively standardized management, large storage capacity, complete variety of grain, and high-quality personnel (see Figure 1 for location information). As national reserves, the low-temperature storage method was used in these five grain storage warehouses with the same management mode. Additionally, the warehouse types were tall, square warehouses. The equipment used for grain storage and transportation were similar, including belt conveyor, grain raking machine, etc. These five grain warehouses are the example for evaluating the hazards of dust during grain storage, as well as the occupational health risks caused by that dust.

Three months of regular site monitoring was carried out from March 2021 to May 2021 due to the increased in-warehousing, out-warehousing, and transporting at the grain storage warehouses in spring, which is convenient for sampling. The sampling plan was designed according to China National Standards GBZ 159-2004 and GBZT 192.1-2007, and the concentration of grain dust was calculated using the filter membrane incremental method. A combination of fixed-point and individual sampling was used at three operating stages for three different grains—maize, wheat, and rice—including in-warehousing, out-warehousing, and transporting. The dust concentration of each warehouse was maintained within a certain range during our sampling process, and there was no systematic reduction in individual warehouses. The sampling height was set to 1.5 m (the breathing height of the workers), and the sampling time was set to 60 min. The gas flow rate was set to 2 L/min. A total of 616 samples were obtained. The detailed information is shown in Table 1.

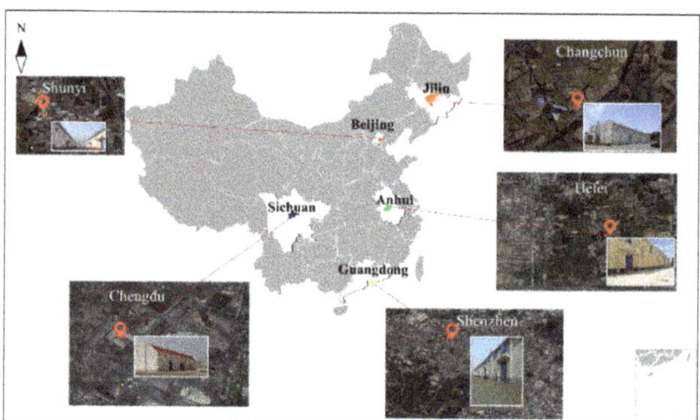

Figure 1. Location of sampling points.

Table 1. Sample information.

Grain	Stage	Sample Size
	Out-warehousing	57
Maize	In-warehousing	62
	Transporting	79
	Out-warehousing	73
Rice	In-warehousing	81
	Transporting	67
	Out-warehousing	83
Wheat	In-warehousing	61
	Transporting	53

The monitoring tool was a dust sampler (AKFC-92 A) made in Changshu, China. The AKFC-92A determines dust concentration, the size of the sampler inlet was 19.63 cm^2, and has the advantages of pulsating airflow, negative pressure, large load capacity, automatic timing sampling, and being explosion proof. The main instrument is shown in Figure 2. The operation procedure was as follows. First, the filter membrane was numbered and weighed to record its quality before sampling. Second, the sampler was fixed horizontally on a tripod platform and placed at the sampling point close to the operator. Third, a trap with the filter membrane was installed on the sampling head, and a certain volume of dusty air was extracted to keep the dust in the filter membrane. Fourth, the filter membrane was removed in a clean area, the dust-receiving surface was folded twice, and then the samples were accurately weighed in the laboratory with an electronic balance (the test accuracy can reach 0.01 mg/m^3) after drying and removing static electricity. Finally, the total dust concentration per unit was calculated from the weight gain of the filter membrane.

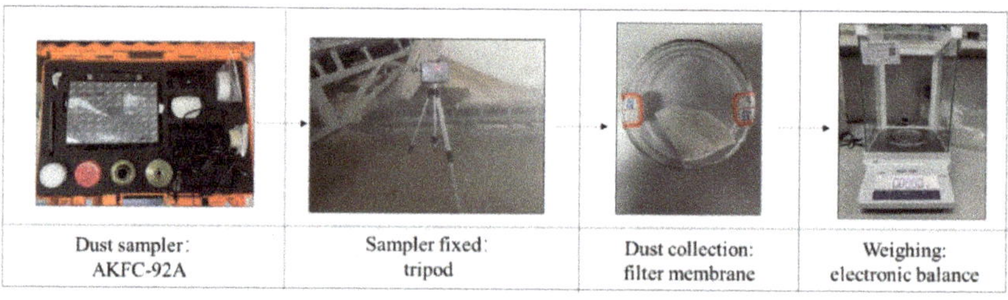

Figure 2. Dust concentration measuring instrument.

2.2. Health Risk Assessment Model

The dust samples used for evaluation in this study came from the warehouses of a large grain storage company to assess the health risks for workers. The inhalation risk assessment model in the *Risk Assessment Guidance for Superfund: Human Health Evaluation Manual, Part F, Vol. 1*, issued in 2009 by the USEPA [42], was selected to evaluate the occupational health risks of the grain storage dust. The key to the model is that the human-absorbed dose is quantified by exposure parameters. Determining the route of exposure is thus a prerequisite for risk assessment. The main exposure pathways for dust that can affect the health of workers are inhalation, which refers to breathing in air containing dust; ingestion, which refers to consumption of food, water, or soil containing dust; and dermal contact, which refers to physical contact of the skin with water or soil containing dust. As an air pollutant, dust mainly enters the human body via inhalation [43], so this study focused on the respiration-based health risks to workers.

An occupational health damage model was constructed with reference to the USEPA modeling principles on health risk and damage quantification. This model included the range determination, dust concentration analysis, health risk characterization, and health damage analysis. The dust generated in rice, wheat, and maize during in-warehousing, out-warehousing, and transporting were the evaluation objects; the collected grain dust data were simulated to select the best distribution; the probability of dust to harm the human health was calculated according to the dose–response relationship; and the dust concentration was transformed into a health risk [44], which was in turn transformed into a life loss caused by disease, expressed in DALY [45]. The dust damage model is shown in Figure 3.

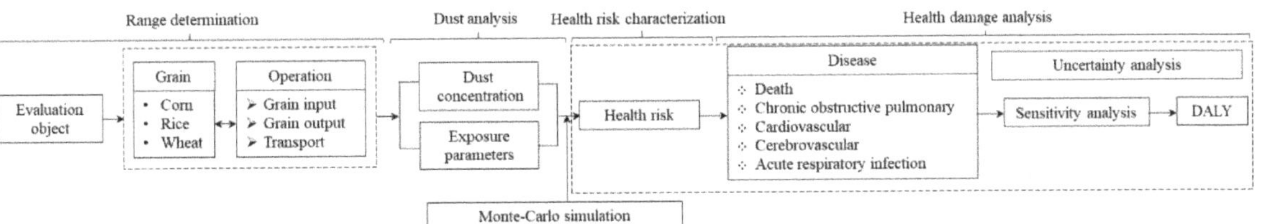

Figure 3. Dust damage model.

2.2.1. Dust Exposure Dose

The dust exposure dose refers to the assessment of individual exposure parameters, which is calculated by monitoring the concentration of dust, exposure duration, and exposure method. However, China has not yet established a complete human exposure parameter database, therefore, due to this lack, the exposure parameter method proposed by the USEPA was used to convert the monitored grain dust concentration into the average daily exposure dose (*ADD*) of the grain warehouse workers. The calculation formula is as follows [46]:

$$ADD = \frac{C \times IR \times ED \times EF \times ET}{BW \times AT}, \quad (1)$$

where *ADD* is the average daily exposure dose of grain storage workers ($mg/kg \cdot d^{-1}$); *C* is the dust concentration on site (obtained by sampling, mg/m^3); *IR* is the inhalation rate of workers (obtained by on-site interview and testing, m^3/h); *ED* is the exposure duration (obtained by on-site interview, a); *EF* is the exposure frequency (obtained by on-site interview, d/a); *ET* is the exposure time (obtained by on-site interview, h/d); *BW* corresponds to the average worker (adult) weight (kg); and *AT* is the average time ($AT = ED \times 365$ days for health risk, d).

2.2.2. Quantizing the Health Risk of Dust

Due to the "threshold" effect, the hazard index R represents the hazard of grain storage dust to the human body. The formula is as follows:

$$R = \frac{ADD}{RfD} \times 10^{-6} \tag{2}$$

where *RfD* is a reference dose of dust and *R* is the health risk of dust. The reference doses for different compounds in the workplace air differ, and the exposure parameter manual issued by the USEPA has been divided. The reference dose standards for different types of dust are different. The *RfD* of silica, cement, wood, and gypsum dust have been calculated as 0.40, 1.20, 1.60, and 3.20, respectively, and a health risk assessment has been completed [47]. However, grain dust contains organic and inorganic substances such as protein, starch, cellulose, and ash, and elements such as Si, Ca, K, Ti, and Cr. The composition is complex, and the *RfD* of grain storage dust is not given. In the present study, dibutyl phthalate (DBP) was selected as a reference for calculation. With reference to the linear relationship between grain dust concentration and DBP in GBZ 2.1 *Occupational Exposure Limits for Hazardous Agents in the Workplace* [48], considering that the standard value of exposure dose and the standard value of environmental concentration have a certain proportional relationship, and the calculation method of Li X.D. [35], it is believed that the linear relationship between DBP and the exposure dose of grain dust is still satisfied, so the *RfD* value of grain dust was calculated to be 1.6 mg/(kg·d).

2.2.3. Dust Health Damage

The DALY model was jointly proposed by Murray and the World Health Organization to quantify the extent of human health damage [49]; it is composed of years of life lost (YLL) and years lived with disability (YLD) [50,51]. The parameters of the DALY are clear, and have been applied to construction, automobile casting, and other industries [52]. Its feasibility and operability have been verified. As dust mainly causes death, chronic obstructive pulmonary disease, cardiovascular disease, cerebrovascular disease, and acute respiratory infections, for this study the damage caused by dust was divided into these five types according to a certain proportion, and the normalized conversion DALY was used to characterize the damage. The formula is as follows:

$$DALY = n \times R \times P \times \sum_i Q_i \times W_i \times L_i \tag{3}$$

where Q_i is the disease risk factor for disease category I; W_i is the effect factor of disease i and takes values between 0 and 1; L_i is the damage factor for disease i (years); P is the number of people affected by specific diseases; and n is the amount of human exposure, namely, the days of operation (days).

The risk and effect factors for the five types of health damage were obtained by referring to the literature [35,37]. The value of the damage factor depends on the evaluated objects. For the five grain warehouses sampled, most of the operators were men from all over China. The life value used in the calculation was derived from the *China Statistical Yearbook* [53]. The values are shown in Table 2.

2.3. *Exposure Parameter Determination*

Exposure parameters are important in health risk assessments, including exposure time, exposure frequency, average time, and exposure duration. Factors such as workers' labor intensity and region have an impact on the assessment results and can interfere with exposure parameters. On-site interviews were conducted with workers in the grain warehouses in five different cities. Additionally, inhalation rate testing was carried out. A total of 45 workers were selected in the in-warehousing operational stage, 49 in the out-warehousing operational stage, and 52 in the transporting operational stage. Their parameter characteristics were obtained, including personal information such as age, gender, height, and weight, as well as work information, such as daily working and rest hours. Crystal Ball 11.1 was used to analyze the survey data in combination with the relevant literature [54–56]. Crystal Ball is a Monte Carlo simulation software launched by Oracle, which can be used for predictive modeling, prediction, simulation, and optimization, random simulation and uncertain risk analysis. It provides a realistic and easy-to-understand uncertainty modeling method, which can achieve the goal and improve the understanding of impact risk by analyzing data and making correct tactical decisions [57]. The results showed that inhalation rate, exposure duration, exposure frequency, exposure time, and average time formed a triangle distribution; body weight followed a normal distribution [58]. The distribution characteristics for the exposure parameters of grain warehouse workers are shown in Table 3.

Table 2. Health damage parameter values.

Disease Endpoints	Q	W	L/a
Death	0.13	1.00	42.2
Chronic obstructive pulmonary disease	0.16	0.15	10
Cardiovascular disease	0.16	0.24	37.2
Cerebrovascular disease	0.20	0.20	37.2
Acute respiratory infections	0.35	0.08	0.04

Table 3. Chinese residents (adults) exposure parameter characteristic.

Exposure Parameters	Abbreviation	Unit	Distribution	Probable Value	Min	Max	SD
Inhalation rate	IR	m^3/h	Triangular	1.8	0.9	2.75	
Body weight	BW	kg	Normal	56.8	42.1	71.6	5.8
Exposure duration	ED	a	Triangular	15	5	25	
Exposure frequency	EF	d/a	Triangular	104.31	98.62	110.54	
Exposure time	ET	h/d	Triangular	5.9	4	6.5	
Average time	AT	d	Triangular	5475	1825	9125	

3. Results and Discussion

3.1. Dust Concentration and Dispersity

After dust sampling, the agglomeration phenomenon in the transportation process was eliminated by ultrasonic treatment, and then the original sample was screened. The particle size distribution of the different grain dust samples was obtained by a laser particle size analyzer. The calculated sample dust concentration data were input into Crystal Ball 11.1 software for fitting, and the results are shown in Table 4.

Table 4. Distribution characteristics of grain dust concentration.

Grain	Dust Concentration			
	Workplace	Distribution	Mean (mg/m^3)	SD
Maize	Out-warehousing	Normal	41.86	14.31
	In-warehousing	Normal	35.92	13.12
	Transporting	Normal	67.25	11.57
Rice	Out-warehousing	Normal	32.54	15.32
	In-warehousing	Normal	30.91	12.69
	Transporting	Normal	59.68	10.67
Wheat	Out-warehousing	Normal	35.29	13.29
	In-warehousing	Normal	26.43	12.67
	Transporting	Normal	55.61	11.12

The dust concentration was at a relatively high level near the conveyor, between 100 and 500 mg/m^3, during the continuous monitoring of the dust concentration. The principle of sampling was to be as close as possible to the working area of the operator, and the position of the sensor was a certain distance from the conveyor, so the concentration of the dust reached 25–70 mg/m^3, which seriously exceeded the standard. Long-term work in such an environment would certainly cause great harm to health. The concentration of maize at each stage was also generally greater than that of rice and wheat, which may be related to the true density.

The three kinds of grain dust have a wide particle size distribution, ranging from 0.6 μm to 950 μm. The deposition sites of the different size particles in respiratory system are also different. Smaller particles can be deposited more deeply and cause greater harm to the human body. It is generally believed that particles with a diameter of less than 2.5 μm have the most impact on human health as they can enter the trachea, bronchus, and alveoli. The proportion of grain dust particles smaller than 2.5 μm exceeded 2%, which could seriously threaten the workers' respiratory system.

3.2. Dust Health Damage Analysis

3.2.1. Health Risks

During risk assessment, uncertainty arose from human parameters and dust concentration. These factors could affect the real reflection of the calculation results to the actual risk values to different degrees, resulting in uncertainty [59,60]. The Monte Carlo simulation effectively solves the problem of uncertainty in risk assessment. Combined with the above formulas, the Crystal Ball 11.1 software was used to simulate and calculate the dust health risk for three kinds of grain and three operation states using the Monte Carlo method. The number of random simulation iterations was set as 10,000, and the confidence level was determined as 95% [61]. The risk results were processed and mapped with OriginPro 9.0. The acceptable risk value recommended by USEPA is 1×10^{-6} [47]. If the risk value for pollutants is less than 1×10^{-6}, it is considered acceptable; if it is greater than 1×10^{-4}, the risk is unacceptable. The results for the dust health risk at different stages for each grain were obtained by this running simulation, as shown in Table 5 and Figures 4 and 5.

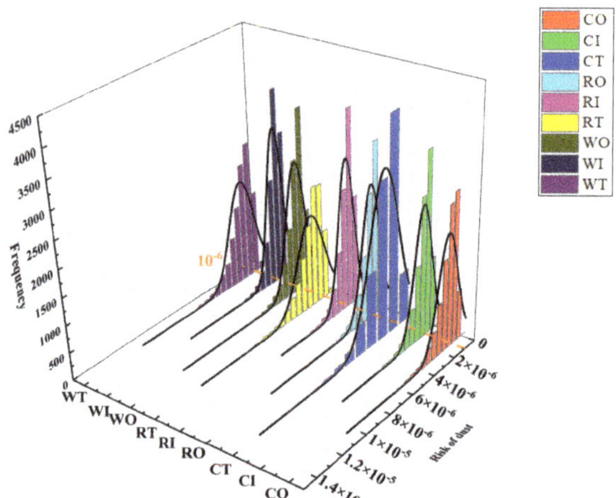

Figure 4. Distribution of health risk values in different storage stages for each grain. Remarks: CO refers to the out-warehousing stage of maize; CI refers to the in-warehousing stage of maize; CT refers to the transporting stage of maize; RO refers to the out-warehousing stage of rice; RI refers to the in-warehousing stage of maize; RT refers to the transporting stage of maize; WO refers to the out-warehousing stage of maize; WI refers to the in-warehousing stage of maize; and WT refers to the transporting stage of maize. The rest of the paper is the same.

Table 5. Statistical values of health risk in different stages of each grain.

Grain	Stage	Health Risk Value		Quantiles				
		Mean	SD	5%	20%	50%	90%	100%
Maize	Out-warehousing	1.42×10^{-6}	9.14×10^{-7}	3.74×10^{-7}	7.09×10^{-7}	1.21×10^{-6}	2.56×10^{-6}	9.63×10^{-6}
	In-warehousing	1.24×10^{-6}	8.27×10^{-7}	3.25×10^{-7}	6.06×10^{-7}	1.56×10^{-6}	2.27×10^{-6}	8.28×10^{-6}
	Transporting	2.34×10^{-6}	1.32×10^{-6}	8.38×10^{-7}	1.30×10^{-6}	2.05×10^{-6}	3.98×10^{-6}	1.24×10^{-5}
Rice	Out-warehousing	1.12×10^{-6}	8.04×10^{-7}	1.80×10^{-7}	4.82×10^{-7}	9.51×10^{-7}	2.14×10^{-6}	9.55×10^{-6}
	In-warehousing	1.05×10^{-6}	7.20×10^{-7}	2.33×10^{-7}	4.94×10^{-7}	8.91×10^{-7}	1.96×10^{-6}	6.89×10^{-6}
	Transporting	2.07×10^{-6}	1.16×10^{-6}	7.41×10^{-7}	1.15×10^{-6}	1.82×10^{-6}	3.54×10^{-6}	1.12×10^{-5}
Wheat	Out-warehousing	1.21×10^{-6}	8.13×10^{-7}	2.92×10^{-7}	5.74×10^{-7}	1.02×10^{-6}	2.24×10^{-6}	1.08×10^{-5}
	In-warehousing	9.03×10^{-6}	6.83×10^{-7}	1.25×10^{-7}	3.84×10^{-7}	7.59×10^{-7}	1.74×10^{-6}	8.09×10^{-6}
	Transporting	1.90×10^{-6}	1.06×10^{-6}	6.89×10^{-7}	1.07×10^{-6}	1.67×10^{-6}	3.23×10^{-6}	1.04×10^{-5}

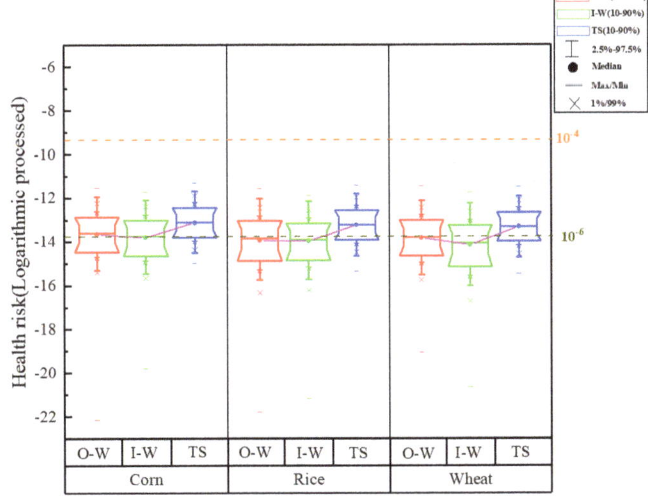

Figure 5. Health risks at different storage stages for each grain. Remarks: O-W refers to the out-warehousing stage; I-W refers to the in-warehousing stage; and TS refers to the transporting stage.

Considering grain type, the health risk of maize in each stage of storage was the largest and followed a lognormal distribution. The out-warehousing stage was $1.42 \times 10^{-6} \pm 9.14 \times 10^{-7}$, the in-warehousing stage was $1.24 \times 10^{-6} \pm 8.27 \times 10^{-7}$, and the transporting stage was $2.34 \times 10^{-6} \pm 1.32 \times 10^{-6}$. The maximum values were 9.63×10^{-6}, 8.28×10^{-6}, and 1.25×10^{-6} for the out-warehousing stage, in-warehousing stage, and transporting stage, respectively, and the average values at each stage were, respectively, 1.42, 1.24, and 2.34 times the acceptable values. The probability of exceeding 10^{-6} was 63%, 53%, and 90%, respectively, but none exceeded the upper limit of the acceptable risk value of $\times 10^{-4}$. The health risk of maize in each storage stage was quite large, especially in the transporting stage, and this indicates the need to take urgent and effective dust reduction measures. The health risk of rice in each storage stage is the second highest and followed a lognormal distribution. The out-warehousing stage was $1.12 \times 10^{-6} \pm 8.04 \times 10^{-7}$, the in-warehousing stage was $1.05 \times 10^{-6} \pm 7.20 \times 10^{-7}$, and the transporting stage was $2.07 \times 10^{-6} \pm 1.16 \times 10^{-6}$. The median values were 9.51×10^{-7}, 8.91×10^{-7}, and 1.82×10^{-6}, respectively. For workers in the storage and transportation of rice, the probability of exceeding value 10^{-6} is 47%, 43%, and 86% in the process of out-warehousing, in-warehousing, and transporting, respectively. Although these values are not more than 1×10^{-4}, it is still very harmful. The health risks of wheat in each storage stage were relatively small, with $1.21 \times 10^{-6} \pm 8.13 \times 10^{-7}$ at the out-warehousing stage, $9.03 \times 10^{-7} \pm 6.83 \times 10^{-7}$ at the in-warehousing stage, and 1.90×10^{-6} at the transporting stage. The average value of the out-warehousing stage was slightly higher than that of rice, which may be due to the smaller standard deviation in the concentration distribution. In long-term wheat storage and transportation operation, the possibility of exceeding the value 10^{-6} was greater than 35%; hence, it also deserves attention.

Based on the storage and transporting stages, the risk of the transporting stage for the three grains is greater than out-warehousing, which is in turn greater than the in-warehousing stage. The health risk value for the maize transporting stage was 1.67 times that of out-warehousing and 1.88 times that of in-warehousing. The maximum, mean, and minimum values were greater than those in the out-warehousing and in-warehousing stage, indicating that workers in the transporting stage are more likely to be more seriously affected by dust. The risk value in the out-warehousing stage was 1.15 times that of in-warehousing, which should also be controlled. The risk value produced by rice transporting was 85% higher than out-warehousing and 97% higher than in-warehousing. The risk value in out-warehousing was seven percentage points higher than in-warehousing. Although the maximum risk value for wheat transporting was slightly less than out-warehousing, the overall risk value was still greater than in the out-warehousing and in-warehousing stages. The main reasons for the high risk in the transfer stage include the use of protective measures (e.g., underground conveyance channels, airtight covers, and dust collectors) and a smaller scale of grain exposure to air during out-warehousing and in-warehousing; the grain also passes through multiple pieces of operation equipment during the transporting stage, and grain dust particles are constantly produced through overturning, flow, impact, and machinery, with particles precipitated under induction, traction, and shear airflow. The risk value in the out-warehousing stage is greater than in-warehousing, mainly as the fumigation and preservation of the grain after entering the warehouse creates pest residues, residual drugs, and powder after pests eat the grain.

3.2.2. Sensitivity Analysis

When the exposure parameters are uncertain, the health risk value may be misleading during decision-making. The sensitivity of each parameter was therefore further analyzed to compare the impact of each parameter on health risk. If the sensitivity analysis is positive, this means that the parameter is positively correlated with risk; if the sensitivity analysis is negative, it means that there is a negative correlation. The correlation is determined by the absolute sensitivity value.

The sensitivity of each exposure parameter for the different grains at different stages is shown in Figure 6. Among the exposure parameters that affect human health, C had the greatest positive impact on maize, rice, and wheat in the out-warehousing and in-warehousing stages, with sensitivities of 57%, 61%, 69%, respectively, in the out-warehousing stage, and 64%, 62%, and 71%, respectively, in the in-warehousing stage; this was followed by ED, with a sensitivity of 46%, 42%, 37%, respectively, in the out-warehousing stage, and 41%, 44%, and 38%, respectively, in the in-warehousing stage. The last was IR, with a sensitivity of about 30%. In the grain transporting stage, the most positive impact is ED, with sensitivities of 55%, 53%, and 52%, respectively, for maize, rice, and wheat. The sensitivity of C and IR differed from ED by about 15 percentage points. ET and EF had relatively small effects on the risk of health damage at each stage, at less than 20% and 5%, respectively. AT and BW had negative sensitivity and were negatively correlated with risk results. The absolute value of the sensitivity of AT for each stage was all greater than 37%, while the absolute value of the sensitivity of BW was less than 19%.

In general, C, ED, IR, and AT are highly sensitive to the health risks of grain storage and transportation, and have a greater impact; BW, ET, and EF are less sensitive, and have only a slight impact on the evaluation results. However, as the same parameters have different effects on different stages of grain production, different measures should be taken to reduce the health risk to workers. At the stages of grain out-warehousing and in-warehousing, dust control should be strengthened to reduce health risks, and the average exposure time to high concentration dust should be reduced. At the grain transporting stage, personnel access to high-concentration dust areas should be strictly monitored and limited, and dust reduction and removal measures should be taken to reduce the average exposure time.

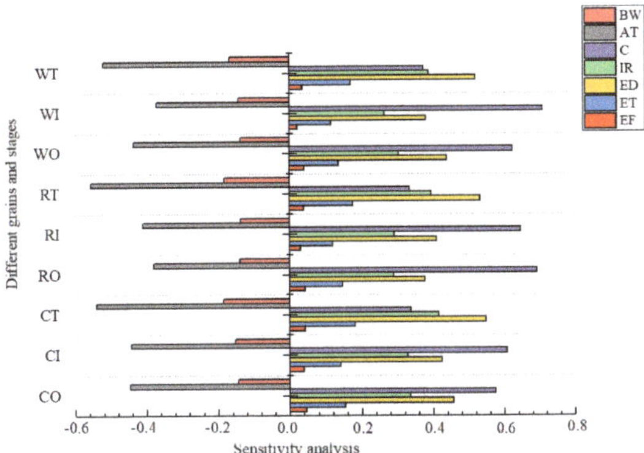

Figure 6. Sensitivity analysis at different stages for each grain.

3.2.3. Health Damages

The health damages and the DALY caused by dust were calculated, as shown in Table 6, and the overall trend was obtained through simulation, as shown in Figures 7–9. Dust caused the greatest damage to health in the maize in-warehousing stage, with an average DALY of 1.1 years. The transporting stage of rice and wheat followed, which were 0.89 and 0.83 years, respectively. The other stages were smaller, all within 1.6 years. At the same time, there were no significant differences in the DALY between different grains and stages. Uncertainty analysis of the DALY showed that the DALY of maize out-warehousing, in-warehousing, and transporting were concentrated in the range of 0.14–1.11, 0.15–0.86, and 0.39–1.57 years, respectively; for rice, they were concentrated in the range of 0.10–0.88, 0.09–0.80, and 0.33–1.43 years, respectively; and for wheat, they were concentrated in range of 0.15–1.01, 0.05–0.69, and 0.28–1.38 years for each stage, respectively.

Table 6. Statistical values of the DALY in different stages of each grain.

Grain	Stage	DALY/a			Quantiles				
		Min	Max	Mean	5%	20%	50%	90%	100%
Maize	Out-warehousing	0.01	2.76	0.63	0.20	0.35	0.57	1.07	2.76
	In-warehousing	0.12	3.28	1.01	0.43	0.64	0.94	1.58	3.28
	Transporting	0.01	2.31	0.54	0.15	0.29	0.49	0.93	2.31
Rice	Out-warehousing	0.01	2.28	0.48	0.09	0.23	0.43	0.89	2.28
	In-warehousing	0.01	2.20	0.46	0.11	0.24	0.42	0.82	2.20
	Transporting	0.14	3.05	0.89	0.38	0.57	0.84	1.40	3.05
Wheat	Out-warehousing	0.01	2.38	0.53	0.15	0.29	0.48	0.94	2.38
	In-warehousing	0.01	2.64	0.40	0.07	0.19	0.36	0.73	2.64
	Transporting	0.11	2.83	0.83	0.35	0.53	0.77	1.31	2.83

According to the data, the DALY caused by dust in the coal mine production process is more than 2 years [62]; for construction engineering, it is more than 3.4 years [37]; and for automobile casting it is more than 3.89 years [52]. By comparison, the health damage caused by grain dust is not as serious as that cased by these three industries. The main reason for this is that the grain storage and transportation process has a certain periodicity, fewer workers, and only a short exposure time. Additionally, the dust of other industries contains a large amount of heavy metal contamination and polycyclic aromatic hydrocarbons, which make the risk value and DALY greater [63]. It should be noted, however, that its dust concentration is comparable to those of the other industries. To reduce the health risk, the generation and diffusion of dust should thus be controlled as much as possible.

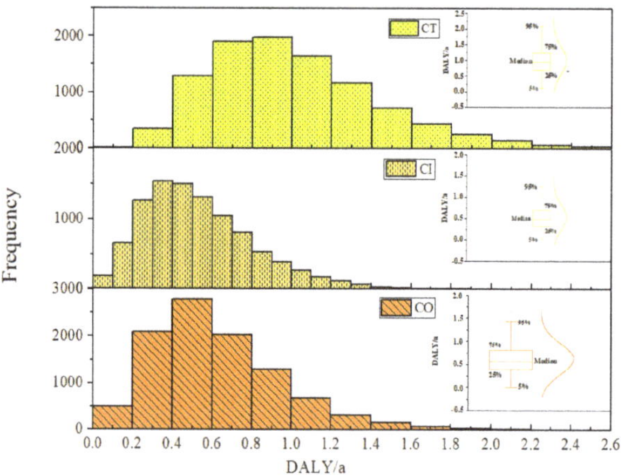

Figure 7. Maize simulation results of the DALY.

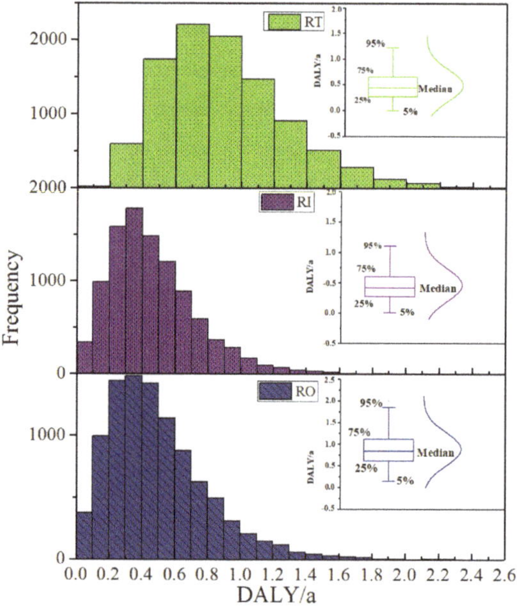

Figure 8. Rice simulation results of the DALY.

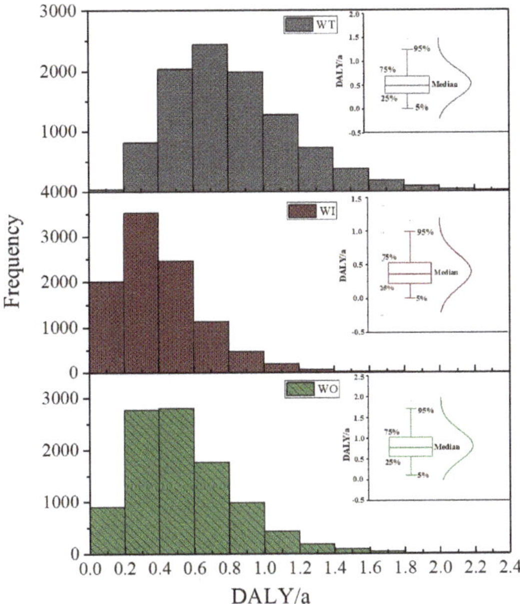

Figure 9. Wheat simulation results of the DALY.

3.2.4. Uncertainty Analysis

The dust concentration level in the grain storage and transportation process is affected by the year, season, region, process, and equipment due to the different equipment, technology, and processes involved in different stages of grain storage and transportation; the sampling time for dust in this health risk assessment was from March to May, so the parameters for dust concentration have certain limitations. However, in this study, a unified calculation of the employee exposure parameters at different stages was carried out, which also creates uncertainty. Exposure parameters such as EF, ED, and ET also refer to the Chinese population exposure parameter manual obtained from the 2011–2012 survey; the use of these data after 10 years may also lead to uncertainties. The *RfD* in the Chinese population exposure parameter manual differs from that provided by USEPA.

3.3. *Implication and Limitations*

The findings of the research presented in this paper have important significance. Previous studies have shown that a large amount of mixed dust is generated during the storage and transportation of grain, and this dust is seriously harmful to human beings. This risk to human health was, however, rarely evaluated further. This study has made a significant contribution to quantifying the health damage caused by grain dust. Based on the analysis of parameter uncertainty, the method of probabilistic risk assessment was used to ensure the comprehensiveness, objectivity, and accuracy of the evaluation result. The Monte Carlo method in the Crystal Ball software was used to deal with uncertainty in the process of risk assessment. The maximum, minimum, standard deviation, quantile, and other numerical risk characteristics were obtained to reflect the harm level of dust from multiple angles. The effects of AT, ED, ET, EF, BW, and C were explored by the sensitivity analysis of the evaluation results, which can effectively help managers choose reasonable prevention and control measures to reduce harm.

The evaluation method used in this study still has some limitations. As mentioned earlier, dust sampling is not universal. The uncertainty caused by the exposure parameter cannot be eliminated. The dermal contact [64,65] and ingestion pathways [66,67] also had an impact on human health. Only the inhalation pathway was considered, which may cause the health risk value presented here to be smaller than the actual risk. In addition, although the researchers tried to select the same grain warehouse for experiment and analysis during the sampling process, system differences were possible between the 616 samples, which will lead to errors in the results.

4. Conclusions

The dust concentration, levels, and sources, as well as the resulting health damage created during transportation and storage at Chinese grain storage companies were thoroughly investigated in this study. First, 616 dust samples from different stages (out-warehousing, in-warehousing, and transporting) and types of grain (maize, wheat, and rice) in five cities were collected and analyzed using the filter membrane method and a laser particle size analyzer. A risk assessment model for grain dust inhalation was established based on the current USEPA health risk assessment system. The health damage of grain dust was quantified by Monte Carlo simulation and Crystal Ball 11.1 software. The DALY was chosen as the final indicator to quantify the health damage. The results showed that the concentration of grain dust ranged from 25 to 70 mg/m^3, and the distribution was normal. The proportion of dust with a particle size less than 10 μm exceeded 10%, which could seriously threaten workers' respiratory system. Based on grain type, the dust risk in each stage followed a lognormal distribution, and the health risk of maize at each stage was the largest, at $1.42 \times 10^{-6} \pm 9.14 \times 10^{-7}$ during out-warehousing, $1.24 \times 10^{-6} \pm 8.27 \times 10^{-7}$ during in-warehousing, and $2.34 \times 10^{-6} \pm 1.32 \times 10^{-6}$ during transport. By stage, the health risk of grain dust can be ranked as follows: the transporting stage > the out-warehousing stage > the in-warehousing stage. The sensitivity analysis indicated that average time (AT), exposure duration (ED), inhalation rate (IR), and dust concentration (C) made the greatest contribution to dust health risk, while AT and body weight exhibited a negative sensitivity. The DALY caused by dust during grain storage was between 0.1 and 3.3 years. The DALY of maize was the largest during the in-warehousing stage, with an average value of 1.01 years, while the DALY of rice and wheat were the largest during the transporting stage, with an average value of 0.89 and 0.83 years, respectively.

The results of this study provide a new perspective for grain storage dust damage assessment. The dust concentration, particle size, and distribution characteristics of three kinds of grain at three storage stages were described, and preventive measures were proposed. The maximum, minimum, standard deviation, and quantile of risk were obtained by the probability risk assessment method to guarantee the comprehensiveness of the results. Furthermore, the DALY can directly reflect the damage of dust to human beings. To resolve the uncertainty of the results caused by various factors and the limitations of the evaluation method, future studies could increase the data sampling and investigation of exposure parameters, and assess the risk for the dermal contact and ingestion pathways to ensure the results are more accurate.

Author Contributions: P.C.: Conceptualization, Writing—Original Draft, Resources, Validation. T.Z.: Methodology, Funding acquisition, Projrct administration; X.C.: Formal analysis, Investigation; X.Y.: Writing—Review & Editing, Data Curation. All authors have read and agreed to the published version of the manuscript.

Funding: This research was funded by the Academy of National Food and Strategic Reserves Administration (No. ZX2006) and National Key R&D Program of China (No. 2018YFD0401403-02).

Institutional Review Board Statement: Not applicable.

Informed Consent Statement: Not applicable.

Acknowledgments: This work was supported by the Academy of National Food and Strategic Reserves Administration (No. ZX2006) and National Key R&D Program of China (No. 2018YFD0401403-02).

Conflicts of Interest: The authors declare no conflict of interest.

References

1. Martin, C.R.; Sauer, D.B. Physical and biological characteristics of grain dust. *Trans. ASAE* **1976**, *19*, 720–723. [CrossRef]
2. Hurst, T.S.; Dosman, J.A. Characterization of health effects of grain dust exposures. *Am. J. Ind. Med.* **1990**, *17*, 27–32. [CrossRef] [PubMed]
3. Parnell, C.B., Jr.; Jones, D.D.; Rutherford, R.D.; Goforth, K.J. Physical properties of five grain dust types. *Environ. Health Perspect.* **1986**, *66*, 183–188. [CrossRef] [PubMed]
4. Spankie, S.; Cherrie, J.W. Exposure to grain dust in Great Britain. *Ann. Occup. Hyg.* **2012**, *56*, 25–36. [PubMed]
5. Chan, M.; Yeung, D.A.E.; Kennedy, S.M. The Impact of Grain Dust on Respiratory Health1-3. *Am Rev. Respir. Dis.* **1992**, *145*, 476–487. [CrossRef] [PubMed]
6. Warren, P.; Cherniack, R.M.; Kam, S.T. Hypersensitivity reactions to grain dust. *J. Allergy Clin. Immunol.* **1974**, *53*, 139–149. [CrossRef]
7. Flaherty, D.; Bhansali, P.; Chavaje, N. Grain fever syndrome induced by inhalation of airborne grain dust. *J. Allergy Clin. Immunol.* **1982**, *69*, 435–443.
8. Davies, R.J.; Green, M.; Schofield, N.M.C. Recurrent ncturnal asthma after exposure to grain dust. *Am. Rev. Respir. Dis.* **1976**, *114*, 1011–1019.
9. Rushton, L. Chronic obstructive pulmonary disease and occupational exposure to silica. *Rev. Environ. Health* **2007**, *22*, 255–272. [CrossRef] [PubMed]
10. Cotton, D.J.; Dosman, J.A. Grain dust and health. III. Environmental factors. *Ann. Intern. Med.* **1978**, *89*, 420–421. [CrossRef]
11. Chan-Yeung, M.; Enarson, D.; Grzybowski, S. Grain dust and respiratory health. *Can. Med. Assoc. J.* **1985**, *133*, 969.
12. Eckhoff, R.K.; Fuhre, K. Dust explosion experiments in a vented 500 m^3 silo cell. *J. Occup. Accid.* **1984**, *6*, 229–240. [CrossRef]
13. Lebecki, K.; Cybulski, K.; Śliz, J.; Dyduch, Z.; Wolanski, P. Large scale grain dust explosions research in Poland. *Shock Waves* **1995**, *5*, 109–114. [CrossRef]
14. Lesikar, B.J.; Parnell, C.B.; Garcia, A. Determination of grain dust explosibility parameters. *Trans. ASAE* **1991**, *34*, 571–0576. [CrossRef]
15. doPico, G.A.; Reddan, W.; Anderson, S.; Dennis, F.; Eugene, S. Acute effects of grain dust exposure during a work shift. *Am. Rev. Respir. Dis.* **1983**, *128*, 399–404. [CrossRef] [PubMed]
16. Krysinska-Traczyk, E.; Perkowski, J.; Dutkiewicz, J. Levels of fungi and mycotoxins in the samples of grain and grain dust collected from five various cereal crops in eastern Poland. *Ann. Agric. Environ. Med.* **2007**, *14*, 159–167. [PubMed]
17. Halstensen, A.S.; Nordby, K.C.; Elen, O.; Eduard, W. Ochratoxin a in grain dust- estimated exposure and relations to agricultural practices in grain production. *Ann. Agric. Environ. Med.* **2004**, *11*, 245–254.
18. Deetz, D.C.; Jagielo, P.J.; Quinn, T.J.; Thorme, P.S.; Bleuer, S.A.; Schwartz, D.A. The kinetics of grain dust-induced inflammation of the lower respiratory tract. *Am. J. Respir. Crit. Care Med.* **1997**, *155*, 254–259. [CrossRef]
19. Von Essen, S.G.; Robbins, R.A.; Thompson, A.B.; Ertl, R.F.; Linder, J.; Rennard, S. Mechanisms of neutrophil recruitment to the lung by grain dust exposure. *Am. Rev. Respir. Dis.* **1998**, *138*, 921–927. [CrossRef] [PubMed]
20. Clapp, W.D.; Becker, S.; Quay, J.; Quay, J.; Watt, J.L.; Thorne, P.S.; Frees, K.L.; Zhang, X.; Koren, H.S.; Lux, C.R.; et al. Grain dust-induced airflow obstruction and inflammation of the lower respiratory tract. *Am. J. Respir. Crit. Care Med.* **1994**, *150*, 611–617. [CrossRef]
21. Behrooz, R.D.; Kaskaoutis, D.G.; Grivas, G.; Mihalopoulos, N. Human health risk assessment for toxic elements in the extreme ambient dust conditions observed in Sistan, Iran. *Chemosphere* **2021**, *262*, 127835. [CrossRef]
22. Guney, M.; Bourges, C.M.J.; Chapuis, R.P.; Zagury, G.J. Lung bioaccessibility of As, Cu, Fe, Mn, Ni, Pb, and Zn in fine fraction (<20 mm) from contaminated soils and mine tailings. *Sci. Total Environ.* **2017**, *579*, 378–386. [CrossRef] [PubMed]
23. Donoghue, A.M. The design of hazard risk assessment matrices for ranking occupational health risks and their application in mining and minerals processing. *Occup. Med.* **2000**, *51*, 118–123. [CrossRef] [PubMed]
24. Lim, H.-S.; Lee, J.-S.; Chon, H.-T.; Sager, M. Heavy metal contamination and health risk assessment in the vicinity of the abandoned Songcheon Au-Ag mine in Korea. *J. Geochem. Explor.* **2008**, *96*, 223–230. [CrossRef]
25. Kan, H.; Chen, B. Particulate air pollution in urban areas of Shanghai, China: Health-based economic assessment. *Sci. Total Environ.* **2004**, *322*, 71–79. [CrossRef]
26. Zhang, M. A health-based-assessment of particulate air pollution in urban areas of Beijing in 2000–2004. *Sci. Total Environ.* **2007**, *376*, 100–108. [CrossRef]
27. Liu, E.; Yan, T.; Birch, G.; Zhu, Y. Pollution and health risk of potentially toxic metals in urban road dust in Nanjing, a mega-city of China. *Sci. Total Environ.* **2014**, *476–477*, 522–531. [CrossRef]
28. Zhang, Y.; Ji, X.; Ku, T.; Li, G.; Sang, N. Heavy metals bound to fine particulate matter from northern China induce season-dependent health risks: A study based on myocardial toxicity. *Environ. Pollut.* **2016**, *216*, 380–390. [CrossRef]
29. Chen, X.; Guo, C.; Song, J.; Wang, X.; Cheng, J. Occupational health risk assessment based on actual dust exposure in a tunnel construction adopting road header in Chongqing, China. *Build. Environ.* **2019**, *165*, 106415. [CrossRef]
30. Zheng, N.; Liu, J.; Wang, Q.; Liang, Z. Health risk assessment of heavy metal exposure to street dust in the zinc smelting district, northeast of China. *Sci. Total Environ.* **2021**, *408*, 726–733. [CrossRef]
31. Tahir, M.W.; Mumtaz, M.W.; Tauseef, S.; Sajjad, M.; Nazeer, A.; Farheen, N.; Iqbal, M. Monitoring of cotton dust and health risk assessment in small-scale weaving industry. *Environ. Monit. Assess.* **2012**, *184*, 4879–4888. [CrossRef] [PubMed]

32. Zazouli, M.A.; Dehbandi, R.; Mohammadyan, M.; Aarabi, M.; Dominguez, A.O.; Kelly, F.J.; Khodabakhshloo, N.; Rahman, M.M.; Naidu, R. Physico-chemical properties and reactive oxygen species generation by respirable coal dust: Implication for human health risk assessment. *J. Hazard. Mater.* **2021**, *405*, 124185. [CrossRef]
33. Li, X.; Zhu, Y.; Zhang, Z. An LCA-based environmental impact assessment model for construction processe. *Build. Environ.* **2010**, *45*, 766–775. [CrossRef]
34. Dong, T.; Li, T.X.; Zhao, X.G.; Cao, S.Z.; Wang, B.B.; Ma, J.; Duan, X.L. Source and Health Risk Assessment of Heavy Metals in Ambient Air PM10 from One Coking Plant. *Environ. Sci.* **2014**, *35*, 1238–1244.
35. Li, X.D.; Su, S.; Huang, T.J. Health damage assessment model for construction dust. *J. Tsinghua Univ. Sci. Technol.* **2015**, *55*, 50–55.
36. Öberg, T.; Bergbäck, B. A review of probabilistic risk assessment of contaminated land (12 pp). *J. Soils Sediments* **2005**, *5*, 213–224. [CrossRef]
37. Zhang, Z.H.; Wu, F. Health impairment due to building construction dust pollution. *J. Tsinghua Univ. Sci. Technol.* **2008**, *48*, 922–925.
38. Peng, C.; Cai, Y.M.; Wang, T.Y.; Xiao, R.B.; Chen, W.P. Regional probabilistic risk assessment of heavy metals in different environmental media and land uses: An urbanization-affected drinking water supply area. *Sci. Rep.* **2016**, *6*, 37084. [CrossRef] [PubMed]
39. Koupaie, E.H.; Eskicioglu, C. Health risk assessment of heavy metals through the consumption of food crops fertilized by biosolids: A probabilistic-based analysis. *J. Hazard. Mater.* **2015**, *300*, 855–865. [CrossRef] [PubMed]
40. Li, F.; Huang, J.H.; Zeng, G.M.; Yuan, X.Z.; Liang, J.; Wang, X.Y. Multimedia health risk assessment: A case study of scenario-uncertainty. *J. Cent. South Univ.* **2012**, *19*, 2901–2909. [CrossRef]
41. Chen, S.C.; Liao, C.M. Health risk assessment on human exposed to environmental polycyclic aromatic hydrocarbons pollution sources. *Sci. Total Environ.* **2006**, *366*, 112–123. [CrossRef]
42. USEPA. *Risk Assessment Guidance for Superfund. Human Health Evaluation Manual. Part F*; USEPA: Washington, DC, USA, 2009; Volume 1.
43. Shen, Z.X.; Cao, J.; Arimoto, R.; Han, Z.; Zhang, R.; Han, Y.; Liu, S.; Okuda, T.; Nakao, S.; Tanako, S. Ionic composition of TSP and PM2.5 during dust storms and air pollution episodes at Xi'an, China. *Atmos. Environ.* **2009**, *43*, 2911–2918. [CrossRef]
44. USEPA. *Appendix A to 40 CFR, Part 423–126 Priority Pollutants*; USEPA: Washington, DC, USA, 2003.
45. Murray, C.J.; Lopez, A.D. Regional patterns of disability-free life expectancy and disability-adjusted life expectancy: Global Burden of Disease Study. *Lancet* **1997**, *349*, 1347–1352. [CrossRef]
46. USEPA. *Exposure Factors Handbook: 2011 Edition EPA/600/R-090/052F*; USEPA: Washington, DC, USA, 2011.
47. Tong, R.P.; Cheng, M.Z.; Zhang, L.; Liu, M.; Yang, X.Y.; Li, X.D.; Yin, W. The construction dust-induced occupational health risk using Monte Carlo simulation. *J. Clean. Prod.* **2018**, *184*, 598–608. [CrossRef]
48. GBZ 2.1. *Occupational Exposure Limits for Hazardous Agents in the Workplace Part 1: Chemical Hazardous Agents*; The Central People's Government of the People's Republic of China: Beijing, China, 2007.
49. Murray, C.J. Quantifying the burden of disease: The technical basis for disability-adjusted life years. *Bull. WHO* **1994**, *72*, 429–445. [PubMed]
50. Soerjomataram, I.; Lortet-Tieulent, J.; Ferlay, J.; Forman, D.; Mathers, C.; Parkin, D.M.; Bray, F. Estimating and validating disability-adjusted life years at the global level: A methodological framework for cancer. *BMC Med. Res. Methodol.* **2012**, *12*, 1. [CrossRef]
51. Pan, S.; An, W.; Li, H.Y.; Su, M.; Zhang, J.L.; Yang, M. Cancer risk assessment on trihalomethanes and haloacetic acids in drinking water of China using disability-adjusted life years. *J. Hazard. Mater.* **2014**, *280*, 288–294. [CrossRef]
52. Tong, R.; Cheng, M.; Ma, X.; Yang, Y.; Liu, Y.; Li, J. Quantitative health risk assessment of inhalation exposure to automobile foundry dust. *Environ. Geochem. Health* **2019**, *41*, 2179–2193. [CrossRef]
53. National Bureau of Statistics. *China Statistical Yearbook 2011*; China Statistics Press: Beijing, China, 2011.
54. Wang, Z.; Duan, X.; Liu, P.; Nie, J.; Zhang, J. Human exposure factors of Chinese people in environmental health risk assessment. *Res. J. Environ. Sci.* **2009**, *22*, 1164–1175.
55. Xiang, H.L.; Yang, J.; Qiu, Z.Z.; Lei, W.X.; Zeng, T.T.; Lan, Z.C. Health Risk Assessment of Tunnel Workers Based on the Investigation and Analysis of Occupational Exposure to PM10. *Environ. Sci.* **2015**, *36*, 2768–2774.
56. Brochu, P.; Ducre-Robitaille, J.F.; Brodeur, J. Physiological daily inhalation rates for free -living individuals aged 1 month to 96 years, using data from doubly labeled water measurements: A proposal for air quality criteria, standard calculations and health risk assessment. *Hum. Ecol. Risk Assess.* **2006**, *12*, 675–701. [CrossRef]
57. Charnes, J. *Financial Modeling with Crystal Ball and Excel, +Website*; John Wiley & Sons: New York, NY, USA, 2012.
58. Hou, J.; Qu, Y.; Ninf, D.; Wang, H. Impact of Human Exposure Factors on Health Risk Assessment for Benzene Contaminated Site. *Environ. Sci. Technol.* **2014**, *37*, 191–195.
59. Qu, C.; Li, B.; Wu, H.; Wang, S.; Giesy, J.P. Multipathway assessment of human health risk posed by polycyclic aromatic hydrocarbons. *Environ. Geochem. Health* **2015**, *37*, 587–601. [CrossRef] [PubMed]
60. Othman, M.; Latif, M.T.; Mohamed, A.F. Health impact assessment from building life cycles and trace metals in coarse particulate matter in urban office environments. *Ecotoxicol. Environ. Saf.* **2018**, *148*, 293–302. [CrossRef]
61. Chiang, K.C.; Chio, C.P.; Chiang, Y.H.; Liao, C.M. Assessing hazardous risks of human exposure to temple airborne polycyclic aromatic hydrocarbons. *J. Hazard. Mater.* **2009**, *166*, 676–685. [CrossRef] [PubMed]

62. Tong, R.P.; Cheng, M.Z.; Ma, X.; Xu, S.R. Ecaluation method of coal dust occupational health damage under uncertainty condition and its application. *China Saf. Sci. J.* **2018**, *28*, 139–144.
63. Martin, C.R. Characterization of grain dust properties. *Trans. ASAE* **1981**, *24*, 738–742. [CrossRef]
64. Tarafdar, A.; Sinha, A. Health risk assessment and source study of PAHs from roadside soil dust of a heavy mining area in India. *Arch. Environ. Occup. Health* **2019**, *74*, 252–262. [CrossRef] [PubMed]
65. Li, F.; Zhang, J.; Huang, J.; Huang, D.; Yang, J.; Song, Y.; Zeng, G. Heavy metals in road dust from Xiandao District, Changsha City, China: Characteristics, health risk assessment, and integrated source identification. *Environ. Sci. Pollut. Res.* **2016**, *23*, 13100–13113. [CrossRef] [PubMed]
66. Ghorbel, M.; Munoz, M.; Courjault-Radé, P.; Destrigneville, C.; de Parseval, P.; Souissi, R.; Souissi, F.; Mammou, A.B.; Abdeljaouad, S. Health risk assessment for human exposure by direct ingestion of Pb, Cd, Zn bearing dust in the former miners' village of Jebel Ressas (NE Tunisia). *Eur. J. Miner.* **2010**, *22*, 639–649. [CrossRef]
67. Liu, Y.; Ma, J.; Yan, H.; Ren, Y.; Wang, B.; Lin, C.; Liu, X. Bioaccessibility and health risk assessment of arsenic in soil and indoor dust in rural and urban areas of Hubei province, China. *Ecotoxicol. Environ. Saf.* **2016**, *126*, 14–22. [CrossRef] [PubMed]

 *atmosphere*

Article

Multi-Media Exposure to Polycyclic Aromatic Hydrocarbons at Lake Chaohu, the Fifth Largest Fresh Water Lake in China: Residual Levels, Sources and Carcinogenic Risk

Ning Qin [1,2], Wei He [2,3], Qishuang He [2,4], Xiangzhen Kong [2,5], Wenxiu Liu [2,6], Qin Wang [7] and Fuliu Xu [2,*]

1 School of Energy and Environmental Engineering, University of Science and Technology Beijing, Beijing 100083, China; qinning@ustb.edu.cn
2 MOE Laboratory for Earth Surface Processes, College of Urban & Environmental Sciences, Peking University, Beijing 100871, China; wei.he@cugb.edu.cn (W.H.); heqs@brcast.org.cn (Q.H.); xzkong@niglas.ac.cn (X.K.); liuwx@craes.org.cn (W.L.)
3 Ministry of Education Key Laboratory Groundwater Circulation & Environmental Evolution, China University Geosciences Beijing, Beijing 100083, China
4 Beijing Municipal Key Lab Agriculture Environment Monitoring, Beijing 100097, China
5 State Key Laboratory of Lake Science and Environment, Nanjing Institute of Geography & Limnology, Chinese Academy of Sciences, Nanjing 210008, China
6 Center for Environmental Health Risk Assessment and Research, Chinese Research Academy of Environmental Sciences, Beijing 100012, China
7 China CDC Key Laboratory of Environment and Population Health, National Institute of Environmental Health, Chinese Center for Disease Control and Prevention, Beijing 100021, China; wangqin@nieh.chinacdc.cn
* Correspondence: author: xufl@urban.pku.edu.cn

Abstract: The residual levels of 16 priority polycyclic aromatic hydrocarbons (PAHs) in environment media and freshwater fish were collected and measured from Lake Chaohu by using Gas chromatography-mass spectrometry. Potential atmospheric sources were identified by molecular diagnostic ratios and the positive matrix factorization (PMF) method. PAH exposure doses through inhalation, intake of water and freshwater fish ingestion were estimated by the assessment model recommended by US EPA. The carcinogenic risks of PAH exposure were evaluated by probabilistic risk assessment and Monte Carlo simulation. The following results were obtained: (1) The PAH_{16} levels in gaseous, particulate phase, water and fish muscles were 59.4 $ng \cdot m^{-3}$, 14.2 $ng \cdot m^{-3}$, 170 $ng \cdot L^{-1}$ and 114 $ng \cdot g^{-1}$, respectively. No significant urban-rural difference was found between two sampling sites except gaseous BaP_{eq}. The relationship between gaseous PAHs and PAH in water was detected by the application of Spearman correlation analysis. (2) Three potential sources were identified by the PMF model. The sources from biomass combustions, coal combustion and vehicle emission accounted for 43.6%, 30.6% and 25.8% of the total PAHs, respectively. (3) Fish intake has the highest lifetime average daily dose (LADD) of 3.01×10^{-6} $mg \cdot kg^{-1} \cdot d^{-1}$, followed by the particle inhalation with LADD of 2.94×10^{-6} $mg \cdot kg^{-1} \cdot d^{-1}$. (4) As a result of probabilistic cancer risk assessment, the median ILCRs were 3.1×10^{-5} to 3.3×10^{-5} in urban and rural residents, which were lower than the suggested serious level but higher than the acceptable level. In summary, the result suggests that potential carcinogenic risk exists among residents around Lake Chaohu. Fish ingestion and inhalation are two major PAH exposure pathways.

Keywords: PAHs; multi-media exposure; health risk; probabilistic risk assessment; Lake Chaohu

Citation: Qin, N.; He, W.; He, Q.; Kong, X.; Liu, W.; Wang, Q.; Xu, F. Multi-Media Exposure to Polycyclic Aromatic Hydrocarbons at Lake Chaohu, the Fifth Largest Fresh Water Lake in China: Residual Levels, Sources and Carcinogenic Risk. *Atmosphere* **2021**, *12*, 1241. https://doi.org/10.3390/atmos12101241

Academic Editor: Alina Barbulescu

Received: 20 July 2021
Accepted: 19 September 2021
Published: 23 September 2021

Publisher's Note: MDPI stays neutral with regard to jurisdictional claims in published maps and institutional affiliations.

Copyright: © 2021 by the authors. Licensee MDPI, Basel, Switzerland. This article is an open access article distributed under the terms and conditions of the Creative Commons Attribution (CC BY) license (https://creativecommons.org/licenses/by/4.0/).

1. Introduction

Polycyclic aromatic hydrocarbons (PAHs) are globally concerned pollutants because of their widespread occurrence, strong persistence and long-range transportation potential [1]. Furthermore, they possess potential toxicity, mutagenicity and carcinogenicity [2–4]. Studies have shown that human cancer causes of skin, lungs and bladder have always been associated with PAHs [5–8], and 16 PAHs are included on the priority pollutants list of the

US EPA. PAHs have a wide variety of sources, including coal combustion, vehicle emission, coking industry and biomass burning [9–11]. After being emitted into the environment, PAHs may redistribute in environmental media and result in people being exposed to these pollutants through multiple pathways, including breathing in polluted air and particles, drinking water, dietary intake and dermal contact with contaminated soil [12,13]. Multi-media distribution and multi-pathway exposure render the assessment of PAH exposure complicated. Therefore, accurately evaluating the contribution of each exposure pathway, characterizing the carcinogenic risk and identifying the sensitive parameters in the exposure process are crucial to the management of PAH emission.

Water bodies act either as a sink [14] or as a source [15] for PAHs in the environment. The atmospheric PAHs can enter water system through wet deposition, dry deposition and gas exchange across the air–water interface [16–18]. Meanwhile, PAHs in water may accumulate in aquatic organisms by direct uptake from water through gills or skin or by the ingestion of suspended particles and contaminated food [19]. Residents living near the lake can be exposed to PAHs by inhaling polluted air and ingesting water and aquatic products.

Lake Chaohu is located in the Anhui Province, which belongs to one of the most developed areas in China, Yangtze River Delta Economic Zone. During the last decades, the PAHs' emission in China, especially in the above-mentioned areas, increased greatly due to the increasing energy demand associated with rapid population growth and economic development and to the low efficiency of energy utilization [20]. Chaohu is famous for its fresh water fish. It is also the drinking water source of large cities such as Hefei and Maanshan. PAH pollution in water system of Chaohu may increase the risk of residents' exposure through fish ingestion and drinking water. Therefore, in recent years, PAH exposure in Lake Chaohu has become a topic of concern. Some studies have been conducted on the PAH residual levels in environment media, source apportionment or environment behaviors [21–25]. Despite the progress in these directions, studies on three issues remain scarce. First, comparison among exposure contribution from different pathways are seldom reported. Second, carcinogenic risk due to total exposure remains unclear. Third, the factors influencing risk assessment are seldom studied. Hence, further studies on PAHs should be performed in order to obtain a comprehensive understanding of the risk profile of PAHs exposure among Chaohu residents.

In this research, EPA priority control PAHs were selected as target chemicals due to their extensive residence in the environment and their threat to public health. The contents of 16 l PAHs in air, particles, lake water and aquatic organisms were measured; physiological and behavior parameters influencing PAH exposure were collected; potential sources were identified; and cancer risks were calculated by US EPA model. The aims of this study were to elucidate the characteristics of multi-media PAH exposure of residents and to provide information for PAH management near Lake Chaohu.

2. Materials and Methods

2.1. Sample Collection and Pretreatment

Two sampling sites were selected near Chaohu City and Zhongmiao Town as urban and rural sites (Figure 1). Atmosphere samples were collected once a month from May 2010 to April 2011 by high-volume samplers. Polyurethane foam (PUF) disk and glass fiber filter (GFF) were used to collect gaseous phase and particulate phase PAHs, respectively. Water was collected from two sampling sites selected near the atmospheric sites. After shaking and mixing, a one liter aliquot of each collected water sample was filtered through a 0.45 μm glass fiber filter using a filtration device consisting of a peristaltic pump (80EL005, Millipore Co., Billerica, MA, USA). Edible aquatic organisms, including spotted steed (Hemibarbus maculatus, HM), carp (Cyprinus carpio, CC), snail (Cipangopaludina chinensis Gray, CCG), topmouth culter (Culter eryropterus, CE), bluntnose black bream (Megalobrama amblycephala, MA), Chinese white prawn (Leander modestus Heller, LMH), whitebait (Hemisalanx prognathus Regan, HPR) and bighead carp (Aristichthys nobilis, AN), were randomly collected in the lake.

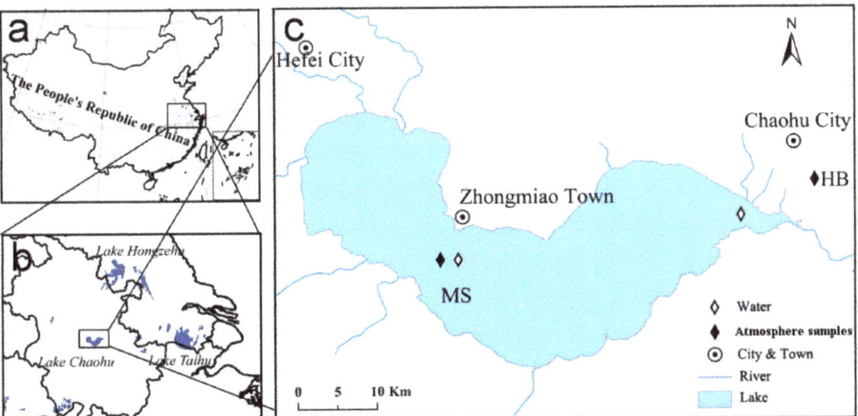

Figure 1. The location of Lake Chaohu and sampling sites: (**a**) People's Republic of China, (**b**) Anhui Province and (**c**) Lake Chaohu.

In the laboratory, PUF and GFF samples were added with surrogate standards of 2-fluoro-1, 10-biphenyl and p-terphenyl-d14 (J&K Scientific, Beijing, China, 2.0 mg mL 1) before measurement. The PUF was Soxhlet extracted with 150 mL 1:1 mixture of n-hexane and acetone for 8 h. GFF was extracted by 25 mL hexane/acetone mixture (1:1) using a microwave accelerated reaction system (CEM Corporation, Matthews, NC, USA). Microwave power was set at 1200 W, and the temperature program was set to the following: ramp up to 100 °C in 10 min and held at 100 °C for another 10 min. Both PUF and GFF extracts were concentrated to 1 mL by rotary evaporation at a temperature below 38 °C and then transferred to a silica/alumina chromatography for cleanup. The elution solution was collected, concentrated, conversed to hexane solution and then added with internal standards (Nap-d8, Ace-d10, Ant-d10, Chr-d12 and Perylene-d12, J&K Scientific Ltd., Beijing, China).

The water samples were extracted by using a solid phase extraction (SPE) system (Supelco, Bellefonte, PA, USA). C18 cartridges (500 mg, 6 mL, Supelco, Bellefonte, PA, USA) were prewashed with dichloromethane (DCM) and conditioned with methanol and deionized water. A 1 L water sample was added with surrogate standards, passed through the SPE system and was extracted. The cartridges were eluted with 10 mL of dichloromethane. The volume of the extracts was reduced by a vacuum rotary evaporator in a water bath and was adjusted to a volume of 1 mL with hexane. Internal standards were added for analysis.

The fish samples were pretreated on the same day after being delivered back to the temporary laboratory. The muscles on both sides of the dorsal and chest were mixed. After obtaining the wet weight, the samples were freeze dried (FDU-830, Tokyo Rikakikai Co., Tokyo, Japan) and grounded into a granular powder with a ball mill (MM400, Retsch GmbH, Haan, Germany). Two gram powder samples were weighed into an extraction tube, and the surrogate standards were added to the samples to indicate recovery. After microwave extraction, the extracts were pressure filtered and concentrated to approximately 1 mL and cleansed by GPC instrument (GPC800+, Lab Tech Ltd., Hongkong, China) with a Bio Beads SX-3 column (300 mm × 20 mm, Bio-Rad Laboratories, Inc., Hercules, CA, USA). Subsequently the concentrate was loaded in a silica gel SPE cartridge (6 mL, 500 mg, Supelco Co., Bellefonte, PA, USA). The cartridge was eluted by hexane and mixed solution of dichloromethane and hexane. The extracts were concentrated to 1 mL, transferred to vials, added with internal standards and sealed for analysis. The details of experiment have been reported in previous research [19,26].

2.2. Instrument Analysis

The samples were analyzed by using Agilent 6890 gas chromatography and a 5976C mass spectrometer detector with a HP-5MS fused silica capillary column (30 m × 0.25 mm × 0.25 μm).

Helium was used as the carrier gas at a flow of 1 mL/min. The samples (1 μL) were injected by the autosampler under a splitless mode at a temperature of 220 °C. The column temperature program was as follows: 50 °C for 2 min, 10 °C/min to 150 °C, 3 °C/min for 240 °C, 240 °C for 5 min, 10 °C/min for 300 °C and 300 °C for 5 min. The ion source temperature of the mass spectrometer was 200 °C, the temperature of the transfer line was 250 °C and the temperature of the quadrupole was 150 °C. The compounds were quantified in the selected ion mode, and the calibration curve was quantified with the internal standard. There were three parallel samples in each species. The method blanks and procedure blanks were prepared following the same procedure.

The quantification was performed by the internal standard method. All of the solvents used were HPLC-grade pure (J&K Chemical, Beijing, China). All of the glassware was cleaned by using an ultrasonic cleaner (KQ-500B, Kun Shan Ultrasonic Instruments Co., Ltd., Kunshan, China) and heated to 400 °C for 6 h. In the sampling process, three parallel samples were been collected from each sample site. The laboratory blanks and sample blanks were analyzed with the true samples. A total of 16 priority control PAHs were measured. The PAH individuals, abbreviations as well as method recoveries in different environment media and aquatic organisms are shown in Table 1.

Table 1. Recoveries and toxicity equivalency factors (TEFs) of 16 PAHs.

Abbreviation	PAHs	Gaseous (%)	Particulate (%)	Water (%)	Aquatic Organisms (%)	TEF
Nap	Naphthalene	46	47	81	115	0.001
Acy	Acenaphthene;	51	48	87	101	0.001
Ace	Acenaphthylene	67	50	81	117	0.001
Flo	Fluorene	75	57	103	105	0.001
Phe	Phenanthrene	83	69	108	107	0.001
Ant	Anthracene	77	71	93	101	0.01
Fla	Fluoranthene	98	87	89	113	0.001
Pyr	Pyrene	124	88	89	122	0.001
BaA	Benzo[a]anthracene	99	97	63	102	0.1
Chr	Chrysene	92	102	62	119	0.01
BbF	Benzo[b]fluoranthene	121	103	43	105	0.1
BkF	Benzo[k]fluorant hene	90	111	44	102	0.1
BaP	Benzo[a]pyrene;	108	103	60	87	1
IcdP	Dibenz[a,h]anthracene	102	119	31	89	0.1
DahA	Indeno [1,2,3-cd]pyrene	127	118	24	93	1
BghiP	Benzo[ghi]perylene	65	115	24	110	0.01

2.3. *Positive Matrix Factorization (PMF)*

In this study, positive matrix factorization method was applied in order to quantitatively identify the major sources. PMF is a useful factorization methodology that can determine source profile and contribution [27,28]. The PMF model can be expressed as follows:

$$X = GF + E \quad (1)$$

where X is the concentration matrix, consisting of n samples and m concentrations of the compounds (n × m); G is the factor contribution matrix; F is the factor profile matrix; and E (n × m) is the residual matrix. The elements of residual matrix are denoted as the following:

$$e_{ij} = x_{ij} - \sum_{k=1}^{p} g_{ik} f_{ki} \quad (2)$$

where x_{ij}, f_{ki} and g_{ik} are the corresponding elements of X, F and G, respectively. Non-negativity constraints are imposed on the contribution and profile matrices, and PMF simultaneously weights individual data points based on uncertainty. $Q(E)$ is an object function and a criterion for the model, defined as the following:

$$Q(E) = \sum_{i=1}^{n} \sum_{j=1}^{m} (e_{ij}/s_{ij})^2 \quad (3)$$

where s_{ij} is the uncertainty of the jth compound in the ith sample.

2.4. Multi-Pathway Exposure and Risk Assessment

The BaP equivalent concentration (BaP_{eq}) and toxicity equivalency factors (TEFs) were used to express the effects of exposure to mixtures of PAHs on health [29]. BaP_{eq} is directly derived from the mass concentrations of different PAHs using TEFs. Therefore, they can be directly compared and contrasted [30]. In order to evaluate the total exposure to dietary PAHs, BaP_{eq} based on BaP toxicity was determined using the following equation:

$$BaP_{eq} = \sum C_i \times TEF \quad (4)$$

where C_i is the concentration of the PAH species in food, and TEF_i is the toxic equivalence factor of the PAH's congener i. (Table 1).

In accordance with the *Exposure Factors Handbook* [31], the lifetime average daily dose (LADD) of PAH exposure through inhalation (air and particle), aquatic product ingestion, and water intake was calculated as follows:

$$LADD = \frac{C \times IR \times EF \times ED}{BW \times AT} \quad (5)$$

where C is the concentration of PAHs in the environment media, and IR is the intake rate of PAHs through inhalation (IR_{inh}, $m^3 \cdot day^{-1}$), water intake rate (IR_{water}, $mL \cdot day^{-1}$) and aquatic product intake rate (IR_{inh}, $g \cdot day^{-1}$). EF is the exposure frequency ($day \cdot year^{-1}$); ED is the exposure duration (year); and BW is body weight (kg). AT is the average lifespan for carcinogens.

High uncertainty exists in risk assessment. Sample measurement errors were inevitable. There are also uncertainties in the parameters and estimates. In probabilistic risk assessment, exposure parameters are considered as random variables. In order to quantify experiment uncertainty and its impact on the estimation of expected risk, a 10,000 times Monte Carlo (MC) technique was used. The Crystal Ball software was employed to implement MC simulation.

3. Results and Discussions

3.1. PAH Residual Levels in Environment Media

The levels of PAHs in the environmental media from Lake Chaohu are presented in Table 2. Sixteen priority PAHs were all detected during both the gaseous phase and particulate phase. In comparison, the detection rates of PAHs with higher than four rings in the water phase were very low due to their poor hydrophilic. The total concentrations of 16 priority PAHs (PAH_{16}) in gaseous and particulate phases were 59.4 ± 51.4 $ng \cdot m^{-3}$ and 14.2 ± 23.5 $ng \cdot m^{-3}$, respectively. The average gaseous PAH_{16} concentrations in urban and rural sites were 3.59 times and 4.95 times higher than in particles. The atmospheric PAH_{16} residual level in this study was lower than the values reported in Guangzhou (337 $ng \cdot m^{-3}$) [32] and in Tianjin (752 $ng \cdot m^{-3}$) [33], but it was greater than the highest level reported in mountain Taishan (9.07 $ng \cdot m^{-3}$) [34]. Compared with data reported abroad, the PAH_{16} level was higher than data reported in Chesapeake Bay (5.31~71.6 $ng \cdot m^{-3}$) [35], Athens (4.8~76 $ng \cdot m^{-3}$) [36] and in southwest Europe (0.32 $ng \cdot m^{-3}$) [37]. PAH_{16} in the muscle of fish from Chaohu Lake (also including snail and shrimp) was also comparable with data reported from other freshwater fish in Hebei (4.76–144 ng/g) [20] and less than data reported in Shanxi (160 ng/g) [38]. Generally speaking, the PAH_{16} contents in environment media and fish indicated a low PAH pollution level in Lake Chaohu. In order to compare the toxicity of difference environment media, the concentrations were converted to BaP_{eq} concentrations. Although the gaseous phase had much higher PAH_{16} content, the BaP_{eq} was much higher in particles. The particulate BaP_{eq} in urban and rural were 11.2 and 5.51 times higher than those in the gaseous phase. The difference between PAH_{16} and BaP_{eq} can be attributed to PAH composition in gas and particles.

Table 2. Residual levels of PAH_{16} and BaP_{eq} in environmental media.

Categories	Media	Unit	Urban			Rural			Kruskal-Wallis Test
			Min	Max	GM	Min	Max	GM ± SD	
PAH_{16}	Gas	$ng{\cdot}m^{-3}$	22.1	186	49.5 ± 46.0	10.9	183	72.3 ± 54.3	$p = 0.157$
	Particle	$ng{\cdot}m^{-3}$	3.41	82.5	13.8 ± 25.6	2.74	69.3	14.6 ± 22.4	$p = 0.773$
	Water	$ng{\cdot}L^{-1}$	57	409	171 ± 119	59.6	779	169 ± 188	$p = 0.544$
	Fish	$ng{\cdot}g^{-1}$	18.5	1029	114 ± 315	18.5	1029	114 ± 315	
BaP_{eq}	Gas	$ng{\cdot}m^{-3}$	0.04	0.38	0.14 ± 0.10	0.06	1.05	0.31 ± 0.26	$p = 0.010$
	Particle	$ng{\cdot}m^{-3}$	0.38	10.1	1.57 ± 2.99	0.31	9.00	1.71 ± 2.74	$p = 0.840$
	Water	$ng{\cdot}L^{-1}$	0.17	1.57	0.46 ± 0.42	0.26	1.36	0.54 ± 0.36	$p = 0.862$
	Fish	$ng{\cdot}g^{-1}$	0.29	20.2	1.75 ± 6.24	0.29	20.2	1.75 ± 6.24	

PAH_{16}: Tthe sum of 16 PAH components; GM: geometric mean; SD: standard deviation.

Spatial difference between urban and rural sites were compared by using the Kruskal–Wallis test. No significant difference was found between two sampling sites except gaseous BaP_{eq}. A $p < 0.05$ significant difference was detected between urban and rural gaseous BaP_{eq} concentrations. The results showed that the concentrations in most environment media were similar in urban and rural area. There may be two reasons accounting for the small spatial difference. First, both Chaohu City and Zhongmiao Town had small populations. No obvious different lifestyle was found between people in urban and rural areas. In particular, there is no obvious heating season in the area around Chaohu. Thus, local emission sources in urban and rural area were not obvious. Second, the city and the town were far away from the local thermal power plant and other industrial pollution sources, resulting in low local pollution levels.

The PAH compositions in environment media were illustrated in Figure 2. It can be observed that the water phase was dominated by the low molecular PAHs. PAHs with less than three rings accounted for 95.0% and 93.3% of total PAHs in water. The same ratios in gaseous phase were 87.1% and 82.9% in urban and rural samples. In the particulate phase, however, the PAHs with more than or equal to four rings contributed to 80.2% and 81.6% of the total PAHs. The high proportion of high molecular weight PAHs results in the increase in toxicity due to the high TEF of high molecular weight PAHs.

Spearman's correlation analysis was used to detect the relationship between content of PAHs in gaseous, particulate phase, water and in aquatic animals. As a result, a significant positive correlation was found between gaseous phase and water phase (Figure 3a). It is reported that wet deposition, dry deposition and gas exchange across the air–water interface are the three major ways that PAHs can enter the water system. The result suggested that gas exchange across the air–water interface is probably an important way for atmospheric PAHs to affect the aquatic system [39]. In contrast, the relationship between particulate and dissolved PAHs was not significant ($p = 0.116$). This can be partly explained by the solubility of different composition in particles and gas phase. The solubility of high molecular weight PAHs was lower than the solubility of those with low molecular weight. On the other hand, high molecular weight PAHs dominated the particle component. Thus, concentration levels of PAHs in particles had a weak correlation with their concentration levels in water. Significant positive correlation was also found between PAH in water and in fish tissues, which indicates the effect of environmental concentration on organisms.

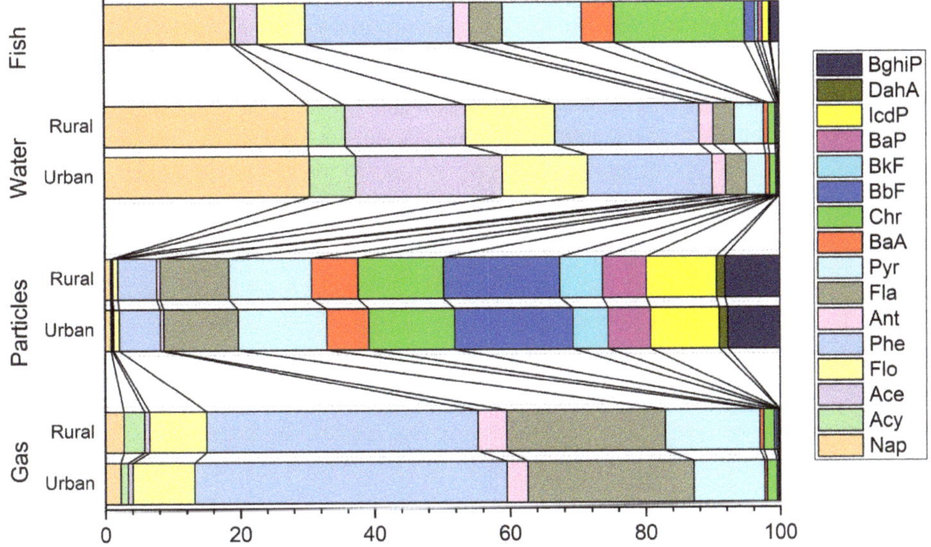

Figure 2. PAHs composition in different environment media.

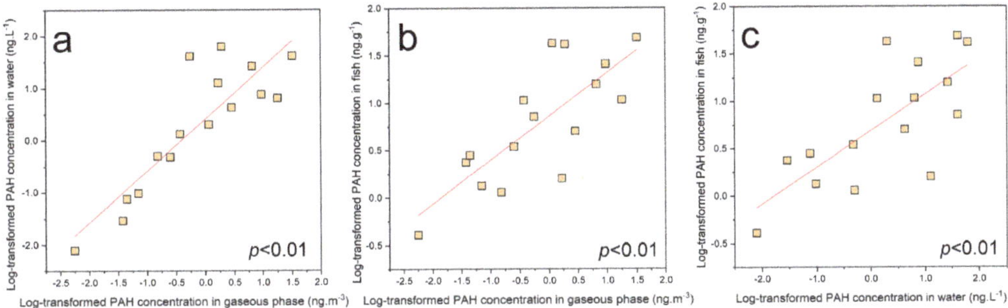

Figure 3. Spearman correlation between PAH levels in environment media (**a**) water and gaseous phase (**b**) and fish tissues and gaseous phase (**c**) water and fish tissues.

3.2. Source Apportionment

3.2.1. Molecular Diagnostic Ratios

PAHs can be formed by multiple anthropogenic activities such as combustion of fossil fuels, or they can be formed naturally in the environment by oil seeps and plant debris and forest and prairie fires. Some methods have been established in order to identify PAH sources: for example, molecular diagnostic ratios (MDRs), the principal component analysis (PCA) method [37,40], the chemical material balance (CMB) model [41], the positive matrix factorization (PMF) method and stable carbon isotopic ratios analysis [42]. In this study, MDRs and PMF were used to identify the major sources and to obtain a reliable conclusion.

The MDR theory [17,43] is based on the hypothesis that some PAH ratios remain constant between the source and the receptor. The MDR method has been widely used in the identification of preliminary sources [44]. Based on the monitoring data, four normally used ratios were employed in this research. Common ratios used include Ant/(Ant + Phe) (mass 178), Fla/(Fla + Pyr) (mass 202), BaA/(BaA +Chr) (mass 228) and IcdP/(IcdP + BghiP) (mass 276).

The ratios of mass 178 increased from January and reached the peak in October (Figure 4). Generally, the ratios varied near the value of 0.1. The lowest value was obtained during the winter, and the highest value was achieved in the summer. For mass 178, a ratio < 0.10 usually is taken as an indication of petroleum, whereas a ratio > 0.10 indicates a dominance of combustion. From November to January, petroleum was the dominant source, and emission from combustion became the greatest contributor during the rest of the year. For mass 202, a ratio of 0.50 is usually defined as the petroleum/combustion transition, and point ratios between 0.40 and 0.50 are more characteristic of liquid fossil fuel (vehicle and crude oil) combustion, whereas ratios > 0.50 are characteristic of grass, wood or coal combustion. The ratios in our study indicated a strong influence of biomass and coal combustion before November, which was consistent with the result of mass 178. For mass 228, BaA/(BaA +Chr), ratios < 0.20 indicate petroleum sources, ratios from 0.20 to 0.35 indicate either petroleum or combustion and those > 0.35 imply combustion. The ratios from May to October were higher than 0.5, which indicated grass, wood or coal combustion. From November to April, the ratios were between 0.40 and 0.50, which was the characteristic of liquid fossil fuel combustion. For IcdP/(IcdP + BghiP), ratios < 0.20 likely indicates petroleum, those between 0.20 and 0.50 imply liquid fossil fuel combustion and ratios > 0.50 imply grass, wood and coal combustion. The ratios of mass 228 in our research indicated a mixture of petroleum or combustion. Combustion was the potential source from July to December. The ratios in our research suggested a potential combustion source [45].

The MDRs indicated that PAHs in Lake Chaohu were mainly from combustion and vehicle emission. We can also found that differences existed when we were using different PAH ratios. The use of PAH MDRs has been criticized in the past due to low accuracy. Overlap areas were reported between commonly applied ratios associated with different types of PAH emissions [44,46]. According to a study of MDRs based on the inventory and monitoring data over 20 years, it was found that the use of MDRs does not respond to known differences in atmospheric emission sources unless the source is strong [47]. due to the limitations of MDRs, a PMF model was also applied to detect the potential sources.

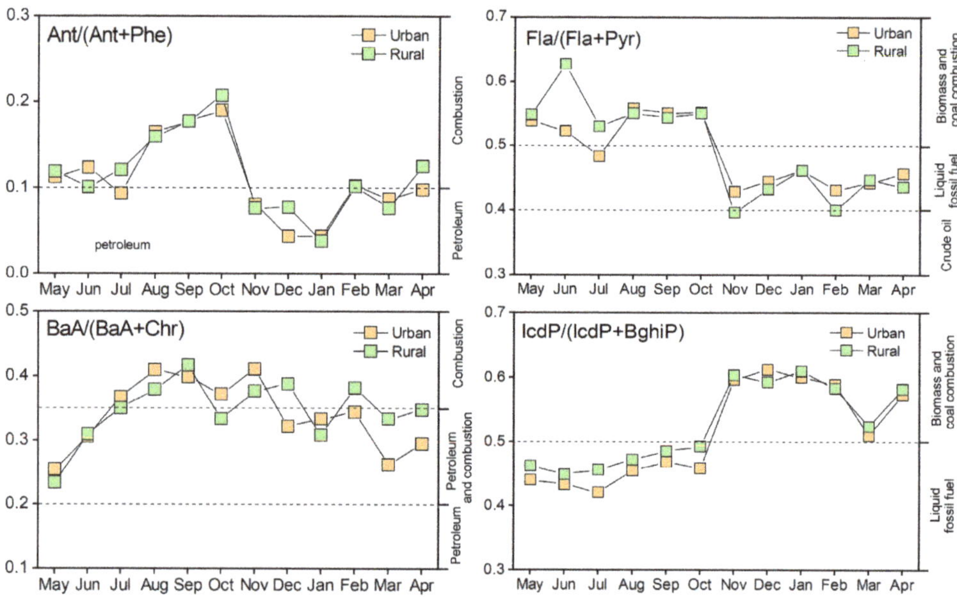

Figure 4. Seasonal and spatial variation of four ratios in atmospheric samples from Lake Chaohu.

3.2.2. PMF Results

When the number of factors for PMF is three, good simulation has been achieved for most PAHs. Therefore, three components were extracted, and the source profiles of three factors are illustrated in Figure 5. It was found that factor one is predominately weighted by low molecular weight PAHs. Factor two is heavily weighted by the middle molecular weight PAHs, and factor three has a higher contribution in the PAHs with more than four rings. The first factor is predominately weighted by Acy, which has been proved to be a tracer of combustion of straws. In addition, factor one has high load on Fla, Pyr and Chr, which are the combustion products of firewood. Therefore, factor one appears to be biomass combustion. Factor two is predominately weighted by Phe and BbF. According to the literature, Flo, Phe and Ant are predominantly considered as coal combustion profiles. The high BbkF and Chr loads are also a typical sign of Chinese domestic coal emissions [2,17,48,49]. Factor two is supposed to be coal combustion emission. For factor three, BghiP has been identified as tracers of auto emissions, and IcdP is considered as a marker of diesel emission. It can be concluded that factor three represents emission from vehicle.

The percentages of the sources from the factors were estimated by PMF. These results showed that the sources from biomass combustions, coal combustion and vehicle emission accounted for 43.6%, 30.6% and 25.8% of the total PAHs, respectively. Compared with the result of MDRs, a similar conclusion was obtained by PMF and MRDs. It can also be observed that coal combustion plays an important part in local PAH emission. Our conclusion is different from some research studies conducted in China. Most domestic studies show that coal is the main source of pollution in China, which is related to the fact that coal is the main energy material in China. China is the largest coal producer and consumer in the world [50]. According to the National Bureau of Statistics, the coal production of China in 2018 was 3.5 billion tons, nearly half of the world's production. Thus, the high coal consumption produces high PAH contribution to the atmosphere. In this study, the atmospheric samples were collected from Mushan island and Chaohu City. The former is a rural area far away from the city, while the latter is a small city with a small population and is a suburb surrounded by rural areas. Compared with coal combustion, firewood combustion is a more important method supplying energy in the research area.

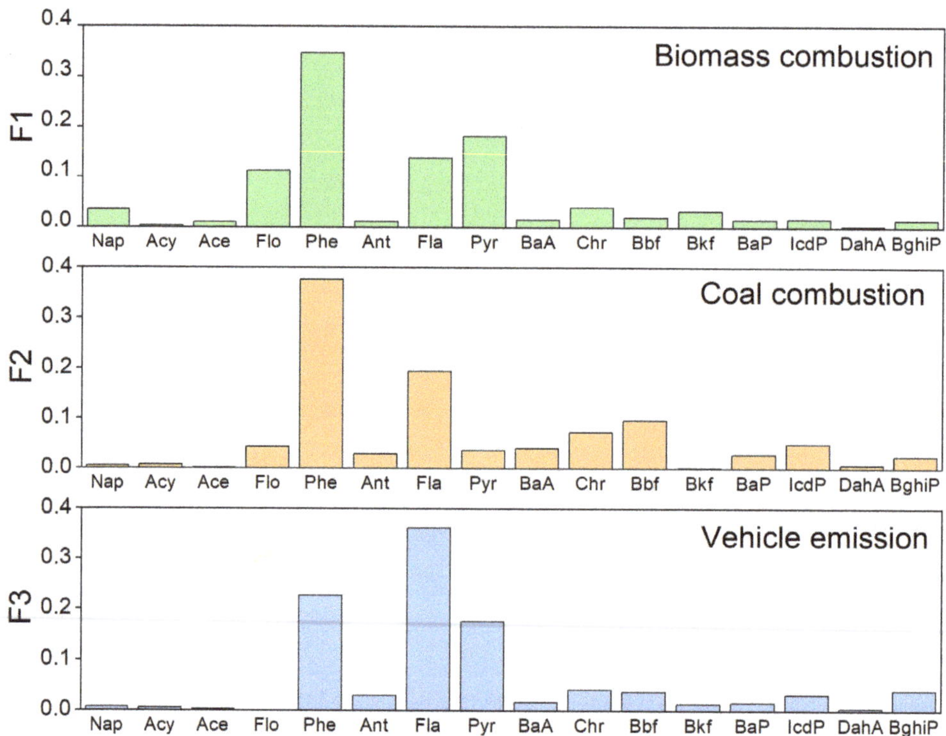

Figure 5. Factors of PMF analysis in Chaohu dustfall.

3.3. Exposure through Different Pathways

3.3.1. Derivation of Exposure Parameters

Three physiological and exposure behavior parameters including bodyweight, water intake rate and inhalation rate were collected from *Exposure Factors Handbook of Chinese Population* [51]. The distribution modes of parameters were fitted by regression models. Specifically, first, we considered a normal distribution for BW and log-normal distributions for the water intake rate and inhalation rate because the normal and log-normal distribution models are the most widely applied in studies on the exposure parameters [52,53]. Second, the 5th, 25th, 50th, 75th and 95th percentiles were collected from the *Exposure Factors Handbook*. The distribution of the parameters was fitted using the Gaussian function. The fit curves and parameters are shown in Figure 6 and Table 3, respectively. It can be observed that relatively good regression results were obtained for all three parameters. In addition, the intake rates of fish consumption in urban and rural were considered as constants. According to a survey conducted between 2010 and 2013, the average daily intakes of freshwater products per capita were 19.0 g/day and 11.1 g/day for the urban and rural populations, respectively [54].

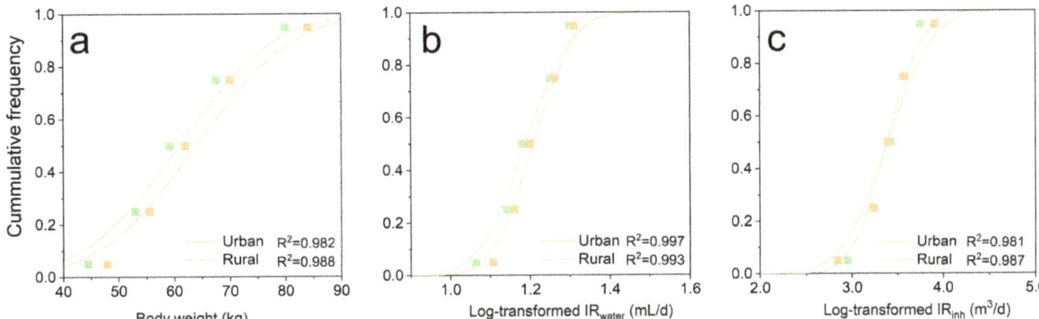

Figure 6. Fitting of (**a**) bodyweight, (**b**) water intake rate and (**c**) inhalation rate of urban and rural population in Anhui Province.

Table 3. Exposure parameters of adults for Monte Carlo simulation.

Media	Unit	Distribution Mode	Urban	Rural	References
Water	ng·L	Log-Normal	LN(−0.77, 0.68)	LN(0.62, 0.56)	Measured
Gas	ng·m^{-3}	Log-Normal	LN(−1.96, 0.67)	LN(−1.18, 0.71)	Measured
Particle	ng·m^{-3}	Log-Normal	LN(0.45, 1.10)	LN(0.54, 1.11)	Measured
Fish	ng·g^{-3}	Log-Normal	LN(0.56, 1.48)	LN(0.56, 1.48)	Measured
BW	Kg	Normal	N(63.90, 13.86)	LN(60.86, 13.64)	[53]
IR(water)	mL·d^{-3}	Log-Normal	LN(7.81, 0.91)	LN(7.80, 0.70)	[53]
IR(inh)	m^3·d^{-1}	Log-Normal	LN(2.78,0.19)	LN(2.74, 0.21)	[53]
IR(fish)	g·d^{-3}	Constant	19.0	11.1	[54]

3.3.2. Estimation of Exposure Doses

The lifetime average daily doses (LADD) of BaP_{eq} exposure through inhalation (air and particle), aquatic product ingestion and water intake were calculated by 10,000 iterations of Monte Carlo simulation. The results were illustrated in Figure 7. BaP_{eq} exposure through fish ingestion had the greatest contribution, with the average value of 3.01 × 10^{-6} (mg·kg^{-1}·d^{-1}). The LADD of residents in urban and rural were 2.97 × 10^{-6} (mg·kg^{-1}·d^{-1}) and 3.07 × 10^{-6} (mg·kg^{-1}·d^{-1}), respectively. Although the fish intake rate in urban residents was 1.71 times higher than that in rural, no obvious difference was observed between two population groups. Exposure through particle inhalation has a comparable contribution of 2.54 × 10^{-6} (mg·kg^{-1}·d^{-1}), which is one and two order of magnitude higher than exposure through gas inhalation and water ingestion, respectively. Our results are consistent with previous research. It has been reported that, for most non-occupationally exposed individuals, diet is the main route of exposure [55,56]. Inhalation of gaseous and particulate also had an ignorable contribution relative to the total BaP_{eq} exposure.

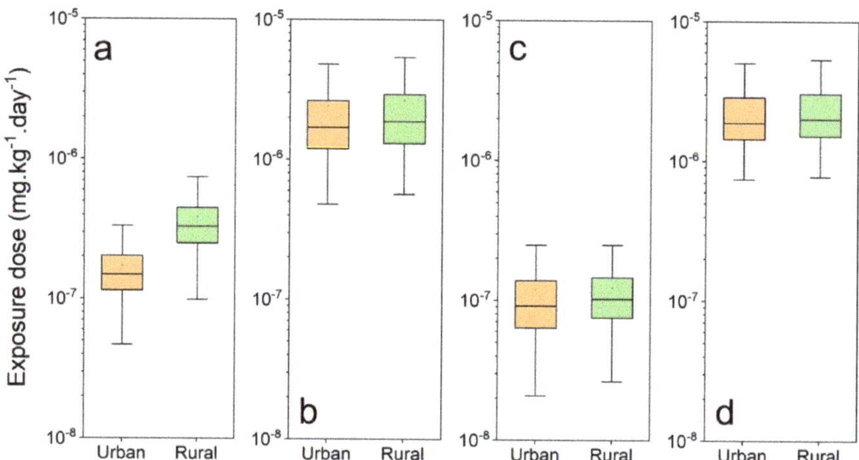

Figure 7. BaP_{eq} exposure through different pathways: (**a**) gaseous phase; (**b**) particulate phase; (**c**) water intake; and (**d**) fish intake.

3.4. Carcinogenic Risk Assessment

The probabilistic cancer risk for urban and rural residents were obtained based on the results of Monte Carlo simulation and presented in Figure 8a,b. The median total PAHs ILCR was 3.1×10^{-5} to 3.3×10^{-5} in urban and rural residents, which showed a slightly higher carcinogenic risk for rural residents than that for urban residents. The difference can be attributed to the variation of exposure behaviors. A one in a million chance of additional human cancer over a 70-year lifetime (ILCR = 10^{-6}) is considered acceptable or inconsequential, and one in ten thousand or greater (ILCR = 10^{-4}) is considered serious (US EPA). Most of the surrogate samples generated by Monte Carlo distributed between the range of 10^{-6} and 10^{-4}, which indicates potential ILCR risk. Due to the different end points, the carcinogenic slopes of inhalation and ingestion are quite different. Therefore, fish ingestion had much higher contribution than particle inhalation. Inhalation (including gas and particles) accounted for 26.5% and 29.1% of the total risk in urban and rural areas, respectively.

Parameter sensitivity was quantitatively assessed by the Spearman's rank correlation coefficient of parameter and risk. The result was shown in Figure 8c,d. As the dominant exposure pathway, fish consumption had the highest risk contribution. PAH concentration in fish had the highest influence on the results of risk assessment. As the most important protection factor, the bodyweight was the second most sensitive parameter. PAH concentration associated with particles was the third most sensitive parameter in this research. It can also be observed that the sensitivity of exposure behavior parameters were relative low compared with media concentrations. This can be attributed to the small difference of these parameters among adults. As the most important behavior parameter, fish consumption rate was not treated as variables in the study due to the limited data support. However, the uncertainty of intake rate of fish should not be ignored due to the high contribution of fish intake on the total risk, which requires further research and more data support in the future.

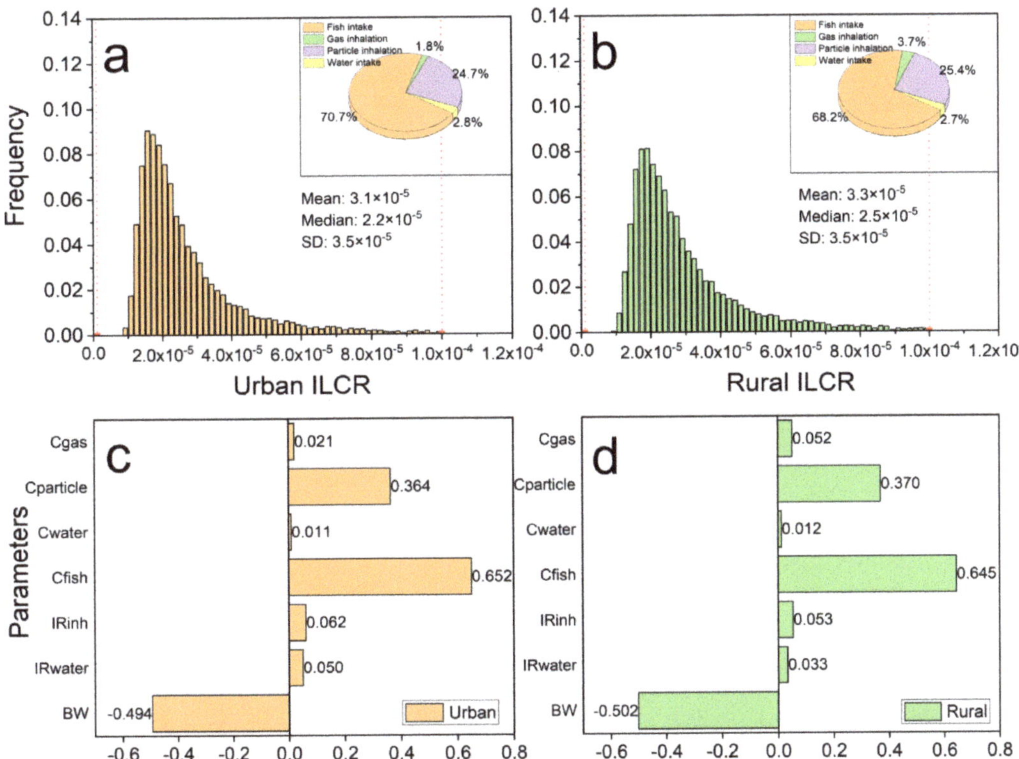

Figure 8. Distributions of incremental lifetime cancer risk (**a**,**b**) and parameter sensitivity for urban and rural populations (**c**,**d**) derived using the Monte Carlo simulation.

4. Conclusions

PAHs in major environment samples were collected in Lake Chao for the entire year. Exposures through four pathways were estimated. ILCRs were characterized. The results found that the gaseous concentration had influence on the PAHs in the water according to the Spearman correlation analysis and may further affect the PAH content in fish tissues. Atmospheric transport is the source of the entire water system. The results of sources apportionment on atmospheric samples indicated the high contribution of biomass combustion. Probabilistic risk assessment suggested that inhalation and fish ingestion are two major pathways of PAH exposure, which are also the key processes in PAHs' risk control for people near Lake Chaohu.

Author Contributions: Investigation, W.H., Q.H., X.K. and W.L.; writing—original draft, N.Q.; writing—review and editing, F.X. and Q.W. All authors have read and agreed to the published version of the manuscript.

Funding: This research was funded by China CDC Key Laboratory of Environment and Population Health, National Institute of Environmental Health, Chinese Center for Disease Control and Prevention (2021-CKL-01), by the Fundamental Research Funds for the Central Universities (FRF-TP-18-071A1) and by the National Natural Science Foundation of China (NSFC) (41977312, 41503104).

Institutional Review Board Statement: Not applicable.

Informed Consent Statement: Not applicable.

Acknowledgments: We also thank the help from National Environmental and Energy Base for International Science and Technology Cooperation.

Conflicts of Interest: All authors declare that they have no conflicts of interest or financial conflicts to disclose.

References

1. Xu, F.-L.; Qin, N.; Zhu, Y.; He, W.; Kong, X.-Z.; Barbour, M.T.; He, Q.-S.; Wang, Y.; Ou-Yang, H.-L.; Tao, S. Multimedia fate modeling of polycyclic aromatic hydrocarbons (PAHs) in Lake Small Baiyangdian, Northern China. *Ecol. Model.* **2013**, *252*, 246–257. [CrossRef]
2. Larsen, R.K.; Baker, J. Source apportionment of polycyclic aromatic hydrocarbons in the urban atmosphere: A comparison of three methods. *Environ. Sci. Technol.* **2003**, *37*, 1873–1881. [CrossRef]
3. Fernandes, M.; Sicre, M.-A.; Boireau, A.; Tronczynski, J. Polyaromatic hydrocarbon (PAH) distributions in the Seine River and its estuary. *Mar. Pollut. Bull.* **1997**, *34*, 857–867. [CrossRef]
4. Man, Y.B.; Mo, W.Y.; Zhang, F.; Wong, M.H. Health risk assessments based on polycyclic aromatic hydrocarbons in freshwater fish cultured using food waste-based diets. *Environ. Pollut.* **2020**, *256*, 113380. [CrossRef] [PubMed]
5. Kim, K.-H.; Jahan, S.A.; Kabir, E.; Brown, R.J.C. A review of airborne polycyclic aromatic hydrocarbons (PAHs) and their human health effects. *Environ. Int.* **2013**, *60*, 71–80. [CrossRef] [PubMed]
6. Miller, B.G.; Doust, E.; Cherrie, J.W.; Hurley, J.F. Lung cancer mortality and exposure to polycyclic aromatic hydrocarbons in British coke oven workers. *BMC Public Health* **2013**, *13*, 962. [CrossRef]
7. Kuang, D.; Zhang, W.; Deng, Q.; Zhang, X.; Huang, K.; Guan, L.; Hu, D.; Wu, T.; Guo, H. Dose-response relationships of polycyclic aromatic hydrocarbons exposure and oxidative damage to DNA and lipid in coke oven workers. *Environ. Sci. Technol.* **2013**, *47*, 7446–7456. [CrossRef] [PubMed]
8. Zhu, Y.; Duan, X.; Qin, N.; Li, J.; Tian, J.; Zhong, Y.; Chen, L.; Fan, R.; Yu, Y.; Wu, G.; et al. Internal biomarkers and external estimation of exposure to polycyclic aromatic hydrocarbons and their relationships with cancer mortality in a high cancer incidence area. *Sci. Total Environ.* **2019**, *688*, 742–750. [CrossRef] [PubMed]
9. Rogge, W.F.; Hildemann, L.M.; Mazurek, M.A.; Cass, G.R.; Simoneit, B.R. Sources of fine organic aerosol.3. Road dust, tire debris, and organonetallic brake lining dust-roads as sources and sinks. *Environ. Sci. Technol.* **1993**, *27*, 1892–1904. [CrossRef]
10. Wang, B.; Li, Z.; Ma, Y.; Qiu, X.; Ren, A. Association of polycyclic aromatic hydrocarbons in housewives' hair with hyper-tension. *Chemosphere* **2016**, *153*, 315–321. [CrossRef]
11. Qin, N.; He, W.; Kong, X.Z.; Liu, W.X.; He, Q.S.; Yang, B.; Ouyang, H.L.; Wang, Q.M.; Xu, F.L. Ecological risk assessment of polycyclic aromatic hydrocarbons (PAHs) in the water from a large Chinese lake based on multiple indicators. *Ecol. Indic.* **2013**, *24*, 599–608. [CrossRef]
12. Andra, S.S.; Austin, C.; Wright, R.O.; Arora, M. Reconstructing pre-natal and early childhood exposure to multi-class organic chemicals using teeth: Towards a retrospective temporal exposome. *Environ. Int.* **2015**, *83*, 137–145. [CrossRef] [PubMed]
13. Yu, Y.; Li, Q.; Wang, H.; Wang, B.; Wang, X.; Ren, A.; Tao, S. Risk of human exposure to polycyclic aromatic hydrocarbons: A case study in Beijing, China. *Environ. Pollut.* **2015**, *205*, 70–77. [CrossRef]
14. Lohmann, R.; Jurado, E.; Pilson, M.E.; Dachs, J. Oceanic deep water formation as a sink of persistent organic pollutants. *Geophys. Res. Lett.* **2006**, *33*, 735. [CrossRef]
15. Zhang, G.; Li, J.; Cheng, H.; Li, X.; Xu, W.; Jones, K.C. Distribution of organochlorine pesticides in the Northern South China Sea: Implications for land outflow and air−sea exchange. *Environ. Sci. Technol.* **2007**, *41*, 3884–3890. [CrossRef]
16. Simonich, S.L.; Hites, R.A. Organic Pollutant Accumulation in Vegetation. *Environ. Sci. Technol.* **1995**, *29*, 2905–2914. [CrossRef] [PubMed]
17. Simcik, M.F.; Eisenreich, S.J.; Lioy, P.J. Source apportionment and source/sink relationships of PAHs in the coastal atmosphere of Chicago and Lake Michigan. *Atmos. Environ.* **1999**, *33*, 5071–5079. [CrossRef]
18. Franz, T.P.; Eisenreich, S.J.; Holsen, T.M. Dry Deposition of Particulate Polychlorinated Biphenyls and Polycyclic Aromatic Hydrocarbons to Lake Michigan. *Environ. Sci. Technol.* **1998**, *32*, 3681–3688. [CrossRef]
19. Qin, N.; He, W.; Liu, W.; Kong, X.; Xu, F.; Giesy, J.P. Tissue distribution, bioaccumulation, and carcinogenic risk of polycyclic aromatic hydrocarbons in aquatic organisms from Lake Chaohu, China. *Sci. Total Environ.* **2020**, *749*, 141577. [CrossRef]
20. Xu, F.-L.; Wu, W.-J.; Wang, J.-J.; Qin, N.; Wang, Y.; He, Q.-S.; He, W.; Tao, S. Residual levels and health risk of polycyclic aromatic hydrocarbons in freshwater fishes from Lake Small Bai-Yang-Dian, Northern China. *Ecol. Model.* **2011**, *222*, 275–286. [CrossRef]
21. He, Y.; Qin, N.; He, W.; Xu, F. The impacts of algae biological pump effect on the occurrence, source apportionment and toxicity of SPM-bound PAHs in lake environment. *Sci. Total Environ.* **2021**, *753*, 141980. [CrossRef] [PubMed]
22. Li, C.; Huo, S.; Yu, Z.; Xi, B.; Zeng, X.; Wu, F. Spatial distribution, potential risk assessment, and source apportionment of polycyclic aromatic hydrocarbons (PAHs) in sediments of Lake Chaohu, China. *Environ. Sci. Pollut. Res.* **2014**, *21*, 12028–12039. [CrossRef] [PubMed]
23. Qin, N.; Kong, X.-Z.; He, W.; He, Q.-S.; Liu, W.-X.; Xu, F.-L. Dustfall-bound polycyclic aromatic hydrocarbons (PAHs) over the fifth largest Chinese lake: Residual levels, source apportionment, and correlations with suspended particulate matter (SPM)-bound PAHs in water. *Environ. Sci. Pollut. Res.* **2021**, 1–13. [CrossRef]

24. Wang, R.; Huang, Q.; Cai, J.; Wang, J. Seasonal variations of atmospheric polycyclic aromatic hydrocarbons (PAHs) sur-rounding Chaohu Lake, China: Source, partitioning behavior, and lung cancer risk. *Atmos. Pollut. Res.* **2021**, *12*. [CrossRef]
25. Zhang, L.; Bai, Y.-S.; Wang, J.-Z.; Peng, S.-C.; Chen, T.-H.; Yin, D.-Q. Identification and determination of the contribution of iron–steel manufacturing industry to sediment-associated polycyclic aromatic hydrocarbons (PAHs) in a large shallow lake of eastern China. *Environ. Sci. Pollut. Res.* **2016**, *23*, 22037–22046. [CrossRef]
26. Qin, N.; He, W.; Kong, X.-Z.; Liu, W.; He, Q.-S.; Yang, B.; Ouyang, H.-L.; Wang, Q.-M.; Xu, F.-L. Atmospheric partitioning and the air–water exchange of polycyclic aromatic hydrocarbons in a large shallow Chinese lake (Lake Chaohu). *Chemosphere* **2013**, *93*, 1685–1693. [CrossRef]
27. Paatero, P. Least squares formulation of robust non-negative factor analysis. *Chemom. Intell. Lab. Syst.* **1997**, *37*, 23–35. [CrossRef]
28. Liu, C.; Tian, F.; Chen, J.; Li, X.; Qiao, X. A comparative study on source apportionment of polycyclic aromatic hydrocarbons in sediments of the Daliao River, China: Positive matrix factorization and factor analysis with non-negative constraints. *Chin. Sci. Bull.* **2010**, *55*, 915–920. [CrossRef]
29. Nisbet, I.C.T.; Lagoy, P.K. Toxic equaivalency factors (TEFs) for Polycyclic Aromatic-Hydrocarbons (PAHs). *Regul. Toxicol. Pharmacol.* **1992**, *16*, 290–300. [CrossRef]
30. Taghvaee, S.; Sowlat, M.H.; Hassanvand, M.S.; Yunesian, M.; Naddafi, K.; Sioutas, C. Source-specific lung cancer risk assessment of ambient PM2.5-bound polycyclic aromatic hydrocarbons (PAHs) in central Tehran. *Environ. Int.* **2018**, *120*, 321–332. [CrossRef]
31. *Exposure Factors Handbook*; EPA/600/P-95/002; US EPA: Washington, DC, USA, 1997.
32. Li, J.; Zhang, G.; Li, X.; Qi, S.H.; Liu, G.Q.; Peng, X.Z. Source seasonality of polycyclic aromatic hydrocarbons (PAHs) in a subtropical city, Guangzhou, South China. *Sci. Total Environ.* **2006**, *355*, 145–155. [CrossRef] [PubMed]
33. Liu, S.; Tao, S.; Liu, W.; Dou, H.; Liu, Y.; Zhao, J.; Little, M.G.; Tian, Z.; Wang, J.; Wang, L.; et al. Seasonal and spatial occurrence and distribution of atmospheric polycyclic aromatic hydrocarbons (PAHs) in rural and urban areas of the North Chinese plain. *Environ. Pollut.* **2008**, *156*, 651–656. [CrossRef] [PubMed]
34. Zhen, Z.; Yin, Y.; Chen, K.; Zhen, X.; Zhang, X.; Jiang, H.; Wang, H.; Kuang, X.; Cui, Y.; Dai, M.; et al. Concentration and atmospheric transport of PM2.5-bound polycyclic aromatic hydrocarbons at Mount Tai, China. *Sci. Total Environ.* **2021**, *786*, 147513. [CrossRef] [PubMed]
35. Gustafson, K.E.; Dickhut, R.M. Gaseous exchange of polycyclic aromatic hydrocarbons across the air–water interface of Southern Chesapeake Bay. *Environ. Sci. Technol.* **1997**, *31*, 1623–1629. [CrossRef]
36. Mandalakis, M.; Tsapakis, M.; Tsoga, A.; Stephanou, E.G. Gas-particle concentrations and distribution of aliphatic hydro-carbons, PAHs, PCBs and PCDD/Fs in the atmosphere of Athens (Greece). *Atmos. Environ.* **2002**, *36*, 4023–4035. [CrossRef]
37. Matos, J.; Silveira, C.; Cerqueira, M. Particle-bound polycyclic aromatic hydrocarbons in a rural background atmosphere of southwestern Europe. *Sci. Total Environ.* **2021**, *87*, 147666. [CrossRef]
38. Xia, Z.; Duan, X.; Qiu, W.; Liu, D.; Wang, B.; Tao, S.; Jiang, Q.; Lu, B.; Song, Y.; Hu, X. Health risk assessment on dietary exposure to polycyclic aromatic hydrocarbons (PAHs) in Taiyuan, China. *Sci. Total Environ.* **2010**, *408*, 5331–5337. [CrossRef]
39. Wu, X.; Wang, Y.; Zhang, Q.; Zhao, H.; Yang, Y.; Zhang, Y.; Xie, Q.; Chen, J. Seasonal variation, air-water exchange, and multivariate source apportionment of polycyclic aromatic hydrocarbons in the coastal area of Dalian, China. *Environ. Pollut.* **2019**, *244*, 405–413. [CrossRef]
40. Křůmal, K.; Mikuška, P.; Horák, J.; Hopan, F.; Kuboňová, L. Influence of boiler output and type on gaseous and particulate emissions from the combustion of coal for residential heating. *Chemosphere* **2021**, *278*, 130402. [CrossRef]
41. Teixeira, E.C.; Agudelo-Castañeda, D.M.; Mattiuzi, C.D.P. Contribution of polycyclic aromatic hydrocarbon (PAH) sources to the urban environment: A comparison of receptor models. *Sci. Total Environ.* **2015**, *538*, 212–219. [CrossRef]
42. Feng, J.; Song, N.; Yu, Y.; Li, Y. Differential analysis of FA-NNC, PCA-MLR, and PMF methods applied in source apportionment of PAHs in street dust. *Environ. Monit. Assess.* **2020**, *192*, 727. [CrossRef]
43. Dickhut, R.M.; Canuel, E.A.; Gustafson, K.E.; Liu, K.; Arzayus, K.M.; Walker, S.E.; Edgecombe, G.; Gaylor, M.O.; Macdonald, E.H. Automotive sources of carcinogenic polycyclic aromatic hydrocarbons associated with particulate matter in the Chesapeake Bay Region. *Environ. Sci. Technol.* **2000**, *34*, 4635–4640. [CrossRef]
44. Davis, E.; Walker, T.R.; Adams, M.; Willis, R.; Norris, G.A.; Henry, R.C. Source apportionment of polycyclic aromatic hydrocarbons (PAHs) in small craft harbor (SCH) surficial sediments in Nova Scotia, Canada. *Sci. Total Environ.* **2019**, *691*, 528–537. [CrossRef]
45. Yunker, M.B.; Macdonald, R.; Vingarzan, R.; Mitchell, R.H.; Goyette, D.; Sylvestre, S. PAHs in the Fraser river basin: A critical appraisal of PAH ratios as indicators of PAH source and composition. *Org. Geochem.* **2002**, *33*, 489–515. [CrossRef]
46. Galarneau, E. Source specificity and atmospheric processing of airborne PAHs: Implications for source apportionment. *Atmospheric Environ.* **2008**, *42*, 8139–8149. [CrossRef]
47. Katsoyiannis, A.; Sweetman, A.; Jones, K.C. PAH molecular diagnostic ratios applied to atmospheric sources: A critical evaluation using two decades of source inventory and air concentration data from the UK. *Environ. Sci. Technol.* **2011**, *45*, 8897–8906. [CrossRef] [PubMed]
48. Harrison, R.M.; Smith, A.D.J.T.; Luhana, L. Source Apportionment of Atmospheric Polycyclic Aromatic Hydrocarbons Collected from an Urban Location in Birmingham, UK. *Environ. Sci. Technol.* **1996**, *30*, 825–832. [CrossRef]
49. Chen, Y.; Sheng, G.; Bi, X.; Feng, Y.; Mai, B.; Fu, J. Emission Factors for Carbonaceous Particles and Polycyclic Aromatic Hydrocarbons from Residential Coal Combustion in China. *Environ. Sci. Technol.* **2005**, *39*, 1861–1867. [CrossRef] [PubMed]

50. Lin, B.-Q.; Liu, J.-H. Estimating coal production peak and trends of coal imports in China. *Energy Policy* **2010**, *38*, 512–519. [CrossRef]
51. Xiuge, Z.; Xiao-li, D. *Exposure Factors Handbook of Chinese Population*; Ministry of Environmental Protection of the People's Republic of China: Beijing, China, 2014.
52. Chen, S.-C.; Liao, C.-M. Health risk assessment on human exposed to environmental polycyclic aromatic hydrocarbons pol-lution sources. *Sci. Total Environ.* **2006**, *366*, 112–123. [CrossRef] [PubMed]
53. Liao, C.-M.; Chio, C.-P.; Chen, W.-Y.; Ju, Y.-R.; Li, W.-H.; Cheng, Y.-H.; Liao, V.; Chen, S.-C.; Ling, M.-P. Lung cancer risk in relation to traffic-related nano/ultrafine particle-bound PAHs exposure: A preliminary probabilistic assessment. *J. Hazard. Mater.* **2011**, *190*, 150–158. [CrossRef] [PubMed]
54. Zhao, L.; He, Y. *Monitoring Report on Nutrition and Health Status of Chinese Residents (2010–2013) Volume I: Dietary and Nutrient Intake*; People's Medical Publishing House Co., LTD: Beijing, China, 2018.
55. Falcó, G.; Domingo, J.L.; Llobet, J.M.; Teixidó, A.; Casas, C.; Müller, L. Polycyclic Aromatic Hydrocarbons in Foods: Human Exposure through the Diet in Catalonia, Spain. *J. Food Prot.* **2003**, *66*, 2325–2331. [CrossRef] [PubMed]
56. Martorell, I.; Nieto, A.; Nadal, M.; Perello, G.; Marce, R.M.; Domingo, J.L. Human exposure to polycyclic aromatic hydro-carbons (PAHs) using data from a duplicate diet study in Catalonia, Spain. *Food Chem. Toxicol.* **2012**, *50*, 4103–4108. [CrossRef] [PubMed]

Article

Unorganized Machines to Estimate the Number of Hospital Admissions Due to Respiratory Diseases Caused by PM_{10} Concentration

Yara de Souza Tadano [1,*], Eduardo Tadeu Bacalhau [2], Luciana Casacio [2], Erickson Puchta [3], Thomas Siqueira Pereira [4], Thiago Antonini Alves [4], Cássia Maria Lie Ugaya [5] and Hugo Valadares Siqueira [3]

1 Department of Mathematics, Federal University of Technology, 330 Doutor Washington Subtil Chueire Street, Ponta Grossa 84017-220, PR, Brazil
2 Center for Marine Studies, Pontal do Paraná Campus, Federal University of Paraná, Beira-mar Avenue, P.O. Box 61, Pontal do Paraná 83255-976, PR, Brazil; bacalhau@ufpr.br (E.T.B.); lucianacasacio@ufpr.br (L.C.)
3 Department of Electric Engineering, Federal University of Technology, 330 Doutor Washington Subtil Chueire Street, Ponta Grossa 84017-220, PR, Brazil; ericksonpuchta@gmail.com (E.P.); hugosiqueira@utfpr.edu.br (H.V.S.)
4 Department of Mechanical Engineering, Federal University of Technology, 330 Doutor Washington Subtil Chueire Street, Ponta Grossa 84017-220, PR, Brazil; thomassiqueira427@gmail.com (T.S.P.); antonini@utfpr.edu.br (T.A.A.)
5 Department of Mechanical, Federal University of Technology, CNPq Fellow, 5000 Dep. Heitor Alencar Furtado Street, Curitiba 81280-340, PR, Brazil; cassiaugaya@utfpr.edu.br
* Correspondence: yaratadano@utfpr.edu.br

Citation: Tadano, Y.d.S.; Bacalhau, E.T.; Casacio, L.; Puchta, E.; Pereira, T.S.; Antonini Alves, T.; Ugaya, C.M.L.; Siqueira, H.V. Unorganized Machines to Estimate the Number of Hospital Admissions Due to Respiratory Diseases Caused by PM_{10} Concentration. *Atmosphere* **2021**, *12*, 1345. https://doi.org/10.3390/atmos12101345

Academic Editors: Hsiao-Chi Chuang and Alina Barbulescu

Received: 17 July 2021
Accepted: 6 October 2021
Published: 14 October 2021

Publisher's Note: MDPI stays neutral with regard to jurisdictional claims in published maps and institutional affiliations.

Copyright: © 2021 by the authors. Licensee MDPI, Basel, Switzerland. This article is an open access article distributed under the terms and conditions of the Creative Commons Attribution (CC BY) license (https://creativecommons.org/licenses/by/4.0/).

Abstract: The particulate matter PM_{10} concentrations have been impacting hospital admissions due to respiratory diseases. The air pollution studies seek to understand how this pollutant affects the health system. Since prediction involves several variables, any disparity causes a disturbance in the overall system, increasing the difficulty of the models' development. Due to the complex nonlinear behavior of the problem and their influencing factors, Artificial Neural Networks are attractive approaches for solving estimations problems. This paper explores two neural network architectures denoted unorganized machines: the echo state networks and the extreme learning machines. Beyond the standard forms, models variations are also proposed: the regularization parameter (RP) to increase the generalization capability, and the Volterra filter to explore nonlinear patterns of the hidden layers. To evaluate the proposed models' performance for the hospital admissions estimation by respiratory diseases, three cities of São Paulo state, Brazil: Cubatão, Campinas and São Paulo, are investigated. Numerical results show the standard models' superior performance for most scenarios. Nevertheless, considering divergent intensity in hospital admissions, the RP models present the best results in terms of data dispersion. Finally, an overall analysis highlights the models' efficiency to assist the hospital admissions management during high air pollution episodes.

Keywords: PM_{10}; health risks; extreme learning machine; echo state network; neural networks

1. Introduction

World Health Organization (WHO) estimates that 91% of the world's population lives in places where air pollution levels exceed the advised limits. This exposure has as a consequence 4.2 million deaths per year due to stroke, heart disease, lung cancer and chronic respiratory illness [1].

In the last decades, the air pollution consequences in the environment and health have been the subject of deep researches [2–4], including the relation between air pollution and human health [5–8] and, specifically, the study of particulate matter (PM) impacts on the respiratory diseases [9–11]. The public health system is currently the main concern for the global governance majority, receiving huge money investments and boosting researches in operational areas. Therefore, several works have been applied to develop mathematical models to improve predicting the diseases caused by PM air concentration.

Generalized Linear Models (GLM) [10–14] and Generalized Additive Models (GAM) [15,16] are statistical regression models usually used to assess air pollution consequences on human health. However, a minimum of data is required to assure that regression models will be able to capture the relationship between the inputs (predictors) and the output (response variable) [17]. For developing countries, as lack of data is a reality, solving the problem using regression models is challenging [18]. For this reason, other models and methods have been applied; since the problem can be seen as a nonlinear mapping task, the Artificial Neural Networks (ANN) approach is the most attractive approach for solving estimation problems. The ANN have been used to solve air pollution mapping tasks [19–22], and they have become increasingly popular over the past decade for predicting the air pollutant's impact on human health [10,17,18,23–25]. Araujo et al. [17] and Kassomenos et al. [24] have shown that the ANN had better performance than linear approaches like the GLM when dealing with nonlinear mapping problems. In this context, Tadano et al. [26] proposed to use two models, known as Unorganized Machines (UM): the echo state networks (ESN) and the extreme learning machines (ELM), to predict hospital admissions. Based on this work, this paper presents a full extension of these models, adding several neural networks variations applied to an enlarged and updated set of instances.

ELM and ESN are ANN architectures used to deal with static nonlinear mapping problems, and are reliable when applied to multiclass classification and, mainly, time series forecasting [27–31]. Thus, the main contribution of this research is an epistemological study that predicts the impact of PM_{10} (particulate matter with an aerodynamic diameter less than 10 μm) daily mean concentrations on hospital admissions due to respiratory diseases using versions of the UM: the addition of regularization parameter applied to increase the generalization capability of the models [32] and the use of the Volterra filter to capture nonlinear patterns of the neural information [33]. To evaluate the performance of the proposed methods, three cities from São Paulo State, Brazil (Campinas, Cubatão and São Paulo city) were considered.

Based on the overall analysis produced, we expect to understand how air pollution affects the health system, especially during global sanitary crises scenarios, avoiding hospital collapse.

This work is organized as follows: Section 2 presents the ELM and ESN standard models, the regularization parameter and nonlinear output layer strategies; Section 3 describes the addressed databases; Section 4 shows the computational results and critical analysis regarding the models' performances; Section 5 presents the main conclusions and future works.

2. Unorganized Machines

Unorganized machines are a designation used as a general term to classify the modern neural network paradigms that unify two kinds of ANN: the echo state networks (ESNs) and the extreme learning machines (ELMs) [27].

In this work, these two architectures are employed to predict the hospital admissions due to respiratory diseases caused by air pollution. Moreover, other models based on the variations and extensions of these models are used [33,34].

2.1. Extreme Learning Machines

The extreme learning machine (ELM) is a feedforward neural network composed of a single hidden layer, similar to the structure of multilayer perceptron (MLP) [28]. Figure 1 illustrates the architecture.

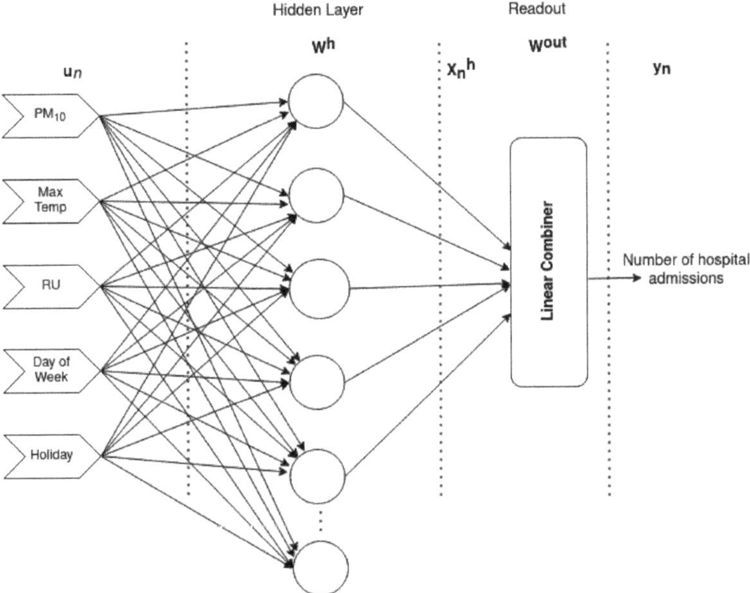

Figure 1. Extreme Learning Machine.

According to Figure 1, the vector $\mathbf{u}_n$ represents all input information: PM_{10} concentration; relative humidity; ambient temperature; the different weekdays; and holidays. This vector $\mathbf{u}_n$ is associated with the matrix $\mathbf{W}^h$ through weights of the hidden layer that can be randomly determined. The unique output layer (readout) $\mathbf{W}^{out}$ is composed of parameters of a linear combiner that are calculated using the Moore-Penrose generalized inverse operator which shall be defined below. Finally, similar to a single-hidden layer multilayer perceptron (MLP), the ELM is also a single hidden layer feedforward neural network, being y_n the output information that indicates the number of hospital admissions.

The activation of the artificial neurons within the hidden layer are given by Equation (1):

$$\mathbf{x}_n^h = f_n^h(\mathbf{W}^h\mathbf{u}_n + \mathbf{b}), \tag{1}$$

being $\mathbf{u}_n = [u_n, u_{n-1}, \ldots, u_{n-K-1}]^T$ the vector that contains the K input signals, $\mathbf{W}^h \in \mathbb{R}^{N\times K}$ the linear input coefficients, $\mathbf{b}$ the vector that represents the biases of the hidden units and $f^h(.) = (f_1^h(.)), f_2^h(.), \ldots, f_N^h(.)$ the activation functions of the hidden neurons. Then, Equation (2) presents the network outputs calculation:

$$y_n = \mathbf{W}^{out}\mathbf{x}_n^h, \tag{2}$$

where $\mathbf{W}^{out}$ is the output matrix.

The output layer (readout) adjustment is the main advantage of ELM models. This strategy is applied only once, considering the error signal [35,36]. Moreover, in dissonance with the traditional feedforward neural networks, when the intermediate activation functions are continuously differentiable, these models can choose the weights of the hidden layer randomly [36–38]. Huang et al. demonstrate that ELMs are universal approximators [39].

These structures are composed of a simple training process, mainly requiring the calculation of the parameters of a linear combiner using the Moore-Penrose generalized inverse operator, as in Equation (3) [36,37,40,41]:

$$\mathbf{W}^{out} = (\mathbf{X}_h^T\mathbf{X}_h)^{-1}\mathbf{X}_h^T\mathbf{d}, \tag{3}$$

where $\mathbf{X}_h \in \mathbb{R}^{T_s \times N}$ is the matrix composed of the intermediate layer outputs and T_s is the training sample numbers, $(\mathbf{X}_h^T\mathbf{X}_h)^{-1}\mathbf{X}_h^T$ is the pseudoinverse of $\mathbf{X}_h$ and $\mathbf{d} \in R^{T_s \times 1}$ is the vector composed of desired outputs.

2.2. Echo State Networks

Echo state networks (ESN) are recurrent neural models known by an effortless training process: the dynamical reservoir (intermediate layer) is fixed, i.e., there is no iterative adjustment. In this sense, the synaptic weights of the reservoir do not use the error function derivatives. Thus, only the output layer is effectively adapted [42]. The adaptation process applies a linear regression scheme similar to the ELM training process, considering that a linear combiner is often applied to the output layer. The neural network structure of ESN can be seen as a general case of ELM because the reservoir presents recurrent loops. Figure 2 illustrates the structure.

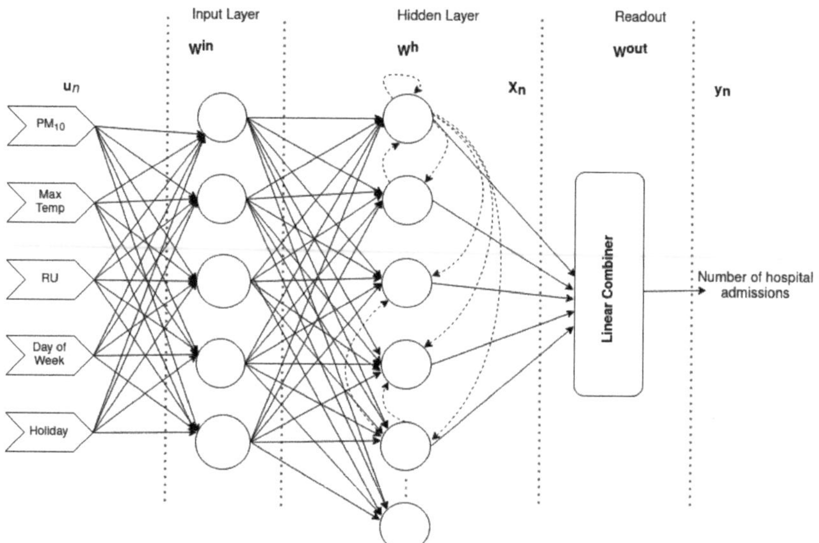

Figure 2. Echo state networks.

Figure 2 shows that the network structure is slightly similar to the ELM model presented in Figure 1, except by the additional input layer ($\mathbf{W}^{in}$), defined as a linear matrix, and feedback loops in the intermediate layer (hidden layer).

Equation (4) expresses the activation of the internal neurons. This activation represents the network states which are influenced by the previous state and the present input:

$$\mathbf{x}_{n+1} = f(\mathbf{W}^{in}\mathbf{u}_{n+1}\mathbf{W}\mathbf{x}_n), \tag{4}$$

where $f(.) = (f_1(.), f_2(.), \ldots, f_N(.))$ gives the activation functions of all neurons within the reservoir, $\mathbf{W}^{in} \in \mathbb{R}^{N \times K}$ is the input weight matrix and $\mathbf{W} \in \mathbb{R}^{N \times N}$ is the recurrent weight matrix.

The linear combinations of the reservoir signals produce the ESN outputs by (5):

$$\mathbf{y}_{n+1} = \mathbf{W}^{out}\mathbf{x}_{n+1}, \tag{5}$$

where $\mathbf{W}^{out} \in \mathbb{R}^{O \times N}$ is the output weight matrix, and O the number of outputs. The parameters of the $\mathbf{W}^{out}$ are determined by Moore-Penrose generalized inverse described in Section 2.1.

Fundamentally, the network model, besides a stable behavior, should present an internal memory that preserves the input signals history formed in the dynamical reservoir [29,35,43]. Both features are contemplated by echo state property (ESP) [29,35,43].

Jaeger et al. suggest in [29] to simplify the weight matrix W, denoting w_{ij} as 0, 0.4 and -0.4 values with probabilities 0.95, 0.025 and 0.025, respectively. On the other hand, Ozturk et al. (2006) suggest a new design for the dynamical reservoir [44] that considers eigenvalues uniformly spreading in the weight matrix. Both approaches are applied in this work.

Having described the unorganized machines in the standard forms, the following subsections describe the variations and extensions which design structures of new models also applied to the proposed problem.

2.3. Regularization Parameter

Primarily proposed by Huang et al. (2011), the regularization strategy aims to improve the model's generalization capability, inducing the solutions obtained by a parameter applied to the Mean Square Error (MSE) cost function. The parameter C is chosen from a validation set of samples, assuming $C = 2^{\lambda}$, with λ discretized in the interval $[-25, 26]$ [32]. The strategy is performed during the interactive process, where all parameters are tested, and only one is selected according to the best MSE validation, via Expression (6):

$$\mathbf{W}^{out} = \left(\frac{\mathbf{I}}{C} + \mathbf{X}_h^T \mathbf{X}_h\right)^{-1} \mathbf{X}_h^T \mathbf{d}, \tag{6}$$

being C the regularization parameter and **I** the identity matrix.

Trying to improve generalization capability given by the parameter C, Kulaif et al. (2013) developed a local search, denoted golden search, to determine better values for the parameter C. The strategy is grounded in two main concepts: significant modifications are obtained in the final solutions if any small parameter variations occur; the function given by each small interval associated with the parameter C and the validation error shall be supposedly quasi-convex [45]. This strategy is also applied in this work.

2.4. Nonlinear Output Layer

Boccato et al. (2011) proposed a variation of nonlinear output layer in ESNs, the Volterra filtering structure [46]. The main concern is to prove the linear dependence between the dynamical echo states, preserving the training process simplicity for the networks. The output signals can be computed through linear combinations of polynomial terms, as in Equation (7) [27]:

$$y_{i,n} = h^0 + \sum_{p=1}^{M} h_p^1 x_{p,n} + \sum_{p-1}^{M} \sum_{q-1}^{M} h_{p,q}^2 x_{p,n} x_{q,n} + \sum_{p-1}^{M} \sum_{q-1}^{M} \sum_{r=1}^{M} h_{p,q,r}^3 x_{p,n} x_{r,n} + \ldots, \tag{7}$$

where $x_{i,n}$ is the output of the $i-th$ neuron of the reservoir (or the $i-th$ echo state) at $n-th$ time instant, h^m the linear combiner coefficient with $m = 1, \ldots, M$, and M the polynomial expansion order.

Similar to Equation (3), the training process simplicity is preserved due to the linear dependence of the outputs regarding the filter parameters. In terms of least squares, Equation (7) guarantee the closed-form solution, allowing the Moore-Penrose inverse operation [47].

However, according to Boccato et al. (2011), the application of a Volterra filter might have as consequence the uncontrollable growth of free parameters and inputs numbers. To prevent these problems, a compression technique known as Principal Component Analysis (PCA) must be applied. Interestingly, the use of PCA is also suitable to avoid the redundancy between echo states [29,48]. In recent years, Chen et al. extended this idea to the ELMs, considering the same premises of the former work [48,49].

All parameters associated with the proposed models: the number of neurons, Volterra Filter orders, the weight values, and the number of simulations, shall be described in Section 4.

3. Case Studies

To evaluate the approach, three cities of São Paulo state, Brazil, with different characteristics, were considered: São Paulo, Campinas and Cubatão. The data set of daily PM_{10} concentration [$\mu g/m^3$], relative humidity [%], and ambient temperature [°C], were obtained on the Environmental Sanitation Technology Company website [50].

The Brazilian National Health System provides data about the daily hospital admissions due to respiratory diseases (RD). The data set considered in this study, available in [51], comprises the International Classification of Diseases 10 (ICD-10)-J00 to J99. In this work, the database was organized as a daily format and separated by the ICD-10 diagnosis.

According to the Brazilian Institute of Geography and Statistics (IBGE) [52], São Paulo City, the largest city in Brazil, has almost 12 million people (data of 2010) in 1500 km^2, which is 7398.26 inhabitants per km^2. The average climate is tropical, about 28 °C in summer and 12 °C in winter [50]. This study considers the period from January 2014 until December 2016. The total number of hospital admissions for respiratory diseases during the studied period, for São Paulo city, was 159,683 occurrences. With regards to the PM_{10} concentration, only four out of twelve air quality monitoring stations had PM_{10} data. In addition, only one station presented less than 100 days of lack of data. To deal with this problem, data from another similar station were used to replace them.

Campinas City is the third most populous city of São Paulo State, with a population of approximately 1,1 million people (data of 2010) spread in 795.7 km^2, a demographic density of 1359.6 inh/km^2 [52]. The climate is tropical with dry winter and rainy summer with an average of 37 °C during summertime. For this city, the data set considered data from January 2017 to December 2019, comprising 15,464 hospital admissions for respiratory diseases. In this case, two of three air quality monitoring stations presented PM_{10} data, however, one had no data for 2019. So, the only station with less missing data (145 days lack) was used.

Cubatão has an estimated 118,720 inhabitants with 142.8 km^2 and 831 inh/km^2 [52]. In the past, it was one of the most global polluted cities because of its large industrial park and for being surrounded by mountains, which makes the air dispersion hard. In the 1980s, the United Nations considered Cubatão the most polluted city in the world. After that, a government, industries and community effort controlled 98% of the air pollutants level in the city [53]. The current experiments considered the data from January 2017 to December 2019, a total of 802 hospital occurrences. For this city, all three air quality monitoring stations had PM_{10} available data. However, only the station with more available data was used, with 158 missing days.

A tendency to decrease hospital admissions on the weekends and holidays is a usual situation. For this reason, the day of the week and holidays were considered as two categorical variables [54]. Thus, in addition to the PM_{10} daily mean concentrations, ambient temperature (T) and relative humidity (RH), the day of the week identifications (1 for Sunday to 7 for Saturday), and a binary flag (h) to recognize if the day is a holiday, were used.

Another important feature is the lag effect of air pollution on human health [10,17,26,55]. A common practice is to consider the effect up to seven days after exposure to air pollution, where lag 0 is the effect on the same day of the exposure, and lag 7 is the effect after seven days of the exposure [54].

Table 1 presents the descriptive statistics for the target (respiratory diseases-RD) and the inputs: PM_{10} concentration, temperature and relative humidity, for each city. All these variables are differed by average, standard deviation and minimum and maximum values.

Table 1. Descriptive statistics for the variables.

City	Variable	Average	S. Deviation	Min.	Max.
São Paulo	RD	144.0	54.7	9.0	409.0
	PM_{10} [$\mu g/m^3$]	28.6	14.0	5.0	97.0
	Temperature [°C]	20.7	3.6	9.9	28.9
	Humidity [%]	48.6	16.1	15.0	93.0
Campinas	RD	16.0	6.0	3.0	37.0
	PM_{10} [$\mu g/m^3$]	21.5	11.3	3.0	84.0
	Temperature [°C]	28.5	3.9	16.6	37.0
	Humidity [%]	42.4	14.4	14.0	90.0
Cubatão	RD	1.0	1.0	0.0	8.0
	PM_{10} [$\mu g/m^3$]	37.6	17.9	11.0	148.0
	Temperature [°C]	27.1	4.3	16.0	40.3
	Humidity [%]	63.5	16.8	19.0	97.0

Note that the cities have different patterns for the target. São Paulo hospitalizations have a wide dispersion, with 9 to 409 daily hospital admissions. Campinas ranges from 3 to 37, while Cubatão, the smallest studied city, has a maximum of eight hospitalizations. It is necessary to highlight that the databases comprise only data from the public health system, not considering data from health insurance and private units.

The maximum daily PM_{10} concentration for Cubatão (148 $\mu g/m^3$) draws attention, because it is almost thrice the WHO 24-hours average limit of 50 $\mu g/m^3$ (Table 1) [56]. Despite that, the hospital admissions are very low (daily maximum of occurrences) since a significant part of the workers of Cubatão live in São Paulo, which is around 63 km far. The hospital admissions might also depend on the air pollutants dispersion pattern and the local population. São Paulo and Campinas maximum daily PM_{10} concentrations are lower than Cubatão, but they are also above the WHO limit of 50 $\mu g/m^3$ (São Paulo-maximum daily of 97 $\mu g/m^3$; Campinas-maximum daily of 84 $\mu g/m^3$) [56].

Since the data set described is large with high variability, it may contain multicollinearity or near-linear dependence among the variables. Multicollinearity occurs when two or more inputs (independent variables) are highly correlated affecting the estimate precision. [57]. To evaluate the data set, the Variance Inflation Factor (VIF) shall be used to diagnose the multicollinearity. VIF is calculated by an inflation of the regression coefficient for a independent variable, assessing its correlation to the dependent variables, and modeling the future relation between them. Then, the VIF for each j_{th} factor can be calculated as:

$$\mathrm{VIF}_j = \frac{1}{1 - R_j^2}, \tag{8}$$

where R_j^2 is the multiple determination coefficient obtained from regressing each independent variable on the others. If VIF exceeds 5, it is an indicator of multicollinearity [57].

In this work, R Studio (R version 4.1.0 (2021-05-18)—"Camp Pontanezen" Copyright (C) 2021 The R Foundation for Statistical Computing Platform: $x86_64 - w64 - mingw32/x64(64 - bit)$) was used to calculate VIF. The results are presented in Table 2, showing no multicollinearity between the inputs of each case study.

Table 2. VIF test results for multicollinearity.

VIF	Cubatão	Campinas	São Paulo
PM_{10}	1.1581	1.5779	1.6365
Relative Humidity	1.9392	2.2771	1.8703
Temperature	1.8825	1.5877	1.2105

In the next section, the proposed models are applied to the presented data, producing a fulfilled analysis of the numerical results obtained.

4. Results and Critical Analysis

The following items describe all models developed to obtain the numerical results in order to evaluate the approach's effectiveness:

- Standard single models: Three versions are developed considering the Standard Models presented in Sections 2.1 and 2.2. The Extreme Learning Machine (ELM), the Echo State Network from Jaeger et al. [29] (ESN J.) and the Echo State Network from Ozturk et al. [44] (ESN O.);
- Regularization Parameter: All standard models are extended, producing three other models through regularization parameter concepts presented in Section 2.3. The ELM with Regularization Parameter (ELM–RP), the ESN J. with Regularization Parameter (ESN J.–RP) and the ESN O. with Regularization Parameter (ESN O.–RP);
- Nonlinear Output Layers: Similarly, three more models are proposed considering the concepts in Section 2.4. The Nonlinear Output Layers strategy is applied to the three single forms creating the ELM with Volterra Filtering Structure (ELM Volt), the ESN J. with Volterra Filtering Structure (ESN J. Volt), and the ESN O. with Volterra Filtering Structure (ESN O. Volt).

The experimental procedure follows the steps summarized in Figure 3:

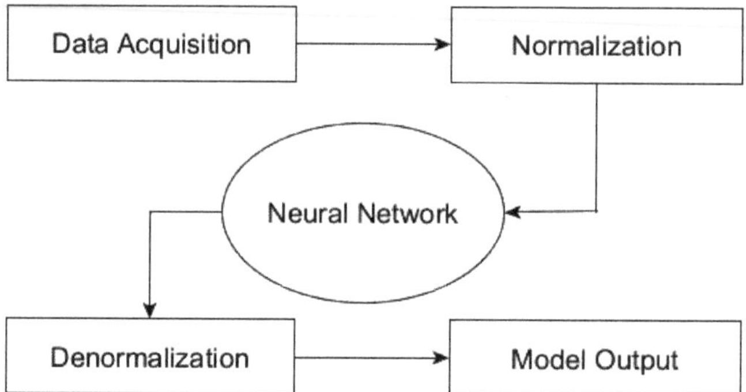

Figure 3. Neural networks appliance steps.

The process begins by collecting the data in the mentioned repositories. Before the insertion of the samples in the neural networks, a normalization procedure is performed due to the limits of the activation function saturation [58]. After the training samples are inserted in the model in order to adjust their free parameters, observing the decrease of the output error. During this process, cross-validation is performed to increase the system generalization capability.

When the training ends, the test samples are inserted in the ANN after the input normalization. The neural response is stored, the normalization is reversed and, finally, the model output is available, which allows the calculation of the models' error. In this work, all models codes were developed in the MATLAB language.

In the training step, the parameters were defined as follows:

- The number of artificial neurons in the hidden layer (or dynamic reservoir) of each model was determined considering a grid search ranging from 3 to 450 neurons;
- The weights were randomly generated in the interval $[-1; +1]$;
- The hyperbolic tangent was addressed as the activation function of the hidden layers;
- The samples were normalized in the interval $[-1; +1]$ before the neural processing;
- The models with RP strategy considered the holdout cross-validation;

- The reservoir designed by Ozturk et al. considered a spectral radius of 0.95 [44];
- The first and the third orders (Equation (7)) of the Volterra filter and the first three principal components of the PCA were considered [48]. These values were defined after empirical tests;
- Before the calculation of the errors, the original domain data was re-scaled.

This work addressed three error metrics to evaluate the solutions quality: Root Mean Square Error (RMSE), Mean Absolute Error (MAE), and Mean Absolute Percentage Error (MAPE), given by (9)–(11), respectively:

$$\text{RMSE} = \sqrt{\frac{1}{N}\sum_{n=1}^{N}(d_n - y_n)^2}, \tag{9}$$

$$\text{MAE} = \frac{1}{N}\sum_{n=1}^{N}|d_n - y_n|, \tag{10}$$

$$\text{MAPE} = \frac{1}{N}\sum_{n=1}^{N}\left|\frac{d_n - y_n}{d_n}\right| \times 100, \tag{11}$$

where d_n is the actual value, y_n is the neural model response and N is the total number of samples.

Tables 3–5 present the computational performances achieved by the nine proposed models for each lag, considering each city. The results present the number of neurons (NN) used in the best performance and the error metrics: RMSE, MAE and MAPE. However, as it can be seen in Table 3 for Cubatão, the error metrics MAPE was not considered due to the expressive number of "zeros" for the actual value (d_n). In the tables, the best results obtained for each error metric and the best model are highlighted in purple. Furthermore, the models highlighted in italic bold with stars are the models which obtained statistically similar results to the best one. This statistical test is described below.

A specific result analysis shows that ELM(RP) had the best results for the all calculated metrics for Cubatão in lag 2. Besides, ELM obtained the best results for Campinas, considering the error metrics RMSE and MAPE in lag 3, but for MAE, ELM(RP) achieved the best results in lag 0. For São Paulo, ELM obtained the smallest error values for different lags: RMSE in lag 2 and MAE in lag 1. Finally, ELM(RP) presented the smallest error metric MAPE in lag3.

Note that the best results obtained by the models sometimes were not replicated for all error metrics. This behavior was evident for São Paulo and Campinas, since the best lag and the best model were not always the same. Similar behavior can be observed in [17,59].

The pairwise Wilcoxon test was applied to evaluate if the results are statistically different considering the RMSE with 30 independent simulations [60]. In Tables 3 and 4, the models highlighted in bold with star tag achieved a p-value higher than 0.05, which means that there is no statistical difference between their results and the best one. For this reason, these models can be considered similar, in terms of performance, to the models that obtained the best results. For Campinas, the standard ELM and all ESNs presented equivalent performances, despite the numerical values being contrasts. For Cubatão, ELM and ELM(RP) results were also similar. At long last, for São Paulo the test did not show any statistical similarity among the models.

Figures 4–6 show the boxplot graphic regarding the RMSE values for each city and the lag associated with the best result.

Table 3. Results for Cubatão (Number of neurons-NN, RMSE and MAE for each model and lag).

		LAG 0			LAG 1	
Model	**NN**	**RMSE**	**MAE**	**NN**	**RMSE**	**MAE**
ELM	250	1.5630	1.1857	300	1.4760	1.1357
ELM(RP)	350	1.5330	1.1643	320	1.4808	1.1357
ELM Volt	350	2.4202	2.0429	450	2.0942	1.7714
ESN J.	320	1.6058	1.2643	450	1.5789	1.1929
ESN J.(RP)	450	1.5879	1.2357	450	1.6345	1.2571
ESN J.Volt	70	2.3815	1.9571	35	1.8323	1.4571
ESN O.	30	1.6257	1.2143	35	1.4904	1.1429
ESN O.(RP)	200	1.6797	1.3357	450	1.7587	1.3929
ESN O.Volt	10	2.7877	2.2286	380	2.7255	2.2429
		LAG 2			**LAG 3**	
Model	**NN**	**RMSE**	**MAE**	**NN**	**RMSE**	**MAE**
*ELM**	420	1.4417	1.1000	350	1.4663	1.1500
ELM(RP)	320	1.4343	1.0714	380	1.4467	1.1286
ELM Volt	420	2.0107	1.6929	100	1.9928	1.6571
ESN J.	300	1.4344	1.1000	380	1.4417	1.1214
ESN J.(RP)	450	1.5142	1.1786	420	1.4541	1.1000
ESN J.Volt	30	2.1827	1.7071	35	2.3664	1.9143
ESN O.	350	1.4760	1.1214	35	1.4880	1.1286
ESN O.(RP)	420	1.6058	1.2357	170	1.5330	1.2071
ESN O.Volt	300	2.4900	2.0429	350	2.6227	2.2143
		LAG 4			**LAG 5**	
Model	**NN**	**RMSE**	**MAE**	**NN**	**RMSE**	**MAE**
ELM	250	1.5071	1.1714	420	1.5353	1.1571
ELM(RP)	250	1.5024	1.1786	400	1.5306	1.1643
ELM Volt	280	2.3890	1.9929	380	2.5114	2.1500
ESN J.	420	1.5142	1.1929	350	1.5561	1.1714
ESN J.(RP)	450	1.5189	1.1643	350	1.5561	1.2214
ESN J.Volt	70	2.2960	1.8714	35	2.5746	2.1500
ESN O.	380	1.6013	1.2786	70	1.5766	1.1571
ESN O.(RP)	380	1.5561	1.2071	250	1.6191	1.2500
ESN O.Volt	50	2.3770	2.0500	30	2.4275	1.9643
		LAG 6			**LAG 7**	
Model	**NN**	**RMSE**	**MAE**	**NN**	**RMSE**	**MAE**
ELM	250	1.4516	1.1500	350	1.5811	1.2286
ELM(RP)	320	1.5515	1.1929	200	1.5376	1.2000
ELM Volt	450	2.4640	2.1429	450	2.4928	2.1643
ESN J.	170	1.6903	1.3429	450	1.5306	1.2214
ESN J.(RP)	420	1.5584	1.2429	420	1.5834	1.2571
ESN J.Volt	40	2.3634	2.0429	70	2.2409	1.8786
ESN O.	40	1.5142	1.1714	35	1.5811	1.2429
ESN O.(RP)	250	1.6410	1.3357	200	1.5969	1.3071
ESN O.Volt	300	2.8322	2.2143	50	2.6390	2.1071

Table 4. Results for Campinas (Number of neurons-NN, RMSE, MAE and MAPE for each model and lag).

	LAG 0				LAG 1			
Model	**NN**	**RMSE**	**MAE**	**MAPE %**	**NN**	**RMSE**	**MAE**	**MAPE %**
ELM	25	6.9017	5.6479	40.2496	3	5.5462	4.4507	36.9895
ELM(RP)	3	5.1094	3.9648	32.7044	25	7.0751	5.7324	40.8537
ELM Volt	25	7.0206	5.2676	37.2186	10	5.3910	4.1338	34.4946
ESN J.	35	6.9394	5.6620	40.0763	50	6.6619	5.4085	41.5375
ESN J.(RP)	3	6.4306	5.0563	50.5484	3	6.4731	5.0845	51.6402
ESN J.Volt	380	6.3540	4.8803	46.8265	3	5.2393	3.9577	33.7701
ESN O.	15	5.8713	4.6127	39.4574	30	6.4878	5.2465	40.5271
ESN O.(RP)	3	6.2473	4.9014	49.1485	7	6.6327	5.2324	52.7208
ESN O.Volt	3	5.6438	4.3873	33.9491	3	5.8743	4.6127	34.6702
	LAG 2				**LAG 3**			
Model	**NN**	**RMSE**	**MAE**	**MAPE %**	**NN**	**RMSE**	**MAE**	**MAPE %**
ELM	15	6.4464	5.1972	38.4853	3	5.0644	4.0282	31.9037
ELM(RP)	3	5.4721	4.3662	33.0808	25	6.6072	5.1761	39.8549
ELM Volt	25	5.9517	4.4507	38.5412	170	6.2020	4.6268	41.3376
*ESN J.**	30	6.4114	5.0915	41.0724	70	6.1260	4.7113	38.2705
*ESN J.(RP)**	3	6.2258	4.7887	48.3773	10	6.2196	4.9296	49.2648
*ESN J.Volt**	3	5.7684	4.4859	35.2526	30	5.7101	4.2817	38.4659
*ESN O.**	25	6.2557	4.9085	39.6942	100	6.2905	4.8310	37.4986
*ESN O.(RP)**	10	6.0630	4.6761	46.9116	5	6.3207	4.9577	49.5265
*ESN O.Volt**	7	5.7648	4.3592	40.2773	450	6.1633	4.8732	42.5624
	LAG 4				**LAG 5**			
Model	**NN**	**RMSE**	**MAE**	**MAPE %**	**NN**	**RMSE**	**MAE**	**MAPE %**
ELM	3	5.7403	4.4648	32.2381	3	5.2928	4.0845	33.6603
ELM(RP)	20	6.6003	5.0986	37.0427	3	5.3200	4.1056	34.3353
ELM Volt	25	5.7885	4.4085	34.7377	30	5.9511	4.7394	37.2746
ESN J.	70	6.2054	4.7746	36.7789	35	6.1070	4.7042	36.9522
ESN J.(RP)	30	6.3184	4.9859	48.8737	3	6.1254	4.7183	46.9130
ESN J.Volt	35	5.2682	4.1056	33.8667	3	5.8934	4.6479	35.1359
ESN O.	120	6.2776	4.8310	36.6999	200	6.3745	4.9859	38.5786
ESN O.(RP)	7	5.9935	4.6831	46.2167	10	6.2377	4.8239	47.9457
ESN O.Volt	3	5.9570	4.3028	36.7693	3	5.4521	4.3732	34.0759
	LAG 6				**LAG 7**			
Model	**NN**	**RMSE**	**MAE**	**MAPE %**	**NN**	**RMSE**	**MAE**	**MAPE %**
ELM	3	5.3068	4.1972	36.8548	35	7.2452	5.6761	42.0235
ELM(RP)	3	5.2474	4.0704	36.8444	35	7.2384	5.6761	42.0908
ELM Volt	25	5.3253	4.2465	35.6237	70	5.1273	4.0930	38.2203
ESN J.	30	6.2360	4.9930	39.2345	50	6.6961	5.1620	41.5200
ESN J.(RP)	5	5.9741	4.6901	44.9676	3	5.8928	4.6549	45.2283
ESN J.Volt	3	5.7873	4.4930	35.0669	3	6.0082	4.8169	34.6062
ESN O.	35	6.3987	5.0563	39.7110	50	6.4579	4.8732	39.7276
ESN O.(RP)	5	5.9487	4.5000	44.5221	5	5.9871	4.6620	45.6153
ESN O.Volt	3	5.6519	4.4014	35.8808	3	5.6687	4.2746	37.4182

Table 5. Results for São Paulo (Number of neurons-NN, RMSE, MAE and MAPE for each model and lag).

	LAG 0				LAG 1			
Model	**NN**	**RMSE**	**MAE**	**MAPE %**	**NN**	**RMSE**	**MAE**	**MAPE %**
ELM	25	61.1156	51.141	43.4189	3	39.7697	30.7821	36.1940
ELM(RP)	15	55.2425	46.4231	41.1963	20	59.1541	48.3205	40.3268
ELM Volt	10	60.8374	48.2179	48.3650	20	72.4500	58.7179	57.4171
ESN J.	420	62.4953	51.3910	42.4775	100	62.1230	50.4167	42.4388
ESN J.(RP)	450	67.7300	55.0449	70.2352	450	67.6519	55.0769	69.8433
ESN J.Volt	400	66.9744	54.0705	64.0419	3	51.2727	40.5833	43.2995
ESN O.	50	58.7160	48.3141	39.7660	50	58.1013	45.9679	40.5099
ESN O.(RP)	380	69.2728	56.6154	71.5779	350	68.8875	55.8526	71.2675
ESN O.Volt	3	57.8689	46.5064	45.7923	3	59.6593	45.7628	50.0376
	LAG 2				LAG 3			
Model	**NN**	**RMSE**	**MAE**	**MAPE %**	**NN**	**RMSE**	**MAE**	**MAPE %**
ELM	3	39.3745	31.2628	36.9582	25	59.9251	48.2308	45.2785
ELM(RP)	20	60.6640	48.2115	41.4840	3	43.9040	35.1026	35.9396
ELM Volt	25	83.6339	69.4423	68.6818	20	71.2603	55.4679	57.0193
ESN J.	70	60.7151	49.1154	42.8314	100	64.7131	52.4423	48.2804
ESN J.(RP)	450	65.7486	53.6795	68.8934	400	65.3756	53.0064	69.0983
ESN J.Volt	3	59.2428	43.6090	46.8035	3	58.6061	44.1795	43.8522
ESN O.	35	58.6986	47.4103	40.7649	15	50.4066	39.3910	41.1334
ESN O.(RP)	420	68.1032	55.3974	70.3439	400	67.3167	54.4936	70.1148
ESN O.Volt	3	55.5113	43.2628	45.1094	5	58.2715	45.4872	46.8127
	LAG 4				LAG 5			
Model	**NN**	**RMSE**	**MAE**	**MAPE %**	**NN**	**RMSE**	**MAE**	**MAPE %**
ELM	15	53.6898	41.4167	43.9006	3	46.4592	37.1474	39.0748
ELM(RP)	3	45.6739	35.9231	42.4408	3	43.6788	33.3077	38.8188
ELM Volt	7	64.8803	49.4487	56.4511	7	49.8634	39.3974	47.2036
ESN J.	450	64.3933	51.3782	47.0148	100	63.9561	51.8141	47.3902
ESN J.(RP)	450	65.9862	52.4808	69.1144	380	66.2419	52.6603	68.9928
ESN J.Volt	380	72.9358	59.3526	64.7092	7	68.9339	55.0513	56.2654
ESN O.	15	55.3579	45.3654	44.0190	25	58.9842	45.2051	44.8766
ESN O.(RP)	380	69.3555	55.5513	72.2436	380	68.0724	54.3590	70.6102
ESN O.Volt	3	50.5135	40.5833	40.8100	3	53.3191	41.0000	48.6203
	LAG 6				LAG 7			
Model	**NN**	**RMSE**	**MAE**	**MAPE %**	**NN**	**RMSE**	**MAE**	**MAPE %**
ELM	20	54.3250	43.9744	43.4307	25	55.3315	43.9744	39.5735
ELM(RP)	3	47.3653	36.9551	39.9839	3	44.0574	35.1923	37.5251
ELM Volt	20	64.5582	46.3654	49.4889	10	75.8567	63.5641	63.1259
ESN J.	150	59.9074	49.0641	45.4045	70	58.9236	47.3141	40.8857
ESN J.(RP)	380	66.0817	52.5705	68.6722	380	65.4964	52.9615	68.4640
ESN J.Volt	3	60.9871	45.6859	47.4705	3	62.1758	47.5641	45.0664
ESN O.	20	54.1340	43.2756	43.7886	35	51.7540	42.4103	39.8727
ESN O.(RP)	420	68.5719	54.7949	70.9594	320	67.4013	54.7564	70.0246
ESN O.Volt	3	57.0886	43.9487	46.1440	3	53.9567	43.3974	44.4320

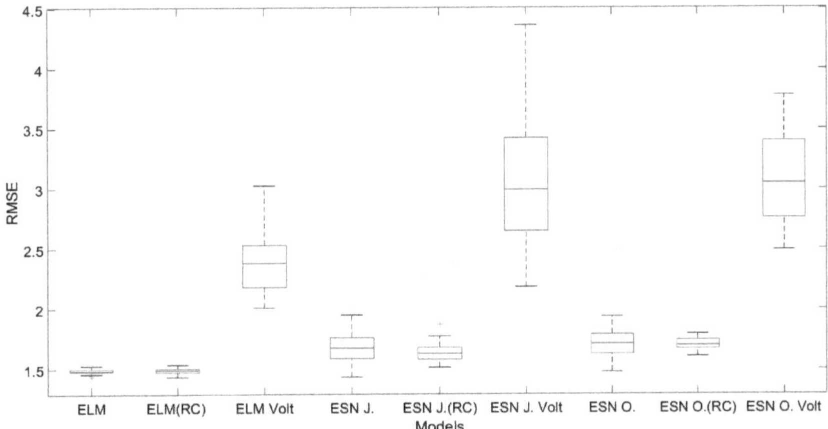

Figure 4. Boxplot graphic regarding the RMSE values for Cubatão-Lag 2.

Considering Cubatão, observe that the smallest dispersion was obtained by ELM (RP) model, which also presented the smallest average value, corroborating the observation from Table 3. The inclusion of the Volterra filter increases the dispersion and the average values for all standard models, representing a significant degree of deterioration in performance.

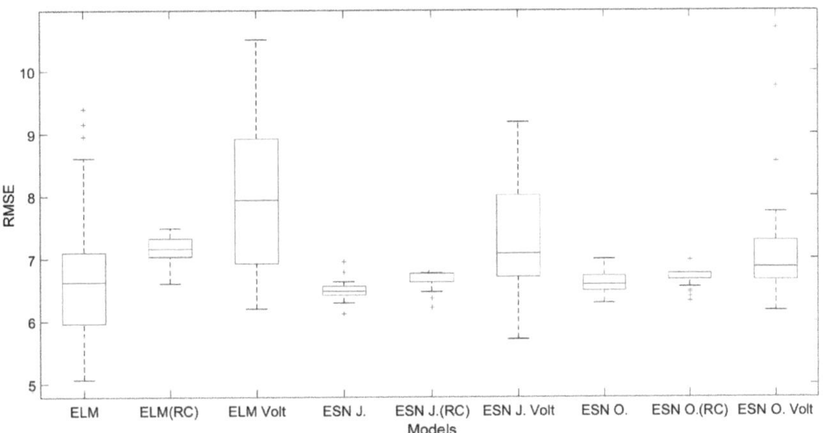

Figure 5. Boxplot graphic regarding the RMSE values for Campinas-Lag 3.

In Campinas' case, only the ELM performances will be considered in this specific analysis since all ESN obtained similar results according to the Wilcoxon test. However, Figure 5 illustrates all models to avoid any curiosity. The RP inclusion decreases the dispersion, while the Volterra filter showed an opposite behavior. Despite that, the best performance in terms of best results regarding 30 simulations was favorable to the use of Volterra filter instead of the RP (note the bottom value in the boxplot). Since the generation of the neurons' weights were random, the algorithms must run at least 30 times, and this fact directly implied a long tail for the boxplots, as can be seen in the Volterra models. Moreover, the best result obtained by ELM does not mean the best performance in terms of dispersion.

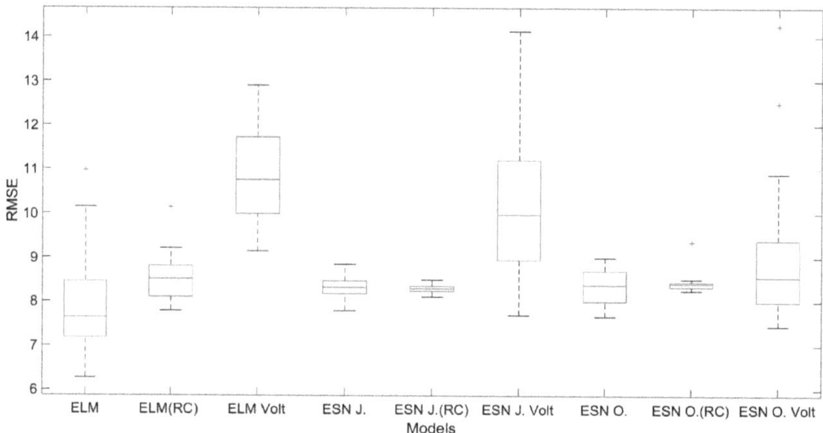

Figure 6. Boxplot graphic regarding the RMSE values for São Paulo-Lag 2.

For São Paulo, the general behavior of the standard models was similar to Campinas. The standard ELM achieved better general errors, even when considered the median value. The inclusion of the RP reduced the dispersion, but it decreased the probability of obtaining better results for the error metrics. On the other hand, the Volterra filter showed a worse performance in terms of dispersion. However, despite the ELM best results for error metric, the model presented a bad dispersion, including an outlier.

Table 6 presents a ranking of best error metric results considering all neural models in ascending order of development. Note that the draws regarding the winners mean that there was no statistical difference between the models. The last column represents the final ranking considering the three cities' results.

Table 6. Ranking of the models' performance.

	Cubatão (lag 2)		Campinas(lag 3)			São Paulo (lag 2)				
Model	RMSE	MAE	RMSE	MAE	MAPE %	RMSE	MAE	MAPE %	Mean	Rank
ELM	1	1	1	1	1	1	1	1	1	1st
ELM(RP)	1	1	9	9	8	5	5	3	5.1	8th
ELM Volt	7	7	8	8	9	9	9	7	8.0	9th
ESN J.	3	3	1	1	1	6	6	4	3.12	3rd
ESN J.(RP)	6	6	1	1	1	7	7	8	4.6	6th
ESN J.Volt	8	8	1	1	1	4	3	6	4.0	5th
ESN O.	4	4	1	1	1	3	4	2	2.5	2nd
ESN O.(RP)	5	5	1	1	1	8	8	9	4.8	7th
ESN O.Volt	9	9	1	1	1	2	2	5	3.8	4th

The standard ELM was the best estimator in all cases as regards the error metric results, but for Cubatão, the results obtained by ELM(RP) were the same. The second and third positions show ESN O. and ESN J. models, respectively. However, despite the main contribution of RP is to increase the models' generalization capability, its use reduced the dispersion of the results, i.e., the models' predictability increased, except for Cubatão. Moreover, the ELM(RP) ranking position was deteriorated by the Campinas results, since all ESNs presented the same statistical performance. Dismissing these aspects, the model could be the second best.

Although the inclusion of the Volterra filter did not improve the performances, the idea of its application was to capture nonlinear patterns among the signals from the hidden layer. Despite the literature presents good performances for this method in correlated tasks [33], its use is not recommended in this case. Similarly, the inclusion of the reservoir designed by Jaeger or Ozturk et al. is not adequate to the problem.

Regarding the number of neurons in the hidden layers (dynamic reservoir), one can see miscellaneous neurons, with a high degree of variation. For Cubatão, the pattern noted was the models used hundreds of neurons in most cases. Interestingly, ESN J. and ESN O. often addressed up to 70 neurons. Moreover, it can be seen in Campinas' case, that the RP models used up to 35 neurons in all cases. Considering São Paulo, the ELM versions tended to use less than 25 neurons, similar to ESN J. Volt, ESN O. and ESN O. Volt. The others models addressed hundreds of neurons. This is a strong indication that a sweep in the neuron amount is needed because a clear pattern regarding this parameter was not found. Even considering the results of the models that presented a p-value large than 0.05, the number of neurons was variable.

In summary, the unorganized machines are particular cases of classic neural models, which the hidden weights are not adjusted. On one hand, the user may lose part of the approximation capability due to this characteristic; on the other hand, there are gains in terms of training effort and stability for the output values during the training, avoiding discrepancies. An important aspect is that these methodologies can be outperformed, depending on the problem. Regarding the use of RP or Volterra filter, the literature indicates that these strategies may increase the mapping capability of the neural models. However, this work showed that in specific cases these approaches did not present efficiency.

Figures 7–9 present the best evolution of the output response in comparison to the actual values.

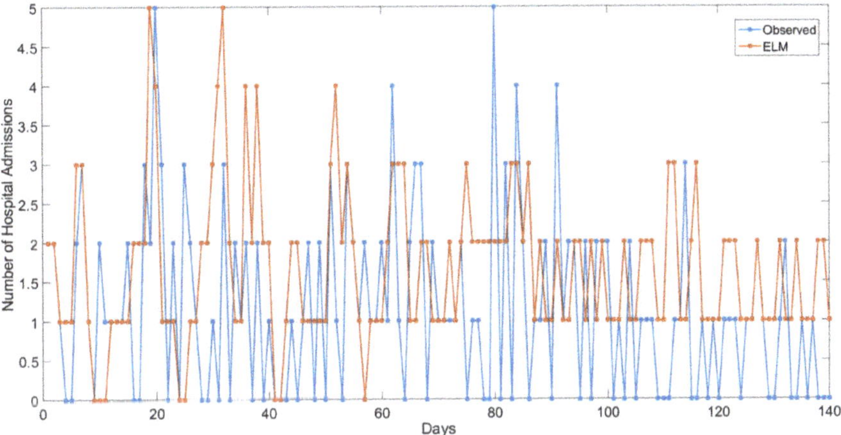

Figure 7. The number of hospital admissions by day of the test set for ELM lag 2-Cubatão (observed versus estimated values).

Figure 7 shows that the prediction task seems to be more difficult when the output has a small range and many "zero" observations. In this case, as the overestimation was small, given that the observed values are zero, it did not interfere in hospital management. Otherwise, in Figure 8, since there were no "zero" observations, the ELM estimations could be considered a suitable performance, except in abrupt cases.

Finally, in Figure 9, ELM reached the smallest RMSE, but comparing with the observed data, it was more difficult to predict the abrupt decrease of hospital admissions occurred around day 70. On the other hand, the ELM(RP) could follow this tendency, but it over and underestimated the number of hospital admissions in many cases. These behaviors are directly related to the number of neurons used by each model since a reduction in this number limited the model approximation capability.

Regarding the best error metric to be used, RMSE seems to be a good strategy, since the error metric was reduced during the neural models training (adjustment) [17,18,61].

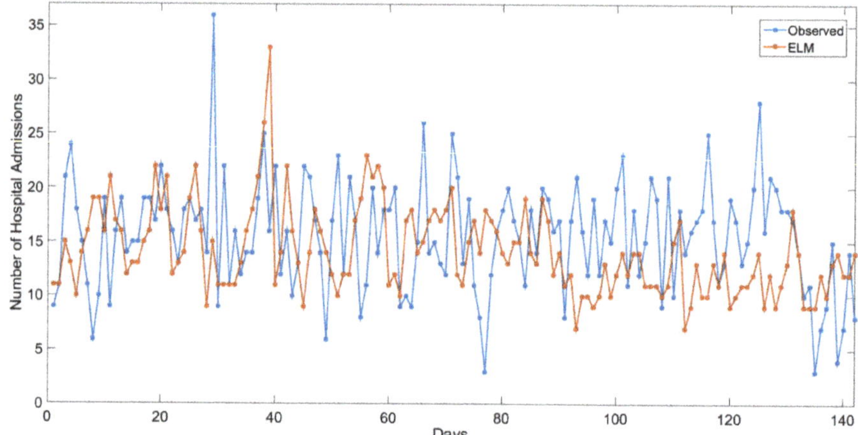

Figure 8. The number of hospital admissions by day of the test set for ELM lag 3-Campinas (observed versus estimated values).

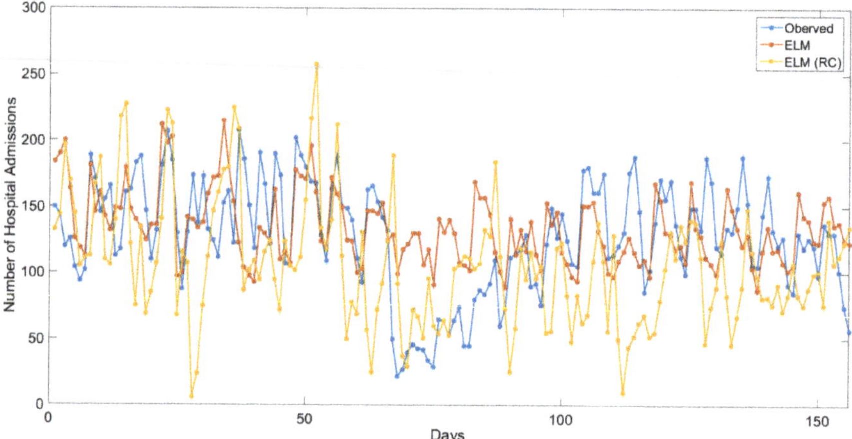

Figure 9. The number of hospital admissions by day of the test set for ELM lag 2-São Paulo (observed versus estimated values).

Table 7 presents a summary of some notable studies showing the association between air pollutants concentration, morbidity (Hospital Admissions or Hospital Emergency) and mortality. This brief description relates the authors, geographic area, considered inputs and predicted variables, the applied methods, metrics, time base, and the best MAPE and RMSE observed for each study. Although these studies present suitable estimations and relevant contributions, they proposed different models, and applied to diverse worldwide places, using specific inputs to predict health effects. For this reason, a comparative analysis of these studies' performances is unfair, as Katri and Tamil [62] previously observed. However, some important aspects can be highlighted.

Two studies [62,63] did not use MAPE or RMSE as error metrics. Khatri and Tamil [62] aimed to compare the performance for peak and non-peak class prediction. The authors used percentage difference in this study and applied MLP, without any consideration about other methods' performance. Shakerkhatibi et al. [64] used other metrics (Delong Method) to compare the predictions using MLP and Conditional Logistic Regression.

Table 7. Summary of studies presenting air pollutant's associations with morbidity and mortality using ANN.

Authors (Year)	Geographic Area of Study	Inputs	Predicted Variable	Methods	Used Metrics	Time Base	Best MAPE	Best RMSE
Kassomenos et al. (2011) [24]	Athens	T, RH, WD, SO_2, black smoke CO, NO_2, NO, O_3	HA for Cardiorespiratory diseases	MLP, GLM	RMSE	daily	NA	0.8950
Moustris et al. (2012) [63]	Athens	T, RH, WS, solar radiation SO_2, PM_{10}, CO, O_3, NO_2 (age subgroups 0–4 years, 5–14 years, 0–14 years)	HA for Asthma	MLP (TLRN)	MBE, RMSE, R^2, IA	daily	NA	3.2
Cengiz and Terzi (2012) [65]	Afyon, Turkey	SO_2, PM_{10}	HA and symptoms (cough, exertional, dyspnea, expectoration) for COPD	MLP, RBF, GLM, GAM	RMSE e MAPE	weekly	4.54	2.38
Shakerkhatibi et al. (2015) [64]	Tabriz, Iran	T. RH, NO, NO_2, NO_X, SO_2, CO, PM_{10}, O_3 (age and gender subgroups)	HA for respiratory and cardiovascular diseases	MLP, CLR	AUC, sensitivity, Specificity and Accuracy (%)	daily	NA	NA
Khatri and Tamil (2017) [62]	Dallas County, Texas, USA	T, RH, WS, CO, O_3, SO_2. NO_2, $PM_{2.5}$	HE for respiratory diseases	MLP	% difference	daily	NA	NA
Tadano et al. (2016) [26]	Campinas city, São Paulo state, Brazil	T, RH, PM_{10}	HA for respiratory diseases	MLP, ESN, ELM	MSE/MAPE	daily	31.2	5.98
Polezer et al. (2018) [10]	Curitiba, Paraná, Brazil	T, RH, $PM_{2.5}$	HA for respiratory diseases	MLP, ESN, ELM	MSE/MAPE	daily	29.87	7.37
Araujo et al. (2020) [17]	Campinas and São Paulo cities, Brazil	T, RH, PM_{10}	HA for respiratory diseases	MLP, GLM, ELM, ESN, RBF, Ensemble	MSE, MAE, MAPE	daily	24.87	3.04
Zhou, Li and Wang (2018) [66]	Hangzhou, Southern part of the Yangtze River Delta, China	T, PM_{10}, $PM_{2.5}$, NO_2, SO_2	Respiratory disease cases	MLP, GAM	AIC, MSE	daily	NA	2.17
Kachba et al. (2020) [18]	São Paulo city, Brazil	CO, NO_x, O_3, SO_2, PM	HA and mortality for respiratory diseases	MLP, ELM, ESN	MSE, MAE, MAPE	monthly	34.53	160.26

WD-Wind Direction; WS-Wind Speed; SO_2-Sulphur Dioxide; CO-Carbon Monoxide; NO_2-Nitrogen Dioxide; NO-Nitrogen Monoxide; O_3-Ozone; NO_x-Nitrogen Oxides; HE-Hospital Emergency; GAM-Generalized Additive Models; TLRN-Time Lagged Recurrent Networks; RBF-Radial Basis Function Network; CLR-Conditional Logistic Regression Modeling; MBE- Mean Bias Error; IA-Index of Agreement; AUC-area under curve.

Considering the variety of applied methods (Table 7), and emphasizing the use of MLP, the performance comparison between ANN and regression models has proved the ANN superior performance. Inspired by these all aspects, the paper's authors believe that this present work, which explores the ELM and ESN models with variations from the RP and the Volterra filter to estimate hospital admissions due to respiratory diseases caused by air pollutants concentration, is a relevant contribution. However, given the harmful effects of PM on human health, and comparing the considered input variables used in the other studies, this work has some limitations, such as the use of only one air pollutant (PM_{10}) and the lack of comparison with a statistical regression modeling.

5. Conclusions

This work predicted the hospital admissions due to respiratory diseases caused by the particulate matter PM_{10} concentrations using the extreme learning machines (ELM) and the echo state networks (ESN) in the standard forms and applying the variations from the regularization parameter (RP) and the Volterra filter. The estimates considered daily PM_{10} concentration, relative humidity, ambient temperature as inputs and predicted the daily hospital admissions for respiratory diseases.

Numerical results indicated the superior performance of the standard models, pointing to ELM as the best predictor for most scenarios. However, regarding Campinas city and the RMSE error metric, a statistical test demonstrated that ESN models were statistically similar when compared to the best one. Besides, a graphic analysis showed that the models with the inclusion of RP strategy presented a reduced dispersion, considering the abrupt variations in hospital admissions, while the Volterra filter showed an opposite behavior, indicating that its application was not suitable for this specific problem. Finally, completing the critical analysis, a ranking of performances classified the models regarding the error metrics for each city. This ranking rewarded the models with statistical similarity rather than models with good dispersion, highlighting the standard models in the first positions.

The application of Unorganized Machines to three different cities was essential to evaluate their good performance in predicting air pollution impacts on human health. An additional graphic analysis of the output response in comparison to the actual values, for the best models, evidenced the good performance of the neural networks to estimate the hospital admissions. This contribution may help governmental bodies and policymakers on the management of hospital planning, mainly during air pollution unfavorable climate periods. Moreover, the good performance of the models confirms the link between all input variables and the output values, verifying that the particulate matter, temperature and relative humidity are fundamental to obtain a good estimation.

A limitation of this study is the lack of large data sets that could bring more uniform performances between the studied cities. As a consequence of the lack of monitoring data, other pollutants variations such as $PM_{2.5}$ cannot be studied.

Considering the continental dimension of Brazil and the characteristics of the different region's climates, it would be paramount to study all regions (states), a hard task due to the lack of monitoring all over the country. Further works shall consider hybrid modeling or ensembles, the use of deseasonalization techniques, and the appliance of other artificial neural networks. Since the ELM is admittedly susceptible to the neurons number changes in the hidden layer and the ESN model is considered robust in this regard, a comparison study should be conducted pointing to the training time required between these models.

Author Contributions: Conceptualization, Y.d.S.T., E.T.B., L.C., and H.V.S.; methodology, H.V.S. and Y.d.S.T.; software, H.V.S., T.S.P., E.P., and T.A.A.; formal analysis, Y.d.S.T., E.T.B., and L.C.; investigation, H.V.S., E.P., T.A.A., and Y.d.S.T.; resources, H.V.S.; data curation, E.T.B. and L.C.; writing—original draft preparation, E.T.B., L.C., and H.V.S.; writing—review and editing, Y.d.S.T. and C.M.L.U.; visualization, H.V.S. and T.A.A.; supervision, Y.d.S.T. and H.V.S.; project administration, H.V.S.; funding acquisition, H.V.S. All authors have read and agreed to the published version of the manuscript.

Funding: This research was funded by the National Council for Scientific and Technological Development (CNPq), grant number 405580/2018-5, and the APC was funded by DIRPPG/UTFPR/PG.

Acknowledgments: The authors thank the Brazilian agencies Coordination for the Improvement of Higher Education Personnel (CAPES)-Financing Code 001, Brazilian National Council for Scientific and Technological Development (CNPq), processes number 40558/2018-5, 315298/2020-0, and Araucaria Foundation, process number 51497, and Federal University of Technology-Parana (UTFPR) for their financial support.

Conflicts of Interest: The authors declare no conflict of interest.

References

1. WHO-World Health Organization. *Ambient Air Pollution: Health Impacts*; WHO: Geneva, Switzerland, 2018.
2. Lelieveld, J.; Evans, J.S.; Fnais, M.; Giannadaki, D.; Pozzer, A. The contribution of outdoor air pollution sources to premature mortality on a global scale. *Nature* **2015**, *525*, 367. [CrossRef] [PubMed]
3. Manisalidis, I.; Stavropoulou, E.; Stavropoulos, A.; Bezirtzoglou, E. Environmental and health impacts of air pollution: A review. *Front. Public Health* **2020**, *8*, 14. [CrossRef] [PubMed]
4. Li, X.; Liu, X. Effects of PM2.5 on chronic airway diesases: A review of research progress. *Atmosphere* **2021**, *12*, 1068. [CrossRef]
5. Ab Manan, N.; Aizuddin, A.N.; Hod, R. Effect of air pollution and hospital admission: A systematic review. *Ann. Glob. Health* **2018**, *84*, 670. [CrossRef] [PubMed]
6. Grigorieva, E.; Lukyanets, A. Combined effect of hot weather and outdoor air pollution on respiratory health: Literature review. *Atmosphere* **2021**, *12*, 790. [CrossRef]
7. Morrissey, K.; Chung, I.; Morse, A.; Parthasarath, S.; Roebuck, M.M.; Tan, M.P.; Wood, A.; Wong, P.F.; Forstick, S.P. The effects of air quality on hospital admissions for chronic respiratory diseases in Petaling Jaya, Malaysia, 2013–2015. *Atmosphere* **2021**, *12*, 1060. [CrossRef]
8. Yitshak-Sade, M.; Nethery, R.; Schwartz, J.D.; Mealli, F.; Dominici, F.; Di, Q.; Awad, Y.A.; Ifergane, G.; Zanobetti, A. PM2.5 and hospital admissions among Medicare enrollees with chronic debilitating brain disorders. *Sci. Total Environ.* **2021**, *755*, 142524. [CrossRef]
9. Anderson, J.O.; Thundiyil, J.G.; Stolbach, A. Clearing the air: A review of the effects of particulate matter air pollution on human health. *J. Med. Toxicol.* **2012**, *8*, 166–175. [CrossRef]
10. Polezer, G.; Tadano, Y.S.; Siqueira, H.V.; Godoi, A.F.; Yamamoto, C.I.; de André, P.A.; Pauliquevis, T.; de Fatima Andrade, M.; Oliveira, A.; Saldiva, P.H.; et al. Assessing the impact of PM 2.5 on respiratory disease using artificial neural networks. *Environ. Pollut.* **2018**, *235*, 394–403. [CrossRef]
11. Ardiles, L.G.; Tadano, Y.S.; Costa, S.; Urbina, V.; Capucim, M.N.; da Silva, I.; Braga, A.; Martins, J.A.; Martins, L.D. Negative binomial regression model for analysis of the relationship between hospitalization and air pollution. *Atmos. Pollut. Res.* **2018**, *9*, 333–341. [CrossRef]
12. McCullagh, P.; Nelder, J.A. *Generalized Linear Models*; Routledge: London, UK, 2019.
13. Belotti, J.T.; Castanho, D.S.; Araujo, L.N.; da Silva, L.V.; Alves, T.A.; Tadano, Y.S.; Stevan, S.L., Jr.; Correa, F.C.; Siqueira, H.V. Air Pollution Epidemiology: A Simplified Generalized Linear Model Approach Optimized by Bio-Inspired Metaheuristics. *Environ. Res.* **2020**, *191*, 110106. [CrossRef] [PubMed]
14. Cromar, K.; Galdson, L.; Palomera, M.J.; Perlmutt, L. Development of a health-based index to indentify the association between air pollution and health effects in Mexico City. *Atmosphere* **2021**, *12*, 372. [CrossRef]
15. Ravindra, K.; Rattan, P.; Mor, S.; Aggarwal, A.N. Generalized additive models: Building evidence of air pollution, climate change and human health. *Environ. Int.* **2019**, *132*, 104987. [CrossRef] [PubMed]
16. Zhou, H.; Geng, H.; Dong, C.; Bai, T. The short-term harvesting effects of ambient particulate matter on mortality in Taiyuan elderly residents: A time-series analysis with a generalized additive distributed lag model. *Ecotoxicol. Environ. Saf.* **2021**, *207*, 111235. [CrossRef]
17. Araujo, L.N.; Belotti, J.T.; Antonini Alves, T.; de Souza Tadano, Y.; Siqueira, H. Ensemble method based on Artificial Neural Networks to estimate air pollution health risks. *Environ. Model. Softw.* **2020**, *123*, 104567. [CrossRef]
18. Kachba, Y.; Chiroli, D.M.d.G.; Belotti, J.T.; Antonini Alves, T.; de Souza Tadano, Y.; Siqueira, H. Artificial Neural Networks to Estimate the Influence of Vehicular Emission Variables on Morbidity and Mortality in the Largest Metropolis in South America. *Sustainability* **2020**, *12*, 2621. [CrossRef]
19. Cabaneros, S.M.; Calautit, J.K.; Hughes, B.R. A review of artificial neural network models for ambient air pollution prediction. *Environ. Model. Softw.* **2019**, *119*, 285–304. [CrossRef]
20. de Mattos Neto, P.S.; Madeiro, F.; Ferreira, T.A.; Cavalcanti, G.D. Hybrid intelligent system for air quality forecasting using phase adjustment. *Eng. Appl. Artif. Intell.* **2014**, *32*, 185–191. [CrossRef]
21. Feng, R.; Zheng, H.J.; Gao, H.; Zhang, A.R.; Huang, C.; Zhang, J.X.; Luo, K.; Fan, J.R. Recurrent Neural Network and random forest for analysis and accurate forecast of atmospheric pollutants: A case study in Hangzhou, China. *J. Clean. Prod.* **2019**, *231*, 1005–1015. [CrossRef]

22. Neto, P.S.D.M.; Firmino, P.R.A.; Siqueira, H.; Tadano, Y.D.S.; Alves, T.A.; De Oliveira, J.F.L.; Marinho, M.H.D.N.; Madeiro, F. Neural-Based Ensembles for Particulate Matter Forecasting. *IEEE Access.* **2021**, *9*, 14470–14490. [CrossRef]
23. Wang, Q.; Liu, Y.; Pan, X. Atmosphere pollutants and mortality rate of respiratory diseases in Beijing. *Sci. Total Environ.* **2008**, *391*, 143–148. [CrossRef]
24. Kassomenos, P.; Petrakis, M.; Sarigiannis, D.; Gotti, A.; Karakitsios, S. Identifying the contribution of physical and chemical stressors to the daily number of hospital admissions implementing an artificial neural network model. *Air Qual. Atmos. Health* **2011**, *4*, 263–272. [CrossRef]
25. Sundaram, N.M.; Sivanandam, S.; Subha, R. Elman neural network mortality predictor for prediction of mortality due to pollution. *Int. J. Appl. Eng. Res* **2016**, *11*, 1835–1840.
26. Tadano, Y.S.; Siqueira, H.V.; Antonini Alves, T. Unorganized machines to predict hospital admissions for respiratory diseases. In Proceedings of the IEEE Latin American Conference on Computational Intelligence (LA-CCI), Cartagena, Colombia, 2–4 November 2016; pp. 1–6.
27. Boccato, L.; Soares, E.S.; Fernandes, M.M.L.P.; Soriano, D.C.; Attux, R. Unorganized Machines: From Turing's Ideas to Modern Connectionist Approaches. *Int. J. Nat. Comput. Res. (IJNCR)* **2011**, *2*, 1–16. [CrossRef]
28. Huang, G.; Huang, G.B.; Song, S.; You, K. Trends in extreme learning machines: A review. *Neural Netw.* **2015**, *61*, 32–48. [CrossRef] [PubMed]
29. Jaeger, H. The "echo state" approach to analysing and training recurrent neural networks-with an erratum note. *Bonn, Ger. Ger. Natl. Res. Cent. Inf. Technol. GMD Tech. Rep.* **2001**, *148*, 13.
30. Jaeger, H. *Short term memory in Echo State Networks*; Technical Report; Fraunhofer Institute for Autonomous Intelligent Systems: Sankt Augustin, Germany, 2001.
31. Siqueira, H.V.; Boccato, L.; Attux, R.; Lyra Filho, C. Echo state networks in seasonal streamflow series prediction. *Learn. Nonlinear Model.* **2012**, *10*, 181–191. [CrossRef]
32. Huang, G.B.; Zhou, H.; Ding, X.; Zhang, R. Extreme learning machine for regression and multiclass classification. *IEEE Trans. Syst. Man, Cybern. Part B Cybern.* **2011**, *42*, 513–529. [CrossRef]
33. Boccato, L.; Lopes, A.; Attux, R.; Von Zuben, F.J. An extended echo state network using Volterra filtering and principal component analysis. *Neural Networks Off. J. Int. Neural Netw. Soc.* **2012**, *32*, 292–302. [CrossRef]
34. Butcher, J.; Verstraeten, D.; Schrauwen, B.; Day, C.; Haycock, P. Extending reservoir computing with random static projections: A hybrid between extreme learning and RC. In Proceedings of the 18th European sSymposium on Artificial Neural Networks, Bruges, Belgium, 28–30 April 2010.
35. Yildiz, I.B.; Jaeger, H.; Kiebel, S. Re-visiting the echo state property. *Neural Netw.* **2012**, *35*, 1–9. [CrossRef]
36. Huang, G.B.; Zhu, Q.Y.; Siew, C.K. Extreme learning machine: Theory and applications. *Neurocomputing* **2006**, *70*, 489–501. [CrossRef]
37. Huang, G.; Chen, L.; Siew, C. Universal approximation using incremental constructive feedforward networks with random hidden nodes. *IEEE Trans. Neural Netw.* **2006**, *17*, 879–92. [CrossRef]
38. Cao, J.; Lin, Z.; Huang, G.B.; Liu, N. Voting based extreme learning machine. *Inf. Sci.* **2012**, *185*, 66–77. [CrossRef]
39. Siqueira, H.; Luna, I. Performance comparison of feedforward neural networks applied to streamflow series forecasting. *Math. Eng. Sci. Aerosp. (MESA)* **2019**, *10*, 41–53.
40. Bartlett, P. The Sample Complexity of Pattern Classification with Neural Networks: The Size of the Weights is More Important than the Size of the Network. *IEEE Trans. Inf. Theory* **1998**, *44*, 525–536. [CrossRef]
41. Liu, X.; Gao, C.; Li, P. A comparative analysis of support vector machines and extreme learning machines. *Neural Netw.* **2012**, *33*, 58–66. [CrossRef] [PubMed]
42. Siqueira, H.; Boccato, L.; Attux, R.; Lyra, C. Echo state networks and extreme learning machines: A comparative study on seasonal streamflow series prediction. In Proceedings of the International Conference on Neural Information Processing, Doha, Qatar, 12–15 November 2012; pp. 491–500.
43. Lukosevicius, M.; Jaeger, H. Reservoir computing approaches to recurrent neural network training. *Comput. Sci. Rev.* **2009**, *3*, 127–149. [CrossRef]
44. Ozturk, M.C.; Xu, D.; Príncipe, J.C. Analysis and design of Echo State Networks. *Neural Comput.* **2007**, *19*, 111–138. [CrossRef]
45. Kulaif, A.C.P.; Von Zuben, F.J. Improved regularization in extreme learning machines. In Proceedings of the 11th Brazilian Congress on Computational Intelligence Porto de Galinhas, Pernambuco, Brazil, 8–11 September 2013; Volume 1, pp. 1–6.
46. Hashem, S. Optimal linear combinations of neural networks. *Neural Netw.* **1997**, *10*, 599–614. [CrossRef]
47. Siqueira, H.; Boccato, L.; Attux, R.; Lyra, C. Unorganized machines for seasonal streamflow series forecasting. *Int. J. Neural Syst.* **2014**, *24*, 1430009. [CrossRef]
48. Joe, H.; Kurowicka, D. *Dependence Modeling: Vine Copula Handbook*; World Scientific Publishing Co. Pte. Ltd.: Singapore, 2011.
49. Toly Chen, Y.C.W. Long-term load forecasting by a collaborative fuzzy-neural approach. *Int. J. Electr. Power Energy Syst.* **2012**, *43*, 454–464. [CrossRef]
50. CETESB-Environmental Sanitation Technology Company. Qualidade do ar no Estado de São Paulo, 2020. Available online: https://cetesb.sp.gov.br/ar/publicacoes-relatorios (accessed on 27 June 2021). (In Portuguese)
51. Datasus-Department of Informatics of the Unique Health System. SIHSUS Reduzida-Ministry of Health, Brazil. Available online: http://www2.datasus.gov.br/DATASUS/index.php?area=0701&item=1&acao=11 (accessed on 1 July 2020).

52. IBGE-Brazilian Institute of Geography and Statistics (in Portuguese: Instituto Brasileiro de Geografia e Estatística. Censo 2010. 2021. Available online: https://censo2010.ibge.gov.br/ (accessed on 27 July 2021).
53. Agrawal, S.B.; Agrawal, M. *Environmental Pollution and Plant Responses*; CRC Press: Boca Raton, FL, USA, 1999.
54. Tadano, Y.S.; Ugaya, C.M.L.; Franco, A.T. Methodology to assess air pollution impact on human health using the generalized linear model with Poisson Regression. In *Air Pollution-Monitoring, Modelling and Health*; InTech: São Paulo, Brazil, 2012.
55. Li, Y.; Ma, Z.; Zheng, C.; Shang, Y. Ambient temperature enhanced acute cardiovascular-respiratory mortality effects of PM 2.5 in Beijing, China. *Int. J. Biometeorol.* **2015**, *59*, 1761–1770. [CrossRef] [PubMed]
56. WHO-World Health Organization. *Air Quality Guidelines for Particulate Matter, Ozone, Nitrogen Dioxide and Sulfur Dioxide-Global Update 2005-Summary of Risk Assessment, 2006*; WHO: Geneva, Switzerland, 2006.
57. Montgomery, D.C.; Peck, E.A.; Vining, G.G. *Introduction to Linear Regression Analysis*; John Wiley & Sons: Hoboken, NJ, USA, 2021.
58. Haykin, S.S. *Neural Networks and Learning Machines*, 3rd ed.; Pearson Education: Upper Saddle River, NJ, USA, 2009.
59. Siqueira, H.; Boccato, L.; Luna, I.; Attux, R.; Lyra, C. Performance analysis of unorganized machines in streamflow forecasting of Brazilian plants. *Appl. Soft Comput.* **2018**, *68*, 494–506. [CrossRef]
60. Cuzick, J. A Wilcoxon-type test for trend. *Stat. Med.* **1985**, *4*, 87–90. [CrossRef] [PubMed]
61. Tadano, Y.S.; Potgieter-Vermaak, S.; Kachba, Y.R.; Chiroli, D.M.; Casacio, L.; Santos-Silva, J.C.; Moreira, C.A.; Machado, V.; Alves, T.A.; Siqueira, H.; et al. Dynamic model to predict the association between air quality, COVID-19 cases, and level of lockdown. *Environ. Pollut.* **2021**, *268*, 115920. [CrossRef]
62. Khatri, K.L.; Tamil, L.S. Early detection of peak demand days of chronic respiratory diseases emergency department visits using artificial neural networks. *IEEE J. Biomed. Health Inform.* **2017**, *22*, 285–290. [CrossRef]
63. Moustris, K.P.; Douros, K.; Nastos, P.T.; Larissi, I.K.; Anthracopoulos, M.B.; Paliatsos, A.G.; Priftis, K.N. Seven-days-ahead forecasting of childhood asthma admissions using artificial neural networks in Athens, Greece. *Int. J. Environ. Health Res.* **2012**, *22*, 93–104. [CrossRef]
64. Shakerkhatibi, M.; Dianat, I.; Jafarabadi, M.A.; Azak, R.; Kousha, A. Air pollution and hospital admissions for cardiorespiratory diseases in Iran: Artificial neural network versus conditional logistic regression. *Int. J. Environ. Sci. Technol.* **2015**, *12*, 3433–3442. [CrossRef]
65. Cengiz, M.A.; Terzi, Y. Comparing models of the effect of air pollutants on hospital admissions and symptoms for chronic obstructive pulmonary disease. *Cent. Eur. J. Public Health* **2012**, *20*, 282. [CrossRef]
66. Zhou, R.; Wu, D.; Li, Y.; Wang, B. Relationship Between Air Pollutants and Outpatient Visits for Respiratory Diseases in Hangzhou. In Proceedings of the 2018 9th International Conference on Information Technology in Medicine and Education (ITME), Hangzhou, China, 19–21 October 2018; pp. 275–280.

Article

Chemical Composition of $PM_{2.5}$ in Wood Fire and LPG Cookstove Homes of Nepali Brick Workers

James D. Johnston [1,*], John D. Beard [1], Emma J. Montague [1], Seshananda Sanjel [2], James H. Lu [1], Haley McBride [1], Frank X. Weber [3] and Ryan T. Chartier [3]

1 Department of Public Health, Brigham Young University, Provo, UT 84602, USA; john_beard@byu.edu (J.D.B.); montagueemma19@gmail.com (E.J.M.); luhuajr@gmail.com (J.H.L.); haleylucille82@gmail.com (H.M.)

2 Department of Community Medicine & Public Health, Karnali Academy of Health Sciences, Jumla 21200, Nepal; seshanandasanjel24@gmail.com

3 RTI International, Research Triangle Park, NC 27709, USA; fxw@rti.org (F.X.W.); rchartier@rti.org (R.T.C.)

* Correspondence: james_johnston@byu.edu; Tel.: +1-801-422-4226

Citation: Johnston, J.D.; Beard, J.D.; Montague, E.J.; Sanjel, S.; Lu, J.H.; McBride, H.; Weber, F.X.; Chartier, R.T. Chemical Composition of $PM_{2.5}$ in Wood Fire and LPG Cookstove Homes of Nepali Brick Workers. *Atmosphere* **2021**, *12*, 911. https://doi.org/10.3390/atmos12070911

Academic Editor: Alina Barbulescu

Received: 25 June 2021
Accepted: 12 July 2021
Published: 15 July 2021

Publisher's Note: MDPI stays neutral with regard to jurisdictional claims in published maps and institutional affiliations.

Copyright: © 2021 by the authors. Licensee MDPI, Basel, Switzerland. This article is an open access article distributed under the terms and conditions of the Creative Commons Attribution (CC BY) license (https://creativecommons.org/licenses/by/4.0/).

Abstract: Household air pollution is a major cause of morbidity and mortality worldwide, largely due to particles $\leq$ 2.5 μm ($PM_{2.5}$). The toxicity of $PM_{2.5}$, however, depends on its physical properties and chemical composition. In this cross-sectional study, we compared the chemical composition of $PM_{2.5}$ in brick workers' homes (n = 16) based on use of wood cooking fire or liquefied petroleum gas (LPG) cookstoves. We collected samples using RTI International particulate matter (PM) exposure monitors (MicroPEMs). We analyzed filters for 33 elements using energy-dispersive X-ray fluorescence and, for black (BC) and brown carbon (BrC), integrating sphere optical transmittance. Wood fire homes had significantly higher concentrations of BC (349 μg/m^3) than LPG homes (6.27 μg/m^3, $p < 0.0001$) or outdoor air (5.36 μg/m^3, $p = 0.002$). Indoor chlorine in wood fire homes averaged 5.86 μg/m^3, which was approximately 34 times the average level in LPG homes (0.17 μg/m^3, $p = 0.0006$). Similarly, potassium in wood fire homes (4.17 μg/m^3) was approximately four times the level in LPG homes (0.98 μg/m^3, $p = 0.001$). In all locations, we found aluminum, calcium, copper, iron, silicon, and titanium in concentrations exceeding those shown to cause respiratory effects in other studies. Our findings suggest the need for multi-faceted interventions to improve air quality for brick workers in Nepal.

Keywords: household air pollution; fine particulate matter; international environmental health; cookstove; respiratory disease; brick worker

1. Introduction

Household air pollution from the indoor burning of solid fuels, such as wood, crop residues, dung, or coal, is associated with 3.8 million deaths annually worldwide [1,2]. Exposure to household air pollution is associated with low birth weight, asthma, chronic obstructive pulmonary disease (COPD), respiratory infections, impaired immune function, coronary heart disease (CHD), stroke, cataracts, and cancers, including lung cancer [3–6]. Among household air pollutants generated from solid fuels, particulate matter (PM) less than or equal to 2.5 microns (μm) in aerodynamic diameter ($PM_{2.5}$), also called fine particulate matter, may be the single largest contributor to this excess disease burden [7,8]. However, the toxicity of $PM_{2.5}$ appears to be partially dependent on its chemical composition, which varies widely based on local emission sources [9–16].

In the Kathmandu Valley, Nepal, there are over 30,000 seasonal brick workers [17]. Most brick workers in Nepal live on-site at the brick kiln [18]. The most common type of housing for these workers is brick huts with tin roofs, often with poor ventilation to the outdoors [19]. Within this population of workers, the two primary methods of cooking are with indoor open wood fires or with liquefied petroleum gas (LPG) cookstoves [18,19].

Previous studies by our group found that Nepali brick workers, in addition to having hazardous work-related respiratory exposures [20], experience significant $PM_{2.5}$ exposures during non-working hours [18,19]. Indoor $PM_{2.5}$ concentrations in brick workers' homes with wood fire and LPG cookstoves were 541.14 and 79.32 $\mu g/m^3$, respectively, and these elevated levels coincided with meal and sleep times [19]. Brick workers suffer a disproportionate burden of respiratory symptoms compared to other workers in the same community [21], and we propose that these symptoms may be partially explained by elevated $PM_{2.5}$ levels in brick workers' homes, particularly among those cooking indoors with open wood fires.

$PM_{2.5}$ generated during wood combustion is composed primarily of elemental (EC) and organic carbon (OC), nitrate and sulfate species, metals, and other elements [22]. However, wood smoke composition depends on the species of wood being burned and the burn temperature [22–24]. Low burn temperatures (300–500 °C), such as during the start-up phase of an open wood fire, generally produce larger particles composed of numerous OC species and low levels of trace elements and metals, while higher burn temperatures (>800 °C), such as during the burn phase of an open wood fire, produce smaller particles composed of higher EC/OC ratios, and higher levels of trace elements and metals [8,23].

The biological mechanisms behind many of the diseases associated with wood smoke inhalation are not well understood, but studies suggest some metals, metalloids, and nonmetal elements may play key roles in air pollution-related diseases [9]. For example, studies of $PM_{2.5}$ constituents in ambient air pollution reported that the elements aluminum (Al), calcium (Ca), chlorine (Cl), iron (Fe), nickel (Ni), titanium (Ti), vanadium (V), and zinc (Zn), as well as black carbon (BC), are associated with increased hospitalizations and mortality, particularly among people $\geq$65 years of age [10,11]. The metals copper (Cu), Fe, potassium (K), and Zn, and the metalloid silicon (Si), are associated with respiratory hospital admissions in children, with the most serious effects seen in those $\leq$5 years of age [12]. The metals Al, Ni, Zn, V, and Ti, and the metalloid Si, are associated with low birth weight [13,14]. Several of these elements are present in wood smoke, in varying concentrations, depending on the species of wood and the burn temperature [22,24–26]. BC is associated with increased morbidity and mortality, primarily from heart and lung diseases [15]. Brown carbon (BrC), another constituent of $PM_{2.5}$ found in areas where the use of solid biomass fuels is high [27], may also influence human health because it can attach to toxic chemicals, such as benzopyrene, and heavy metals [28].

Understanding the chemical composition of $PM_{2.5}$ among specific populations may help elucidate relationships between exposure and disease. Our previous study measured the chemical composition of $PM_{2.5}$ in brick workers' homes during daytime hours when most home occupants were working, and thus did not capture pollutants generated during non-working hour activities such as cooking [18]. The purpose of this study, therefore, was to measure the chemical composition of $PM_{2.5}$ over a full day in order to characterize non-working hour exposures.

2. Materials and Methods

2.1. Study Design

We collected $PM_{2.5}$ samples using both filter-based and real-time nephelometer methods. Our previous study reported the $PM_{2.5}$ total mass and nephelometer trend analyses [19]. For this study, we analyzed the 25 mm 3.0 µm PTFE filters (Zefon International, Ocala, FL, USA) for 35 chemical constituents. The methods for home selection, measures of housing characteristics, and air filter handling and sampling strategy, described briefly here, are described in full in our previously published paper [19]. We used a cross-sectional study design to measure $PM_{2.5}$ constituents in brick workers' homes (n = 17) from a single brick kiln in Bhaktapur, Nepal. We recruited homes by convenience sampling, and we classified them as either wood cooking fire or LPG cookstove homes. The typical construction of the homes sampled in this study was detailed previously [19]. We collected samples from 30 April to 3 May 2019 for approximately 21 h (median: 21.21; interquartile range: 2.21) in each home. We administered an extant questionnaire [18], by means of an interpreter, to measure housing factors, including number of people in the home, number of children in the home, primary fuel used for cooking, presence of smokers in the home, and the number of smokers in the home. We also measured the living area of the home and calculated the occupant density as the number of occupants divided by the home area in m^2. Prior to data collection, Brigham Young University's (BYU) Institutional Review Board (IRB) determined that this study did not meet the definition of human subject research, per 45 CFR 46 [29], based on the fact the unit of study was the home rather than the individual.

2.2. Indoor and Outdoor $PM_{2.5}$ Measurements

We collected $PM_{2.5}$ samples using MicroPEM V.3.2A personal exposure monitors (RTI International, Research Triangle Park, NC, USA), which we placed indoors, on a tripod, approximately 1.2 m from the floor. Simultaneous daily outdoor samples were collected on-site at the brick kiln in a centralized location. Detailed methods describing MicroPEM preparation, placement, and filter handling were described previously [19].

2.3. Elemental Analysis

RTI International performed the analysis of the 25 mm filters for 33 elements following the IO3.3 compendium method [30], which was modified for use with the Thermo (Thermo Fisher Scientific, Waltham, MA, USA) ARL energy-dispersive X-ray fluorescence instrument equipped with a silicon drift detector. This instrument configuration was used because it could produce enough spectral counts to fully quantify each element, while collimating the beam. The instrument was calibrated with thin-film standards (Micromatter Technologies Inc., Surrey, BC, Canada) that approximated PM deposition on a filter and the unknown samples were analyzed under identical excitation conditions. The samples were analyzed under vacuum with five different energy conditions to achieve maximum sensitivity, while avoiding overlapping spectra. A camera system within the instrument chamber was used to ensure the beam was focused on the exposed area of the filter to accurately quantify the elements of concern. A multi-element thin film standard was analyzed with each tray of samples to ensure there was acceptable instrument performance across the mass range and to assess instrument drift.

2.4. Carbon Analysis

Following gravimetric analysis, all sample filters were shipped to RTI International for optical analysis using RTI International's integrating sphere optical transmittance technique [31]. The optical transmittance through the filter and the deposited PM sample were measured at seven wavelengths, ranging from near-infrared (940 nm) to blue (430 nm). All the sample filter transmittance data were adjusted using the mean transmittance of 10 blank filters from the same manufacturer's lot. An empirically-derived algorithm used the measured wavelength-dependent transmittance values to quantify the BC and lightly absorbing BrC contributions to the total PM collected on the sample filter. This technique, and similar optical methods, have been used in numerous PM exposure studies as a low-cost and non-destructive means of obtaining basic PM compositional data from sample filters [32–36].

2.5. Statistical Analyses

All statistical analyses were conducted using SAS version 9.4 (SAS Institute, Inc., Cary, NC, USA). Although we collected 20 total $PM_{2.5}$ samples, the filter of one sample tore and could not be analyzed for $PM_{2.5}$ chemical components. Thus, we excluded that sample from all statistical analyses. We used $\alpha = 0.05$ as the significance level for all analyses.

We calculated the frequencies and percentages for categorical characteristics of homes at the brick kiln and arithmetic means, standard deviations, minimums, first quartiles, medians, third quartiles, and maximums for the continuous characteristics of the homes at the brick kiln. For $PM_{2.5}$ chemical components, we calculated the frequency and percentage of samples that had concentrations below the lower detection limits (LDL), at or between the LDLs and upper detection limits (UDL), and above the UDLs. We also calculated the geometric means (GM), 95% confidence intervals (CI), minimums, and maximums for concentrations of $PM_{2.5}$ chemical components. We used GMs because the distributions of the concentrations of almost all $PM_{2.5}$ chemical components were right-skewed. For individual $PM_{2.5}$ chemical components that had all concentrations at or between the LDLs and UDLs, we used separate intercept-only linear regression models, with the natural logarithm of concentrations of individual $PM_{2.5}$ chemical components as the dependent variables, and then exponentiated intercept coefficients to calculate GMs and 95% CIs. For individual $PM_{2.5}$ chemical components with some concentrations below the LDLs or above the UDLs, we used separate intercept-only Tobit regression models, with the natural logarithm of concentrations of individual $PM_{2.5}$ chemical components as the dependent variables, and then exponentiated intercept coefficients to calculate GMs and 95% CIs.

We used decision rules that were similar to those of Beard at al. [37], who based their decision rules on information from Lubin et al. [38], to determine the appropriate types of regression models to use for the analyses of individual $PM_{2.5}$ chemical components with varying proportions of concentrations at or between the LDLs and UDLs. For the individual $PM_{2.5}$ chemical components that had all concentrations at or between the LDLs and UDLs, we estimated the p-values and unadjusted associations between the individual characteristics of homes at the brick kiln and the concentrations of individual $PM_{2.5}$ chemical components, using separate simple linear regression models, with the natural logarithm of the concentrations of individual $PM_{2.5}$ chemical components as the dependent variables. For the individual $PM_{2.5}$ chemical components that had >30–99% of concentrations at or between the LDLs and UDLs, we estimated the p-values and unadjusted associations between the individual characteristics of homes at the brick kiln and the concentrations of individual $PM_{2.5}$ chemical components, using separate simple Tobit regression models, with the natural logarithm of the concentrations of individual $PM_{2.5}$ chemical components as the dependent variables. For individual $PM_{2.5}$ chemical components that had >0–30% of concentrations at or between the LDLs and UDLs, we estimated the p-values and unadjusted associations between individual characteristics of homes at the brick kiln and the concentrations of individual $PM_{2.5}$ chemical components, using separate simple exact unconditional logistic regression models, with dichotomous indicator variables (i.e., one if the concentration was $\geq$LDL and zero if the concentration was <LDL) as dependent variables. For each of the three types of regression models, we exponentiated slope coefficients to calculate GMs, geometric mean ratios (GMR), or exact odds ratios (OR) and 95% CIs.

We considered several versions (e.g., linear; linear and quadratic; linear, quadratic, and cubic; natural logarithm; and categorical) of continuous characteristics of homes at the brick kilns and used the versions that had the lowest values of the Akaike Information Criterion (AIC) [39,40]. Where appropriate, we conducted pairwise comparisons of the GMs of concentrations of $PM_{2.5}$ chemical components for each category of home area and fuel type and location and used the Tukey (linear regression models) or Tukey-Kramer (Tobit regression models) method to adjust the p-values for multiple comparisons. We estimated multivariable linear or Tobit regression models when more than one characteristic of homes at the brick kilns were statistically significantly associated with concentrations of a particular $PM_{2.5}$ chemical component. For sensitivity analyses, we repeated analyses using home volume instead of home area.

3. Results

3.1. Characteristics of Homes of Brick Workers

For the 16 homes that we collected $PM_{2.5}$ samples from (i.e., excluding the one home for which the filter tore), the home area was 5.41–9.50 m^2 for 38%, >9.50–10.67 m^2 for 31%, and >10.67–31.40 m^2 for 31% (Table 1). The mean number of people in the home was 3.31 and the mean occupant density was 33.54 people per 100 m^2. Sixty-three percent of homes had 0–1 children and 38% had 2–3 children. Fifty percent of homes had smokers and the median number of smokers in the home was 0.5. Sixty-nine percent of homes used LPG for fuel and 31% used wood.

Table 1. Characteristics of homes [a] at a brick kiln in Bhaktapur, Nepal (May 2019).

Characteristic	Homes, n (%)	Mean	SD	Min	Q1	Median	Q3	Max
Total	16 (100)							
Home area [b], m^2								
5.41–9.50	6 (38)							
>9.50–10.67	5 (31)							
>10.67–31.40	5 (31)							
Number of people in home		3.31	1.54	1.00	2.00	3.00	4.00	7.00
Occupant density, number of people/100 m^2		33.54	17.00	9.55	22.47	30.11	38.55	73.96
Number of children in home								
0–1	10 (63)							
2–3	6 (38)							
Smokers in home								
No	8 (50)							
Yes	8 (50)							
Number of smokers in home		0.75	1.06	0.00	0.00	0.50	1.00	4.00
Fuel type								
LPG	11 (69)							
Wood	5 (31)							

Abbreviations: LPG, liquefied petroleum gas; Max, maximum; Min, minimum; Q1, first quartile; Q3, third quartile; and SD, standard deviation. [a] The filter of one sample tore and could not be analyzed, and, thus, was excluded from analyses. [b] Categories based on tertiles.

3.2. Summary Statistics for $PM_{2.5}$ Chemical Component Concentrations

Six $PM_{2.5}$ chemical components had all concentrations below the LDLs (Table 2). For the other 29 $PM_{2.5}$ chemical components, GMs of the concentrations ranged from 0.000042 $\mu g/m^3$ for $PM_{2.5}$ cerium to 16.09 $\mu g/m^3$ for $PM_{2.5}$ BC, with a median GM of 0.016 $\mu g/m^3$ for $PM_{2.5}$ barium (Ba).

Table 2. Summary statistics for the mean of samples inside or outside homes [a] at a brick kiln in Bhaktapur, Nepal (May 2019).

					Between LDL and UDL					
$PM_{2.5}$ Chemical Component, $\mu g/m^3$	LDL Mass (μg)	LDL Concentration Range	Missing, n	Below LDL, n (%)	n (%)	GM [b]	95% CI [b]	Min [c]	Max [c]	Above UDL
Al	0.012	0.018, 0.052		2 (11)	17 (89)	0.30	0.14, 0.66	0.031	2.19	0 (0)
Sb	0.24	0.34, 1.01		19 (100)	0 (0)	NA	NA	NA	NA	0 (0)
As	0.0024	0.0034, 0.010		17 (89)	2 (11)	0.0031	0.0025, 0.0039	0.0039	0.0039	0 (0)
Ba	0.0047	0.0068, 0.020		7 (37)	12 (63)	0.016	0.0083, 0.032	0.010	0.10	0 (0)
BC	0.50	0.73, 2.15		1 (5)	14 (74)	16.09	5.82, 44.52	1.84	107.36	4 (21) [d]
Br	0.0021	0.0030, 0.0089		0 (0)	19 (100)	0.022 [e]	0.016, 0.030 [e]	0.0078	0.061	0 (0)
BrC	0.50	0.73, 2.15	4	0 (0)	15 (100)	10.56 [e]	7.89, 14.13 [e]	2.34	17.88	0 (0)
Cd	0.082	0.12, 0.35		19 (100)	0 (0)	NA	NA	NA	NA	0 (0)
Cs	0.0024	0.0034, 0.010		6 (32)	13 (68)	0.0077	0.0050, 0.012	0.0048	0.024	0 (0)
Ca	0.0022	0.0033, 0.0096		0 (0)	19 (100)	0.18 [e]	0.075, 0.42 [e]	0.0060	1.33	0 (0)
Ce	0.0024	0.0034, 0.010		18 (95)	1 (5)	0.000042	0.0000000052, 0.35	0.015	0.015	0 (0)
Cl	0.0019	0.0028, 0.0082		0 (0)	19 (100)	0.38 [e]	0.14, 1.07 [e]	0.023	17.13	0 (0)
Cr	0.0013	0.0019, 0.0055		10 (53)	9 (47)	0.0020	0.0013, 0.0032	0.0019	0.0086	0 (0)
Co	0.00096	0.0014, 0.0041		15 (79)	4 (21)	0.00093	0.00047, 0.0019	0.0015	0.0032	0 (0)
Cu	0.0016	0.0023, 0.0067		6 (32)	13 (68)	0.0043	0.0025, 0.0076	0.0025	0.048	0 (0)
In	0.12	0.18, 0.52		19 (100)	0 (0)	NA	NA	NA	NA	0 (0)
Fe	0.0016	0.0024, 0.0071		0 (0)	19 (100)	0.26 [e]	0.12, 0.57 [e]	0.022	1.69	0 (0)
Pb	0.0049	0.0071, 0.021		6 (32)	13 (68)	0.014	0.0088, 0.021	0.0099	0.11	0 (0)
Mg	0.0050	0.0072, 0.021		5 (26)	14 (74)	0.032	0.016, 0.065	0.011	0.20	0 (0)
Mn	0.0018	0.0025, 0.0075		6 (32)	13 (68)	0.0083	0.0040, 0.017	0.0034	0.054	0 (0)
Mo	0.012	0.017, 0.051		19 (100)	0 (0)	NA	NA	NA	NA	0 (0)
Ni	0.0010	0.0015, 0.0043		10 (53)	9 (47)	0.0017	0.0013, 0.0023	0.0020	0.0038	0 (0)

Table 2. *Cont.*

$PM_{2.5}$ Chemical Component, $\mu g/m^3$	LDL Mass (μg)	LDL Concentration Range	Missing, n	Below LDL, n (%)	Between LDL and UDL n (%)	GM [b]	95% CI [b]	Min [c]	Max [c]	Above UDL
P	0.0024	0.0036, 0.011		5 (26)	14 (74)	0.011	0.0061, 0.021	0.0045	0.064	0 (0)
K	0.0019	0.0028, 0.0083		0 (0)	19 (100)	1.44 [e]	0.94, 2.18 [e]	0.38	5.86	0 (0)
Rb	0.0023	0.0034, 0.010		9 (47)	10 (53)	0.0043	0.0022, 0.0081	0.0043	0.028	0 (0)
Se	0.0022	0.0033, 0.0097		12 (63)	7 (37)	0.0033	0.0026, 0.0041	0.0039	0.0070	0 (0)
Si	0.0064	0.0093, 0.027		0 (0)	19 (100)	1.08 [e]	0.58, 2.00 [e]	0.14	5.15	0 (0)
Ag	0.055	0.079, 0.24		19 (100)	0 (0)	NA	NA	NA	NA	0 (0)
Na	0.010	0.015, 0.045		2 (11)	17 (89)	0.13	0.088, 0.19	0.034	0.37	0 (0)
Sr	0.0030	0.0043, 0.013		17 (89)	2 (11)	0.0015	0.00025, 0.0088	0.0075	0.0092	0 (0)
S	0.0026	0.0038, 0.011		0 (0)	19 (100)	2.77 [e]	2.20, 3.48 [e]	0.99	4.79	0 (0)
Sn	0.18	0.26, 0.76		19 (100)	0 (0)	NA	NA	NA	NA	0 (0)
Ti	0.00085	0.0012, 0.0036		2 (11)	17 (89)	0.021	0.0088, 0.050	0.0018	0.19	0 (0)
V	0.0011	0.0015, 0.0046		9 (47)	10 (53)	0.0021	0.0012, 0.0039	0.0020	0.013	0 (0)
Zn	0.0015	0.0022, 0.0066		0 (0)	19 (100)	0.059 [e]	0.039, 0.089 [e]	0.019	0.39	0 (0)

Abbreviations: Al, aluminum; Sb, antimony; As, arsenic; Ba, barium; BC, black carbon; Br, bromine; BrC, brown carbon; Cd, cadmium; Cs, cesium; Ca, calcium; Ce, cerium; Cl, chlorine; Cr, chromium; Co, cobalt; CI, confidence interval; Cu, copper; GM, geometric mean; In, indium; Fe, iron; Pb, lead; LDL, lower detection limit; Mg, magnesium; Mn, manganese; Max, maximum; Min, minimum; Mo, molybdenum; Ni, nickel; NA, not applicable; $PM_{2.5}$, particulate matter with an aerodynamic diameter less than 2.5 µm; P, phosphorus; K, potassium; Rb, rubidium; Se, selenium; Si, silicon; Ag, silver; Na, sodium; Sr, strontium; S, sulfur; Sn, tin; Ti, titanium; UDL, upper detection limit; V, vanadium; and Zn, zinc. [a] The filter of one inside sample tore and could not be analyzed, and so that home was excluded from analyses. Of the remaining 19 samples, 16 were from inside and three were from outside the homes. [b] Estimated via simple (i.e., unadjusted), intercept only Tobit regression models of the natural logarithm transformed values. [c] Calculated from samples that had values at or between the LDL and UDL. [d] The UDL mass was 80 µg and the UDL concentration range was 116.08 to 344.21 $\mu g/m^3$. [e] Estimated via simple (i.e., unadjusted), intercept only linear regression models of the natural logarithm transformed values.

3.3. Associations between Characteristics of Homes of Brick Workers and $PM_{2.5}$ Chemical Component Concentrations

Home area was significantly associated with concentrations of $PM_{2.5}$ Cl ($p = 0.03$) and $PM_{2.5}$ Cu ($p = 0.005$; Supplementary Materials, Table S1). For $PM_{2.5}$ Cl, pairwise comparisons indicated significant differences between 5.41–9.50 m^2 (GM = 2.95 $\mu g/m^3$) and >10.67–31.40 m^2 (GM = 0.14 $\mu g/m^3$; $p = 0.04$), but not between 5.41–9.50 m^2 and >9.50–10.67 m^2 (GM = 0.25 $\mu g/m^3$; $p = 0.10$) or >9.50–10.67 m^2 and >10.67–31.40 m^2 ($p = 0.85$). For $PM_{2.5}$ Cu, pairwise comparisons indicated significant differences between 5.41–9.50 m^2 (GM = 0.013 $\mu g/m^3$) and >9.50–10.67 m^2 (GM = 0.0024 $\mu g/m^3$; $p = 0.01$), and between 5.41–9.50 m^2 and >10.67–31.40 m^2 (GM = 0.0035 $\mu g/m^3$; $p = 0.03$), but not between >9.50–10.67 m^2 and >10.67–31.40 m^2 ($p = 0.82$). The number of people in the home, occupant density, and the number of children in the home were not significantly associated with concentrations of any $PM_{2.5}$ chemical component (Supplementary Materials, Tables S1 and S2). The presence of smokers in the home was significantly associated with concentrations of 22 of 29 (76%) $PM_{2.5}$ chemical components, and the GMs of concentrations were higher in homes with smokers than in homes without smokers for all 22 significant associations (Table 3). Similarly, the number of smokers in the home was significantly associated with concentrations of 20 (69%) $PM_{2.5}$ chemical components and the GMRs were greater than one for all 20 significant associations (i.e., the GMs of the concentrations of those 20 $PM_{2.5}$ chemical components increased as the number of smokers in the home increased). Fuel type and location was significantly associated with concentrations of 22 (76%) $PM_{2.5}$ chemical components (Table 4). Pairwise comparisons indicated significant differences between LPG, indoor, and wood, indoor, for 21 (95%) of the significant associations, and the GMs of concentrations were higher for wood, indoor, than for LPG, indoor, for all 21 significant differences. Pairwise comparisons indicated significant differences between LPG, indoor, and outdoor for one (5%) of the significant associations (i.e., $PM_{2.5}$ BrC) and the GM of concentrations was higher for LPG, indoor, than for outdoor. Pairwise comparisons indicated a significant difference between wood, indoor, and outdoor for six (27%) of the significant associations, and the GMs of concentrations were higher for wood, indoor, than for outdoor for all six significant differences.

Table 3. Associations between the mean of samples inside homes [a] and smokers in the home and the number of smokers in the home at a brick kiln in Bhaktapur, Nepal (May 2019).

	Smokers in Home					Number of Smokers in Home		
	No		Yes					
$PM_{2.5}$ Chemical Component, $\mu g/m^3$	GM [b]	95% CI [b]	GM [b]	95% CI [b]	*p*-Value [b]	GMR [b,c]	95% CI [b,c]	*p*-Value [b]
Al	0.10	0.043, 0.23	0.88	0.38, 2.04	0.0003	2.50	1.33, 4.71	0.005
As	1.00	Reference	2.66 [d,e]	0.30, ∞ [d,e]	NA	4.21 [d]	0.80, 235.33 [d]	NA
Ba	0.0064	0.0026, 0.016	0.037	0.018, 0.076	0.003	2.19	1.31, 3.67	0.003
BC	4.81	1.65, 13.98	80.22	24.43, 263.43	0.0006	5.67	1.55, 20.75	0.009
Br	0.015 [f]	0.010, 0.021 [f]	0.038 [f]	0.027, 0.054 [f]	0.001 [f]	1.41 [f]	1.05, 1.90 [f]	0.03 [f]
BrC	11.32 [f]	8.99, 14.25 [f]	14.92 [f]	10.78, 20.66 [f]	0.15 [f]	1.25 [f]	0.93, 1.66 [f]	0.12 [f]
Cs	0.0045	0.0025, 0.0080	0.012	0.0068, 0.019	0.02	1.59	1.12, 2.26	0.009
Ca	0.060 [f]	0.021, 0.17 [f]	0.53 [f]	0.19, 1.49 [f]	0.006 [f]	2.59 [f]	1.22, 5.50 [f]	0.02 [f]
Ce	1.00	Reference	1.00 [d,e]	0.053, ∞ [d,e]	NA	1.21 [d]	0.034, 6.51 [d]	NA
Cl	0.083 [f]	0.035, 0.20 [f]	3.26 [f]	1.38, 7.75 [f]	<0.0001 [f]	3.86 [f]	1.58, 9.42 [f]	0.006 [f]
Cr	0.00097	0.00045, 0.0021	0.0032	0.0022, 0.0048	0.005	1.53	1.06, 2.21	0.02
Co	1.00	Reference	2.66 [d,e]	0.30, ∞ [d,e]	NA	4.21 [d]	0.80, 235.33 [d]	NA
Cu	0.0022	0.0013, 0.0038	0.012	0.0076, 0.019	<0.0001	1.66	1.01, 2.73	0.05
Fe	0.086 [f]	0.034, 0.21 [f]	0.72 [f]	0.29, 1.79 [f]	0.003 [f]	2.49 [f]	1.26, 4.94 [f]	0.01 [f]
Pb	0.0069	0.0035, 0.013	0.026	0.015, 0.045	0.002	1.49	0.93, 2.38	0.10
Mg	0.014	0.0059, 0.034	0.067	0.029, 0.15	0.01	1.90	1.03, 3.50	0.04
Mn	0.0031	0.0013, 0.0074	0.021	0.0097, 0.045	0.001	2.35	1.34, 4.13	0.003
Ni	0.0011	0.00069, 0.0018	0.0023	0.0017, 0.0031	0.01	1.28	1.01, 1.61	0.04
P	0.0042	0.0022, 0.0081	0.029	0.016, 0.052	<0.0001	2.21	1.36, 3.58	0.001
K	0.91 [f]	0.52, 1.60 [f]	2.62 [f]	1.49, 4.61 [f]	0.01 [f]	1.62 [f]	1.08, 2.42 [f]	0.02 [f]
Rb	0.0017	0.00060, 0.0051	0.0095	0.0045, 0.020	0.008	2.02	1.14, 3.57	0.02
Se	0.0023	0.0014, 0.0040	0.0037	0.0027, 0.0051	0.11	1.07	0.83, 1.39	0.59
Si	0.48 [f]	0.23, 0.96 [f]	2.32 [f]	1.15, 4.71 [f]	0.004 [f]	1.98 [f]	1.17, 3.35 [f]	0.01 [f]
Na	0.14	0.078, 0.27	0.096	0.051, 0.18	0.36	1.01	0.66, 1.55	0.96
Sr	1.00	Reference	1.00 [d,e]	0.053, ∞ [d,e]	NA	2.28 [d,e]	0.93, ∞ [d,e]	NA
S	2.21 [f]	1.69, 2.88 [f]	3.64 [f]	2.79, 4.76 [f]	0.01 [f]	1.25 [f]	1.04, 1.52 [f]	0.02 [f]
Ti	0.0063	0.0024, 0.016	0.066	0.026, 0.17	0.0005	2.76	1.38, 5.52	0.004
V	0.00077	0.00029, 0.0020	0.0048	0.0028, 0.0083	0.001	2.06	1.32, 3.20	0.002
Zn	0.035 [f]	0.021, 0.061 [f]	0.10 [f]	0.060, 0.18 [f]	0.01 [f]	1.46 [f]	0.95, 2.23 [f]	0.08 [f]

Abbreviations: Al, aluminum; As, arsenic; Ba, barium; BC, black carbon; Br, bromine; BrC, brown carbon; Cs, cesium; Ca, calcium; Ce, cerium, Cl, chlorine; Cr, chromium; Co, cobalt; CI, confidence interval; Cu, copper; GM, geometric mean; GMR, geometric mean ratio; Fe, iron; Pb, lead; Mg, magnesium; Mn, manganese; Ni, nickel; NA, not applicable; $PM_{2.5}$, particulate matter with an aerodynamic diameter less than 2.5 μm; P, phosphorus; K, potassium; Rb, rubidium; Se, selenium; Si, silicon; Na, sodium; Sr, strontium; S, sulfur; Ti, titanium; V, vanadium; and Zn, zinc. [a] The filter of one sample tore and could not be analyzed, so that home was excluded from analyses. [b] Estimated via simple (i.e., unadjusted) Tobit regression models of the natural logarithm transformed values. [c] Exponentiated regression coefficient and 95% CI (i.e., GM $PM_{2.5}$ chemical component concentration ratio for a specified change in the independent variable or $\exp(\beta) - 1$ = percent change in GM $PM_{2.5}$ chemical component concentration for a specified change in the independent variable). [d] Exact odds ratio and 95% CI; estimated via simple (i.e., unadjusted) exact unconditional logistic regression models. [e] Median unbiased estimate. [f] Estimated via simple (i.e., unadjusted) linear regression models of the natural logarithm transformed values.

Table 4. Associations between the mean of samples inside or outside homes [a] and fuel type and location at a brick kiln in Bhaktapur, Nepal (May 2019).

	Fuel Type and Location									
	LPG, Indoor		Wood, Indoor		Outdoor			LPG, Indoor vs. Wood, Indoor	LPG, Indoor vs. Outdoor	Wood, Indoor vs. Outdoor
$PM_{2.5}$ Chemical Component, μg/m^3	GM [b]	95% CI [b]	GM [b]	95% CI [b]	GM [b]	95% CI [b]	*p*-Value [b]	*p*-Value [b,c]	*p*-Value [b,c]	*p*-Value [b,c]
Al	0.14	0.062, 0.33	1.48	0.44, 4.98	0.37	0.074, 1.85	0.007	0.005	0.56	0.37
As	1.00	Reference	6.36 [d,e]	0.70, ∞ [d,e]	[f]	[f]	NA	NA	NA	NA
Ba	0.0088	0.0047, 0.016	0.062	0.028, 0.14	0.031	0.011, 0.089	0.0005	0.0004	0.11	0.56
BC	6.27	2.73, 14.41	349.04	64.46, 1,889.91	5.36	1.02, 28.21	0.0001	<0.0001	0.98	0.002
Br	0.018 [g]	0.013, 0.025 [g]	0.043 [g]	0.026, 0.072 [g]	0.013 [g]	0.0069, 0.025 [g]	0.01 [g]	0.02 [g]	0.67 [g]	0.02 [g]
BrC	12.03 [g]	9.09, 15.91 [g]	17.52 [g]	6.93, 44.32 [g]	5.54 [g]	3.24, 9.47 [g]	0.03 [g]	0.68 [g]	0.04 [g]	0.09 [g]
Cs	0.0052	0.0033, 0.0081	0.015	0.0085, 0.028	0.012	0.0055, 0.028	0.009	0.01	0.15	0.91
Ca	0.081 [g]	0.032, 0.21 [g]	1.01 [g]	0.25, 4.11 [g]	0.17 [g]	0.028, 1.06 [g]	0.02 [g]	0.02 [g]	0.72 [g]	0.26 [g]
Ce	1.00	Reference	2.20 [d,e]	0.12, ∞ [d,e]	[f]	[f]	NA	NA	NA	NA
Cl	0.17 [g]	0.072, 0.42 [g]	5.86 [g]	1.59, 21.52 [g]	0.071 [g]	0.013, 0.38 [g]	0.0003 [g]	0.0006 [g]	0.59 [g]	0.001 [g]
Cr	0.0013	0.00071, 0.0023	0.0039	0.0023, 0.0066	0.0034	0.0016, 0.0071	0.01	0.01	0.09	0.95
Co	1.00	Reference	6.36 [d,e]	0.70, ∞ [d,e]	12.47 [d,e]	1.31, ∞ [d,e]	NA	NA	NA	NA
Cu	0.0032	0.0019, 0.0054	0.015	0.0075, 0.030	0.0020	0.00064, 0.0064	0.0006	0.001	0.74	0.009
Fe	0.12 [g]	0.051, 0.28 [g]	1.25 [g]	0.35. 4.44 [g]	0.33 [g]	0.064, 1.69 [g]	0.02 [g]	0.01 [g]	0.48 [g]	0.38 [g]
Pb	0.0095	0.0061, 0.015	0.036	0.020, 0.064	0.014	0.0061, 0.031	0.002	0.001	0.70	0.15
Mg	0.017	0.0083, 0.036	0.11	0.040, 0.30	0.054	0.014, 0.21	0.01	0.009	0.30	0.68
Mn	0.0042	0.0022, 0.0077	0.040	0.017, 0.090	0.015	0.0048, 0.044	<0.0001	<0.0001	0.12	0.33
Ni	0.0015	0.0011, 0.0021	0.0021	0.0014, 0.0033	0.0028	0.0016, 0.0047	0.11	NA	NA	NA
P	0.0061	0.0036, 0.010	0.045	0.022, 0.092	0.017	0.0065, 0.045	<0.0001	<0.0001	0.15	0.25
K	0.98 [g]	0.67, 1.45 [g]	4.17 [g]	2.34, 7.43 [g]	0.98 [g]	0.46, 2.06 [g]	0.001 [g]	0.001 [g]	>0.99 [g]	0.01 [g]
Rb	0.0027	0.0016, 0.0046	0.018	0.011, 0.031	0.0048	0.0022, 0.010	<0.0001	<0.0001	0.46	0.01
Se	0.0031	0.0023, 0.0042	0.0032	0.0022, 0.0048	0.0039	0.0025, 0.0062	0.71	NA	NA	NA
Si	0.60 [g]	0.30, 1.20 [g]	3.58 [g]	1.28, 9.97 [g]	1.23 [g]	0.33, 4.62 [g]	0.03 [g]	0.02 [g]	0.58 [g]	0.39 [g]
Na	0.13	0.079, 0.21	0.097	0.046, 0.20	0.21	0.083, 0.54	0.44	NA	NA	NA
Sr	1.00	Reference	2.20 [d,e]	0.12, ∞ [d,e]	3.67 [d,e]	0.19, ∞ [d,e]	NA	NA	NA	NA
S	2.35 [g]	1.80, 3.06 [g]	4.29 [g]	2.90, 6.35 [g]	2.45 [g]	1.47, 4.06 [g]	0.04 [g]	0.04 [g]	0.99 [g]	0.18 [g]
Ti	0.0088	0.0037, 0.021	0.13	0.036, 0.46	0.028	0.0052, 0.15	0.003	0.002	0.46	0.34
V	0.0013	0.00079, 0.0022	0.0075	0.0043, 0.013	0.0035	0.0017, 0.0075	<0.0001	<0.0001	0.08	0.25
Zn	0.039 [g]	0.026, 0.060 [g]	0.15 [g]	0.084, 0.28 [g]	0.053 [g]	0.024, 0.12 [g]	0.004 [g]	0.003 [g]	0.75 [g]	0.09 [g]

Abbreviations: Al, aluminum; As, arsenic; Ba, barium; BC, black carbon; Br, bromine; BrC, brown carbon; Cs, cesium; Ca, calcium; Ce, cerium; Cl, chlorine; Cr, chromium; Co, cobalt; CI, confidence interval; Cu, copper; GM, geometric mean; Fe, iron; Pb, lead; LPG, liquefied petroleum gas; Mg, magnesium; Mn, manganese; Ni, nickel; NA, not applicable; $PM_{2.5}$, particulate matter with an aerodynamic diameter less than 2.5 μm; P, phosphorus; K, potassium; Rb, rubidium; Se, selenium; Si, silicon; Na, sodium; Sr, strontium; S, sulfur; Ti, titanium; V, vanadium; and Zn, zinc. [a] The filter of one inside sample tore and could not be analyzed, so that home was excluded from analyses. Of the remaining 19 samples, 16 were from inside and three were from outside the homes. [b] Estimated via simple (i.e., unadjusted) Tobit regression models of the natural logarithm transformed values. [c] Used the Tukey-Kramer (for Tobit regression models) or Tukey (for linear regression models) methods to adjust for multiple comparisons. [d] Exact odds ratio and 95% CI; estimated via simple (i.e., unadjusted) exact unconditional logistic regression models. [e] Median unbiased estimate. [f] Degenerate; unable to estimate. [g] Estimated via simple (i.e., unadjusted) linear regression models of the natural logarithm transformed values.

3.4. Adjusted Associations between Characteristics of Homes of Brick Workers and $PM_{2.5}$ Chemical Component Concentrations

Smokers in the home was most consistently significantly associated with $PM_{2.5}$ Cl and $PM_{2.5}$ Cu when some combination of home area, smokers in the home or number of smokers in the home, and fuel type were included as independent variables in the multivariable linear regression models (Supplementary Materials, Table S3). Smokers in the home was significantly associated with concentrations of six (24%) $PM_{2.5}$ chemical components, and fuel type was significantly associated with concentrations of nine (36%)

$PM_{2.5}$ chemical components when smokers in the home and fuel type were included as independent variables in the multivariable linear or Tobit regression models (Table 5). The number of smokers in the home was not significantly associated with concentrations of any $PM_{2.5}$ chemical component, but fuel type was significantly associated with concentrations of 14 (56%) $PM_{2.5}$ chemical components when the number of smokers in the home and fuel type were included as independent variables in the multivariable linear or Tobit regression models (Table 6).

Table 5. Associations between the mean of samples inside homes [a] and smokers in home and fuel type, mutually adjusted for each other at a brick kiln in Bhaktapur, Nepal (May 2019).

	Smokers in Home	Fuel Type
$PM_{2.5}$ Chemical Component	***p*-Value [b]**	***p*-Value [b]**
Al	0.08	0.09
Ba	0.23	0.03
BC	0.22	0.002
Br	0.04 [c]	0.30 [c]
BrC	0.30 [c]	0.55 [c]
Cs	0.39	0.12
Ca	0.20 [c]	0.08 [c]
Cl	0.002 [c]	0.06 [c]
Cr	0.10	0.20
Cu	0.003	0.22
Fe	0.14 [c]	0.09 [c]
Pb	0.13	0.10
Mg	0.39	0.09
Mn	0.22	0.008
Ni	0.02	0.62
P	0.02	0.02
K	0.50 [c]	0.02 [c]
Rb	0.68	0.002
Se	0.05	0.24
Si	0.16 [c]	0.09 [c]
Na	0.50	0.99
S	0.32 [c]	0.09 [c]
Ti	0.12	0.04
V	0.07	0.004
Zn	0.37 [c]	0.03 [c]

Abbreviations: Al, aluminum; Ba, barium; BC, black carbon; Br, bromine; BrC, brown carbon; Cs, cesium; Ca, calcium; Cl, chlorine; Cr, chromium; Cu, copper; Fe, iron; Pb, lead; Mg, magnesium; Mn, manganese; Ni, nickel; $PM_{2.5}$, particulate matter with an aerodynamic diameter less than 2.5 μm; P, phosphorus; K, potassium; Rb, rubidium; Se, selenium; Si, silicon; Na, sodium; S, sulfur; Ti, titanium; V, vanadium; and Zn, zinc. [a] The filter of one sample tore and could not be analyzed, so that home was excluded from analyses. [b] Estimated via multivariable Tobit regression models of the natural logarithm transformed values adjusted for smokers in the home and fuel type. [c] Estimated via multivariable linear regression models of the natural logarithm transformed values adjusted for smokers in the home and fuel type.

Table 6. Associations between the mean of samples inside homes [a] and the number of smokers in the home and fuel type, mutually adjusted for each other at a brick kiln in Bhaktapur, Nepal (May 2019).

	Number of Smokers in Home	Fuel Type
$PM_{2.5}$ Chemical Component	***p*-Value [b]**	***p*-Value [b]**
Al	0.35	0.04
Ba	0.30	0.02
BC	0.72	0.002
Br	0.48 [c]	0.11 [c]
BrC	0.30 [c]	0.99 [c]

Table 6. *Cont.*

	Number of Smokers in Home	Fuel Type
$PM_{2.5}$ Chemical Component	p-Value [b]	p-Value [b]
Cs	0.30	0.14
Ca	0.46 [c]	0.06 [c]
Cl	0.32 [c]	0.02 [c]
Cr	0.50	0.07
Cu	0.98	0.01
Fe	0.39 [c]	0.06 [c]
Pb	0.65	0.004
Mg	0.77	0.04
Mn	0.39	0.005
Ni	0.09	0.82
P	0.27	0.006
K	0.75 [c]	0.01 [c]
Rb	0.92	0.0007
Se	0.57	0.82
Si	0.43 [c]	0.06 [c]
Na	0.51	0.37
S	0.51 [c]	0.07 [c]
Ti	0.39	0.02
V	0.20	0.002
Zn	0.70 [c]	0.008 [c]

Abbreviations: Al, aluminum; Ba, barium; BC, black carbon; Br, bromine; BrC, brown carbon; Cs, cesium; Ca, calcium; Cl, chlorine; Cr, chromium; Cu, copper; Fe, iron; Pb, lead; Mg, magnesium; Mn, manganese; Ni, nickel; $PM_{2.5}$, particulate matter with an aerodynamic diameter less than 2.5 μm; P, phosphorus; K, potassium; Rb, rubidium; Se, selenium; Si, silicon; Na, sodium; S, sulfur; Ti, titanium; V, vanadium; and Zn, zinc. [a] The filter of one sample tore and could not be analyzed, so that home was excluded from analyses. [b] Estimated via multivariable Tobit regression models of the natural logarithm transformed values adjusted for the number of smokers in the home and fuel type. [c] Estimated via multivariable linear regression models of the natural logarithm transformed values adjusted for the number of smokers in the home and fuel type.

3.5. Sensitivity Analyses

The results were almost identical when we repeated analyses using home volume instead of home area (not shown), with the one exception being that home volume was not significantly associated with concentrations of $PM_{2.5}$ Cl (8.52–16.00 m^3: GM = 1.23; 95% CI: 0.26, 5.94 $\mu g/m^3$; and >16.00–53.44 m^3: GM = 0.22; 95% CI: 0.046, 1.06 $\mu g/m^3$; $p = 0.12$). In other words, home volume was significantly associated with concentrations of $PM_{2.5}$ Cu (8.52–16.00 m^3: GM = 0.0086; 95% CI: 0.0043, 0.017 $\mu g/m^3$; and >16.00–53.44 m^3: GM = 0.0030; 95% CI: 0.0014, 0.0063 $\mu g/m^3$; $p = 0.04$), but not with concentrations of any other $PM_{2.5}$ chemical component (not shown).

4. Discussion

This research was conducted as a follow-up to a previous study we conducted in 2018 [18]. In our previous study, we collected $PM_{2.5}$ samples in on-site brick workers' homes in Bhaktapur, Nepal, but the sampling time was limited to approximately seven hours during the middle of the day when most workers were not at home. Thus, we were not able to measure $PM_{2.5}$ generated during cooking and other household activities during non-working hours. The longer sampling time (approximately 21 h in each home) used in the current study allowed us to characterize $PM_{2.5}$ constituents across both working and non-working hours. Using the seven-hour samples in our previous study, we found no difference in the chemical composition of indoor vs. outdoor air, except for Cl, which was higher indoors. Our previous study also found that the primary fuel used for cooking was significantly associated with only two $PM_{2.5}$ chemical components, Cl and K, which were both higher in wood fuel homes. In contrast, in this study we found significant differences for 22 chemical components based on cooking fuel type and location (LPG, indoor vs. wood, indoor vs. outdoor). Pairwise comparisons indicated fuel type was the primary source of these significant differences. We attributed these differences in results among studies to non-working hour activities in the home that were not captured in our previous study, but were captured in our current study.

The major elemental aerosol-phase tracers of wood smoke are Cl and K, both of which are commonly found in $PM_{2.5}$ generated from wood combustion [41]. Like our previous study, we found significantly higher levels of both elements in homes where wood fires were used for cooking. For wood fire homes, the indoor Cl level averaged 5.86 $\mu g/m^3$, which was approximately 34 times the average level in LPG homes (0.17 $\mu g/m^3$). Similarly, K levels in wood fire homes (4.17 $\mu g/m^3$) were approximately four times the levels in LPG homes (0.98 $\mu g/m^3$). Furthermore, our results showed significant differences in Cl and K levels between wood fire homes and outdoor air, but not between LPG homes and outdoor air, suggesting the high levels of Cl and K in our study originated from cooking indoors over wood fires. Our findings are consistent with previous studies conducted in homes in West Africa and India. In higher-income homes in Accra, Ghana, where residents tend to cook indoors with LPG cookstoves, Zhou et al. reported average indoor Cl and K levels of 0.34 and 1.08 $\mu g/m^3$, respectively [42]. By comparison, average Cl and K levels in enclosed cookhouses using firewood in The Gambia were 7.90 and 10.75 $\mu g/m^3$, respectively [42]. In unventilated, low-income homes in India where solid biomass fuels were used for cooking, annual Cl and K levels averaged 5.8 and 7.6 $\mu g/m^3$, respectively [43].

Among the 35 analytes, BC accounted for the highest concentration in wood fire homes (349.04 $\mu g/m^3$), where levels were 56 and 65 times the levels in LPG homes and outdoor air, respectively. BC is released into the air as a result of incomplete combustion of fuels, and prolonged or extreme exposure is associated with increased morbidity and mortality, primarily from cardiac and respiratory illnesses [15]. When BC acts as a carrier for polycyclic aromatic hydrocarbons, it is linked to adverse health effects, including cancer and severe immune, reproductive, and pulmonary damage [15,44]. Several additional elemental species identified in our study were previously shown to be associated with burning wood. For example, we found concentrations of Al, Ca, magnesium (Mg), phosphorus (P), and Si were significantly higher in homes with wood cooking fires than in homes with LPG cookstoves. All of these elements were shown in previous studies to be associated with high temperature burning of wood or wood pellets in stoves [23,45].

Respiratory illnesses are common among brick workers in Nepal [21], and occupational exposures likely play an important role in this finding [20]. However, previous studies of urban ambient $PM_{2.5}$ constituents found that several metals and other elements were associated with respiratory disease in adults and children [10,12] at much lower concentrations than those found in our study. For example, Bell et al. found associations between respiratory hospital admissions in adults $\geq$65 years of age and $PM_{2.5}$ constituents Al, Ca, Cl, BC, Ni, Si, Ti, and V [10]. In our study, all of these constituents, with the exception of Ni and V, were found in higher concentrations than those reported by Bell et al. in all sampled locations (wood indoor, LPG indoor, and outdoor air), and V concentrations in our study were higher in wood fire homes. Ostro et al. found associations between respiratory hospital admissions in children and concentrations of Cu, Fe, K, Si, and Zn in ambient air [12]. Again, we found each of these constituents in our samples, and in most cases at higher concentrations than those reported by Ostro et al. Differences in concentrations were most pronounced in wood fire homes, where element concentrations ranged from 1.9–35 times the ambient concentrations reported by Ostro et al. We propose that repeated exposure to the high concentrations of metals and other elements in both indoor and outdoor air may contribute significantly to the respiratory symptoms seen among brick workers in Nepal.

One of the most noticeable differences between this study and our previous one was the number of chemical components that had significantly higher concentrations in homes with smokers. Depending on the tobacco source, cigarette smoke contains varying levels of several metals that are associated with deleterious health effects, including Al, arsenic (As), Ba, beryllium (Be), cadmium (Cd), cobalt (Co), chromium (Cr), Cu, Fe, lead (Pb), manganese (Mn), mercury (Hg), Ni, selenium (Se), Si, V, and Zn [46,47]. Our current study found these metals in higher concentrations in homes with smokers compared to homes with non-smokers, with the exception of Cd, which had all sample concentrations below the LDL, and Be and Hg, which we did not test for. Toxicologically, these metals are associated with allergic sensitization and inflammation, COPD, cancer, asthma, immune system suppression [47], vascular endothelium damage, and the development of atherosclerosis [46].

In our previous study, we discussed concerns about small, overcrowded housing among brick workers in Nepal, and specifically regarding the potential for indoor pollution to concentrate in smaller, poorly ventilated homes [18]. The finding that home area was significantly associated with $PM_{2.5}$ Cl and Cu appears to support this concern. In the cases of both Cl and Cu, smaller homes (i.e., 5.41–9.50 m^2) had the highest concentrations (Supplementary Materials, Table S1). Although not statistically significant, smaller homes also had the highest GM concentrations for 22 other chemical components. Smaller home area appears to contribute to a build-up in air pollution concentrations. The small sample size and reduced statistical power in this study may have contributed to our inability to detect significant associations between home area and concentrations of $PM_{2.5}$ constituents for elements other than Cl and Cu.

We used multivariable linear or Tobit regression models that included two or three characteristics of homes at the brick kilns (i.e., smoking, fuel type, and home area) as independent variables to determine whether significant associations between these characteristics and concentrations of $PM_{2.5}$ chemical components, found using simple (unadjusted) regression models, remained statistically significant when we adjusted for the other characteristic(s). As stated previously, Cu, Ni, and Se were previously found in cigarette smoke [46,47] and all three $PM_{2.5}$ chemical components were significantly associated with smokers in the home in our study when we adjusted for fuel type. In addition, Cu was significantly associated with smokers in the home when we adjusted for home area and fuel type. Al, BC, Cl, Mg, P, and K were previously found in wood smoke [15,23,41,45] and all six $PM_{2.5}$ chemical components were significantly associated with fuel type in our study when we adjusted for smokers in the home and/or number of smokers in the home. However, Cl was not significantly associated with fuel type when we adjusted for home area and smokers in the home or number of smokers in the home. Cl was instead significantly associated with smokers in the home when we adjusted for home area and fuel type. Ba, Pb, Mn, V, and Zn were previously found in cigarette smoke [46,47], but none of these $PM_{2.5}$ chemical components were significantly associated with smokers in the home or number of smokers in the home in our study when we adjusted for fuel type. All five $PM_{2.5}$ chemical components were instead significantly associated with fuel type when we adjusted for smokers in the home and/or number of smokers in the home. The reasons for these discrepancies between our results and those of previous studies are unknown, but our small sample size and the fact that all five homes that used wood for fuel also had smokers in the home may have contributed.

Although we did not have a sufficient sample size to conduct principal component analysis in this study, we can make some conjecture about possible pollution sources. Of the 29 analytes that had at least one sample concentration above the LDL, only one (BrC) was significantly different between LPG homes and outdoor air. This finding may be explained by stir-fry cooking within the home [48], or possibly by activities such as burning candles or smoking indoors during non-working hours. Non-significant differences in the remaining 28 analytes may be largely explained by infiltration of ambient air pollution through gaps in brick workers' homes, as discussed previously [18,19]. There are currently over 100 operating brick kilns in the Kathmandu Valley, most of which are coal fired [17,49,50]. Several analytes from our samples are known to originate from coal burning, such as Al, As, Ba, Ca, Fe, K, Mg, Mn, P, Se, Si, and Ti, depending on the source of the coal [51]. In addition, the kiln from which our samples were collected is located near the Araniko Highway, a major roadway through Bhaktapur. Vehicle exhaust is a source of several metals that we found in ambient air, as well as in participant homes, including Cr, Cu, Fe, Ni, Pb, and Zn [52–54]. Tire fading and brake wear may be responsible for Zn and Cd, and Cu and Zn, respectively [52,55], although Cd concentrations were below the LDL for all samples in our study.

One limitation in understanding the contribution of wood burning to elemental composition in our study is that we did not measure burn temperature, which greatly affects the chemical composition of particles [24]. In future studies, we may also consider using the EC/OC ratio or measuring methyl chloride levels as more definitive markers of wood smoke in our study homes, as well as looking more closely at the bioavailability of PM-bound metals to understand the toxicological properties of $PM_{2.5}$ in brick workers' homes. We were unable to obtain measurements for BrC for four samples because the amount of BC on the filters surpassed the UDL, which rendered the optical transmittance method unfeasible. This study was also limited because samples were obtained from homes at a single brick kiln and we had a relatively small sample size. A larger sample size would have allowed for the use of principal component analysis or related methods, such as positive matrix factorization, which was used in other studies [42], to determine the sources of pollution. Other limitations of this study (e.g., unmeasured confounding by temporal factors, lack of health data, etc.) were discussed previously [19].

5. Conclusions

Based on the findings of this and other studies [18,19], we suggest a multi-faceted approach is needed to protect brick workers in the Kathmandu Valley from the adverse health effects associated with poor air quality. The atmospheric pressure, wind direction and velocity, humidity, and the bowl-shaped topography of Kathmandu Valley add to the air pollution problems [56]. As air pollution remains a major issue, it is of paramount importance to educate the general population regarding the detrimental effects of air pollution and preventative measures to inhibit extreme outcomes [57]. The government of Nepal has to take primary responsibility to address the consequences of this problem by developing policies and action plans to reduce ambient air pollution and, ultimately, its consequences [57]. As the primary source of indoor air pollution in Nepal is the burning of solid fuels for cooking, improved stoves, smoke hoods, vented or chimney stoves, and clean fuel replacements would reduce the disease burden due to indoor air pollution exposure [58]. Considering 50% of homes in this study had at least one smoker, and that smoking is a significant predictor of respiratory illness among brick workers in Nepal [21], future interventions to improve indoor air quality in brick workers' homes should also include smoking cessation programs [59,60].

Supplementary Materials: The following are available online at https://www.mdpi.com/article/10.3390/atmos12070911/s1, Table S1: Associations between the mean of samples inside homes and home area and number of people in home at a brick kiln in Bhaktapur, Nepal, May 2019, Table S2: Associations between the mean of samples inside homes and occupant density and number of children in home at a brick kiln in Bhaktapur, Nepal, May 2019, and Table S3: Associations between the mean of samples inside homes and home area, smokers in home, number of smokers in home, and fuel type mutually adjusted for each other at a brick kiln in Bhaktapur, Nepal, May 2019.

Author Contributions: Conceptualization, J.D.J., J.D.B., F.X.W. and R.T.C.; methodology, J.D.J., J.D.B., F.X.W. and R.T.C.; validation, F.X.W. and R.T.C.; formal analysis, J.D.B., F.X.W., R.T.C. and J.D.J.; investigation, J.D.J.; resources, J.D.J., F.X.W. and R.T.C.; data curation, J.D.J., J.D.B., F.X.W. and R.T.C.; writing—original draft preparation, J.D.J., J.D.B., E.J.M., S.S., J.H.L., H.M., F.X.W. and R.T.C.; writing—review and editing, J.D.J., J.D.B., E.J.M., S.S., J.H.L., H.M., F.X.W. and R.T.C.; supervision, J.D.J. and J.D.B.; project administration, J.D.J. All authors have read and agreed to the published version of the manuscript.

Funding: This research received no external funding.

Institutional Review Board Statement: Brigham Young University's (BYU) Institutional Review Board (IRB) determined this study did not meet the definition of human subject research, per 45 CFR 46 [29], based on the fact that the unit of study was the home rather than the individual.

Informed Consent Statement: Not Applicable.

Data Availability Statement: The data presented in this study are available upon request from the corresponding author.

Acknowledgments: We thank the brick kiln manager and staff, Sirish, our interpreter, and especially the brick workers and their families for providing support during sample collection.

Conflicts of Interest: The authors declare no conflict of interest.

References

1. World Health Organization. Household Air Pollution and Health. 2018. Available online: https://www.who.int/news-room/fact-sheets/detail/household-air-pollution-and-health (accessed on 21 May 2021).
2. Landrigan, P.J.; Fuller, R.; Acosta, N.J.R.; Adeyi, O.; Arnold, R.; Basu, N.; Baldé, A.B.; Bertollini, R.; Bose-O'Reilly, S.; Boufford, J.I.; et al. The Lancet Commission on pollution and health. *Lancet* **2018**, *391*, 462–512. [CrossRef]
3. Fullerton, D.G.; Bruce, N.; Gordon, S. Indoor air pollution from biomass fuel smoke is a major health concern in the developing world. *Trans. R. Soc. Trop. Med. Hyg.* **2008**, *102*, 843–851. [CrossRef]
4. Kim, K.-H.; Jahan, S.A.; Kabir, E. A review of diseases associated with household air pollution due to the use of biomass fuels. *J. Hazard. Mater.* **2011**, *192*, 425–431. [CrossRef]
5. Zhang, J.; Smith, K.R. Household Air Pollution from Coal and Biomass Fuels in China: Measurements, Health Impacts, and Interventions. *Environ. Health Perspect.* **2007**, *115*, 848–855. [CrossRef]
6. Apte, K.; Salvi, S. Household air pollution and its effects on health. *F1000Research* **2016**, *5*, 2593. [CrossRef]
7. Mannucci, P.M.; Harari, S.; Martinelli, I.; Franchini, M. Effects on health of air pollution: A narrative review. *Intern. Emerg. Med.* **2015**, *10*, 657–662. [CrossRef]
8. GBD 2013 Risk Factors Collaborators; Forouzanfar, M.H.; Alexander, L.; Anderson, H.R.; Bachman, V.F.; Biryukov, S.; Brauer, M.; Burnett, R.; Casey, D.; Coates, M.M.; et al. Global, regional, and national comparative risk assessment of 79 behavioural, environmental and occupational, and metabolic risks or clusters of risks in 188 countries, 1990–2013: A systematic analysis for the Global Burden of Disease Study 2013. *Lancet* **2015**, *386*, 2287–2323. [CrossRef]
9. Schwarze, P.E.; Øvrevik, J.; Låg, M.; Refsnes, M.; Nafstad, P.; Hetland, R.B.; Dybing, E. Particulate matter properties and health effects: Consistency of epidemiological and toxicological studies. *Hum. Exp. Toxicol.* **2006**, *25*, 559–579. [CrossRef] [PubMed]
10. Bell, M.L.; Ebisu, K.; Leaderer, B.; Gent, J.F.; Lee, H.J.; Koutrakis, P.; Wang, Y.; Dominici, F.; Peng, R.D. Associations of PM2.5 Constituents and Sources with Hospital Admissions: Analysis of Four Counties in Connecticut and Massachusetts (USA) for Persons $\geq$65 Years of Age. *Environ. Health Perspect.* **2014**, *122*, 138–144. [CrossRef]
11. Burnett, R.T.; Book, J.; Dann, T.; Delocla, C.; Philips, O.; Cakmak, S.; Vincent, R.; Goldberg, M.S.; Krewski, D. Association between particulate- and gas-phase components of urban air pollution and daily mortality in eight Canadian cities. *Inhal. Toxicol.* **2000**, *12*, 15–39. [CrossRef] [PubMed]
12. Ostro, B.; Roth, L.; Malig, B.; Marty, M. The Effects of Fine Particle Components on Respiratory Hospital Admissions in Children. *Environ. Health Perspect.* **2009**, *117*, 475–480. [CrossRef]
13. Ebisu, K.; Bell, M. Airborne PM 2.5 Chemical Components and Low Birth Weight in the Northeastern and Mid-Atlantic Regions of the United States. *Environ. Health Perspect.* **2012**, *120*, 1746–1752. [CrossRef] [PubMed]
14. Bell, M.L.; Belanger, K.; Ebisu, K.; Gent, J.F.; Lee, H.J.; Koutrakis, P.; Leaderer, B.P. Prenatal exposure to fine particulate matter and birth weight: Variations by particulate constituents and sources. *Epidemiology* **2010**, *21*, 884. [CrossRef]
15. Ali, M.U.; Siyi, L.; Yousaf, B.; Abbas, Q.; Hameed, R.; Zheng, C.; Kuang, X.; Wong, M.H. Emission sources and full spectrum of health impacts of black carbon associated polycyclic aromatic hydrocarbons (PAHs) in urban environment: A review. *Crit. Rev. Environ. Sci. Technol.* **2021**, *51*, 857–896. [CrossRef]
16. Laden, F.; Neas, L.M.; Dockery, D.W.; Schwartz, J. Association of fine particulate matter from different sources with daily mortality in six U.S. cities. *Environ. Health Perspect.* **2000**, *108*, 941–947. [CrossRef] [PubMed]
17. Haack, B.N.; Khatiwada, G. Rice and Bricks: Environmental Issues and Mapping of the Unusual Crop Rotation Pattern in the Kathmandu Valley, Nepal. *Environ. Manag.* **2007**, *39*, 774–782. [CrossRef]
18. Thygerson, S.M.; Beard, J.D.; House, M.J.; Smith, R.L.; Burbidge, H.C.; Andrus, K.N.; Weber, F.; Chartier, R.; Johnston, J.D. Air-Quality Assessment of On-Site Brick-Kiln Worker Housing in Bhaktapur, Nepal: Chemical Speciation of Indoor and Outdoor PM2.5 Pollution. *Int. J. Environ. Res. Public Health* **2019**, *16*, 4114. [CrossRef] [PubMed]
19. Johnston, J.D.; Hawks, M.E.; Johnston, H.B.; Johnson, L.A.; Beard, J.D. Comparison of Liquefied Petroleum Gas Cookstoves and Wood Cooking Fires on $PM_{2.5}$ Trends in Brick Workers' Homes in Nepal. *Int. J. Environ. Res. Public Health* **2020**, *17*, 5681. [CrossRef]
20. Sanjel, S.; Khanal, S.N.; Thygerson, S.M.; Carter, W.; Johnston, J.D.; Joshi, S.K. Exposure to respirable silica among clay brick workers in Kathmandu valley, Nepal. *Arch. Environ. Occup. Health* **2018**, *73*, 347–350. [CrossRef]

21. Sanjel, S.; Khanal, S.N.; Thygerson, S.M.; Carter, W.S.; Johnston, J.D.; Joshi, S.K. Respiratory symptoms and illnesses related to the concentration of airborne particulate matter among brick kiln workers in Kathmandu valley, Nepal. *Ann. Occup. Environ. Med.* **2017**, *29*, 9. [CrossRef]
22. Fine, P.M.; Cass, G.R.; Simoneit, B.R. Chemical characterization of fine particle emissions from fireplace combustion of woods grown in the northeastern United States. *Environ. Sci. Technol.* **2001**, *35*, 2665–2675. [CrossRef]
23. Bolling, A.K.; Pagels, J.; Yttri, K.E.; Barregard, L.; Sallsten, G.; Schwarze, P.E.; Boman, C. Health effects of residential wood smoke particles: The importance of combustion conditions and physicochemical particle properties. *Part. Fibre Toxicol.* **2009**, *6*, 29. [CrossRef]
24. Rau, J.A. Composition and Size Distribution of Residential Wood Smoke Particles. *Aerosol Sci. Technol.* **1989**, *10*, 181–192. [CrossRef]
25. Kleeman, M.J.; Schauer, A.J.J.; Cass, G.R. Size and Composition Distribution of Fine Particulate Matter Emitted from Wood Burning, Meat Charbroiling, and Cigarettes. *Environ. Sci. Technol.* **1999**, *33*, 3516–3523. [CrossRef]
26. Larson, T.V.; Koenig, J.Q. Wood smoke: Emissions and noncancer respiratory effects. *Ann. Rev. Public Health* **1994**, *15*, 133–156. [CrossRef]
27. Laskin, A.; Laskin, J.; Nizkorodov, S. Chemistry of Atmospheric Brown Carbon. *Chem. Rev.* **2015**, *115*, 4335–4382. [CrossRef] [PubMed]
28. Yan, J.; Wang, X.; Gong, P.; Wang, C.; Cong, Z. Review of brown carbon aerosols: Recent progress and perspectives. *Sci. Total Environ.* **2018**, *634*, 1475–1485. [CrossRef] [PubMed]
29. US Department of Health and Human Services. Code of Federal Regulations, Title 45, Part 46, Protection of Human Subjects, Part 46.102. Available online: https://www.hhs.gov/ohrp/regulations-and-policy/regulations/45-cfr-46/revised-common-rule-regulatory-text/index.html (accessed on 13 July 2021).
30. Kellog, B.; Winberry, W.T. Determination of metals in ambient particulate matter using x-ray fluorescence (XRF) spectroscopy. In *Compendium of Methods for the Determination of Inorganic Compounds in Ambient Air*; US EPA: Cincinatti, OH, USA, 1999.
31. Lawless, P.A.; Rodes, C.E.; Ensor, D.S. Multiwavelength absorbance of filter deposits for determination of environmental tobacco smoke and black carbon. *Atmos. Environ.* **2004**, *38*, 3373–3383. [CrossRef]
32. Williams, R.; Rea, A.; Vette, A.; Croghan, C.; Whitaker, D.; Stevens, C.; McDow, S.; Fortmann, R.; Sheldon, L.; Wilson, H.; et al. The design and field implementation of the Detroit Exposure and Aerosol Research Study. *J. Expo. Sci. Environ. Epidemiol.* **2008**, *19*, 643–659. [CrossRef]
33. Rodes, C.E.; Lawless, P.A.; Thornburg, J.; Williams, R.W.; Croghan, C.W. DEARS particulate matter relationships for personal, indoor, outdoor, and central site settings for a general population. *Atmos. Environ.* **2010**, *44*, 1386–1399. [CrossRef]
34. Zhao, W.; Hopke, P.K.; Gelfand, E.W.; Rabinovitch, N. Use of an expanded receptor model for personal exposure analysis in schoolchildren with asthma. *Atmos. Environ.* **2007**, *41*, 4084–4096. [CrossRef]
35. Williams, R.; Jones, P.; Croghan, C.; Thornburg, J.; Rodes, C. The influence of human and environmental exposure factors on personal NO_2 exposures. *J. Expo. Sci. Environ. Epidemiol.* **2011**, *22*, 109–115. [CrossRef]
36. Sloan, C.D.; Weber, F.; Bradshaw, R.K.; Philipp, T.J.; Barber, W.B.; Palmer, V.L.; Graul, R.J.; Tuttle, S.C.; Chartier, R.T.; Johnston, J.D. Elemental analysis of infant airborne particulate exposures. *J. Expo. Sci. Environ. Epidemiol.* **2016**, *27*, 526–534. [CrossRef]
37. Beard, J.D.; Erdely, A.; Dahm, M.M.; de Perio, M.A.; Birch, M.E.; Evans, D.E.; Fernback, J.E.; Eye, T.; Kodali, V.; Mercer, R.R.; et al. Carbon nanotube and nanofiber exposure and sputum and blood biomarkers of early effect among U.S. workers. *Environ. Int.* **2018**, *116*, 214–228. [CrossRef]
38. Lubin, J.H.; Colt, J.S.; Camann, D.; Davis, S.; Cerhan, J.; Severson, R.K.; Bernstein, L.; Hartge, P. Epidemiologic Evaluation of Measurement Data in the Presence of Detection Limits. *Environ. Health Perspect.* **2004**, *112*, 1691–1696. [CrossRef]
39. Akaike, H. A new look at the statistical model identification. *IEEE Trans. Autom. Control.* **1974**, *19*, 716–723. [CrossRef]
40. Howe, C.J.; Cole, S.R.; Westreich, D.; Greenland, S.; Napravnik, S.; Eron, J.J. Splines for Trend Analysis and Continuous Confounder Control. *Epidemiology* **2011**, *22*, 874–875. [CrossRef] [PubMed]
41. Khalil, M.; Rasmussen, R. Tracers of wood smoke. *Atmospheric Environ.* **2003**, *37*, 1211–1222. [CrossRef]
42. Zhou, Z.; Dionisio, K.L.; Verissimo, T.G.; Kerr, A.S.; Coull, B.; Howie, S.; Arku, R.E.; Koutrakis, P.; Spengler, J.D.; Fornace, K.; et al. Chemical Characterization and Source Apportionment of Household Fine Particulate Matter in Rural, Peri-urban, and Urban West Africa. *Environ. Sci. Technol.* **2014**, *48*, 1343–1351. [CrossRef] [PubMed]
43. Matawle, J.L.; Pervez, S.; Shrivastava, A.; Tiwari, S.; Pant, P.; Deb, M.K.; Bisht, D.S.; Pervez, Y.F. $PM_{2.5}$ pollution from household solid fuel burning practices in central India: 1. Impact on indoor air quality and associated health risks. *Environ. Geochem. Health* **2017**, *39*, 1045–1058. [CrossRef] [PubMed]
44. Breton, C.V.; Marutani, A.N. Air Pollution and Epigenetics: Recent Findings. *Curr. Environ. Health Rep.* **2014**, *1*, 35–45. [CrossRef]
45. Lind, T.; Vaimari, T.; Kauppinen, E.; Nilsson, K.; Sfiris, G.; Maenhaut, W. ASH formation mechanisms during combustion of wood in circulating fluidized beds. *Proc. Combust. Inst.* **2000**, *28*, 2287–2295. [CrossRef]
46. Bernhard, D.; Rossmann, A.; Wick, G. Metals in cigarette smoke. *IUBMB Life* **2005**, *57*, 805–809. [CrossRef]
47. Pappas, R.S. Toxic elements in tobacco and in cigarette smoke: Inflammation and sensitization. *Metallomics* **2011**, *3*, 1181–1198. [CrossRef]
48. Sankhyan, S. Indoor black and brown carbon from cooking activities and outdoor penetration: Insights from the HOMEChem Study. Ph.D. Thesis, University of Colorado at Boulder, Boulder, CO, USA, 2019.

49. ENPHO. *A Study on Status of Brick Industry in the Kathmandu Valley*; ENPHO: Kathmandu, Nepal, 2001.
50. Raut, A. Brick Kilns in Kathmandu Valley: Current status, environmental impacts and future options. *Himal. J. Sci.* **2003**, *1*, 59–61. [CrossRef]
51. Watson, J.G.; Chow, J.C.; E Houck, J. PM2.5 chemical source profiles for vehicle exhaust, vegetative burning, geological material, and coal burning in Northwestern Colorado during 1995. *Chemosphere* **2001**, *43*, 1141–1151. [CrossRef]
52. Hong, N.; Zhu, P.; Liu, A.; Zhao, X.; Guan, Y. Using an innovative flag element ratio approach to tracking potential sources of heavy metals on urban road surfaces. *Environ. Pollut.* **2018**, *243*, 410–417. [CrossRef] [PubMed]
53. Kumari, S.; Jain, M.K.; Elumalai, S.P. Assessment of Pollution and Health Risks of Heavy Metals in Particulate Matter and Road Dust Along the Road Network of Dhanbad, India. *J. Health Pollut.* **2021**, *11*, 210305.
54. Bilos, C.; Colombo, J.C.; Skorupka, C.N.; Presa, M.J.R. Sources, distribution and variability of airborne trace metals in La Plata City area, Argentina. *Environ. Pollut.* **2001**, *111*, 149–158. [CrossRef]
55. Hjortenkrans, D.S.T.; Bergbäck, B.G.; Häggerud, A.V. Metal Emissions from Brake Linings and Tires: Case Studies of Stockholm, Sweden 1995/1998 and 2005. *Environ. Sci. Technol.* **2007**, *41*, 5224–5230. [CrossRef] [PubMed]
56. Giri, D.; Krishna, M.V.; Adhikary, P.R. The influence of meteorological conditions on PM10 concentrations in Kathmandu Valley. *Int. J. Environ. Res.* **2008**, *2*, 49–60.
57. SSaud, B.; Paudel, G. The Threat of Ambient Air Pollution in Kathmandu, Nepal. *J. Environ. Public Health* **2018**, *2018*, 1504591. [CrossRef]
58. Malla, M.B.; Bruce, N.; Bates, E.; Rehfuess, E. Applying global cost-benefit analysis methods to indoor air pollution mitigation interventions in Nepal, Kenya and Sudan: Insights and challenges. *Energy Policy* **2011**, *39*, 7518–7529. [CrossRef]
59. Van den Brand, F.; Anagelhout, G.; Winkens, B.; Chavannes, N.H.; Van Schayck, O.C.P. Effect of a workplace-based group training programme combined with financial incentives on smoking cessation: A cluster-randomised controlled trial. *Lancet Public Health* **2018**, *3*, e536–e544. [CrossRef]
60. Cahill, K.; Lancaster, T.R. Workplace interventions for smoking cessation. *Cochrane Database Syst. Rev.* **2014**. [CrossRef] [PubMed]

MDPI
St. Alban-Anlage 66
4052 Basel
Switzerland
Tel. +41 61 683 77 34
Fax +41 61 302 89 18
www.mdpi.com

Atmosphere Editorial Office
E-mail: atmosphere@mdpi.com
www.mdpi.com/journal/atmosphere

www.ingramcontent.com/pod-product-compliance
Lightning Source LLC
LaVergne TN
LVHW070928170726
843515LV00005B/897